THE ELEMENTS

THE ELEMENTS

■

Written and compiled by

JOHN EMSLEY

Reader in Inorganic Chemistry,
Department of Chemistry,
King's College, London

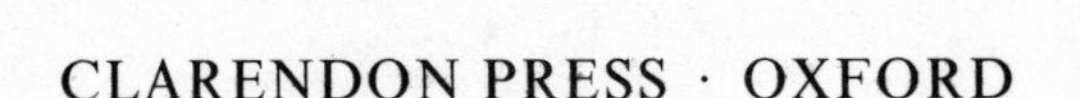

CLARENDON PRESS · OXFORD

Oxford University Press, Walton Street, Oxford OX2 6DP
Oxford New York Toronto
Delhi Bombay Calcutta Madras Karachi
Petaling Jaya Singapore Hong Kong Tokyo
Nairobi Dar es Salaam Cape Town
Melbourne Auckland
and associated companies in
Berlin Ibadan

Oxford is a trade mark of Oxford University Press

Published in the United States
by Oxford University Press, New York

First published 1989
Reprinted (with corrections) 1989

British Library Cataloguing in Publication Data
Emsley, John
The elements.
1. Chemical elements, chemical compounds—Technical data—For schools
I. Title
540'.212
ISBN 0-19-855238-6
ISBN 0-19-855237-8 (pbk)

Library of Congress Cataloging in Publication Data
Emsley, J. (John)
The elements/written and compiled by John Emsley.
p. cm. Includes index.
1. Chemical elements—Handbooks, manuals, etc. I. Title.
QD466.E48 1988 546—dc19 88-19011
ISBN 0-19-855238-6
ISBN 0-19-855237-8 (pbk)

Typeset by Cotswold Typesetting Ltd, Gloucester
Printed in Great Britain
by Bookcraft (Bath) Ltd
Midsomer Norton, Avon

Preface

MOST people are aware that everything we see around us is composed of a limited number of chemical elements. Most scientists occasionally need to find information about individual elements, and it comes as a surprise to them to discover that such information is not always easily obtained. Even chemists may have difficulty in tracking down certain pieces of data. Chemists have at their disposal several excellent collections of numerical facts, but we tend to view our world through the eyes of analytical, organic, inorganic, or physical chemists, and compile our data accordingly.

Curiously, a handbook about the elements themselves may be seen as the province of all, and yet be the responsibility of no one. This may explain why it has remained neglected for so long, but hopefully the book you now hold fills this gap. It should not be seen as just a chemistry book, but as a general book to which all scientists could refer. That there is a widespread interest in the chemical elements was made clear to me a few years ago.

1984 was the 150th anniversary of the birth of Dimitri Mendeleyev, who devised the Periodic Table of the elements in 1869. To commemorate the event I wrote a short article for the popular science magazine *New Scientist*. In this, I also asked for readers' comments on the controversy over how the groups of the periodic table should be numbered. The response was remarkable—letters arrived by every post for the next six months. Most agreed that there was a need to change from the Roman numbers I–VIII (with the sub-divisions A and B) which were then in use, because of the conflicting ways in which these group numbers were used in different parts of the world.

However there was very little agreement on what should replace them. What agreement there was supported the use of Arabic numerals 1–18, and this was the system adopted for the *New Scientist* Periodic Table produced that year. Happily this was also the system favoured by the American Chemical Society and it had the support of the International Union of Pure and Applied Chemists. It was first proposed by the Swedish chemist Arne Ölander in 1958.

While the periodic table is universally recognized as the hallmark of chemistry, it generally conveys only the barest minimum of information about each element. The *New Scientist* version, slightly modified and reproduced on the inside front cover, contains only the chemical symbol of each element, its atomic number, and relative atomic mass. Other versions have been produced that crowded each box with up to ten or more pieces of data such as melting and boiling points, densities, crystal structures, electron arrangement, etc. Some companies produce special versions for physicists, crystallographers, spectroscopists, and even bird fanciers!

Many years ago I began my own collection of data about the elements. To start with I simply bought a large note book with over 100 pages and wrote the name of a different element at the top of each page. Over the years my collection of numerical information about each element grew until the book finally burst its binding as I stuck in more and more leaves. It is an edited version of this book that you now have in your hands. Not all the information about every element is here, nor could it be, since I have deliberately kept within the format of a double page for each element, although I have tried to include as much data as possible within this framework.

For each element I have grouped together properties under the headings chemical, physical, nuclear, and electronic, as well as giving the history and derivation of the element's name. I have also included something on its environmental and biological importance. All the data are given in SI units in the main tables, but there are full details of how to convert this information into other, commonly used, units in the Key chapter.

The final part of the book consists of a series of double tables of properties, arranged first in order of the elements and then in order of the property itself. Again the choice is partly my own but these tables were mainly determined by the need to include as many elements as possible for each property, and there are only a certain number of properties that are common to all, or most, of the elements.

Finally I make no apologies for including tables of such 'obvious' data as the elements in alphabetical order with their chemical symbols, and chemical symbols in alphabetical order with their full names (see p. 251 ff.). The need for these came home to me as I was preparing the *New Scientist* version of the periodic table. Opinion among other scientists is that chemists assume too much, and while we know without thinking that W is tungsten and Sn is tin, many of our fellow scientists need these symbols confirmed when they encounter them.

Hopefully this book will find its way on to the shelves of all scientists, and not just chemists. In particular I have in mind those scientists working in the media who have an uphill struggle to make chemistry interesting. With *The Elements* as a source book they have at their fingertips background information about an element with which they can back up their stories.

London J.E.
December 1987

The key to *The Elements*

THE curious thing about numerical information is the way it varies slightly from book to book. Usually these variations are of the order of ±2 per cent, small but irritating. The need to standardize data has been recognized by several organizations, such as the International Union of Pure and Applied Chemists (IUPAC) and the National Bureau of Standards (NBS) in Washington DC. When such bodies set up committees to assess data and decide on the most reliable values, the job of an author like myself becomes much easier. For example, the table of thermodynamic data of the elements comes from an NBS book while the standard reduction potentials come from an IUPAC publication.

Rather than quote alternative values or a range of values for certain properties I have assumed that well established specialists in the collecting of data, such as those of the *CRC handbook of chemistry and physics* (reference 1) and *Lange's handbook of chemistry* (reference 2) provide reliable information, even though it may not be in SI units. Other compilations with extensive chemical data are references 3–6. In the sections which follow I give the sources of my information, explain the SI units used, and convert the data to other commonly used units.

Note: n.a. used in the tables means not available.

The elements: discovery, names, formulae, and relative atomic masses

IUPAC is the official body responsible for approving the names and formulae given to elements, and authenticating their relative atomic masses. The US names of aluminium (aluminum) and caesium (cesium) are so near to the recommended names that they present no problem, except that the latter name leads to a slightly different position in a list of elements arranged alphabetically. Changes are reported in *Pure and Applied Chemistry*, IUPAC's official journal. The relative atomic masses given here are taken from *Pure and Applied Chemistry*, 1986, **58**, 1677.

The most comprehensive volume on the discovery and history of the elements is M. E. Weeks and H. M. Leicester's, *Discovery of the elements*, published by the *Journal of Chemical Education*, Easton (PA), 1968. An outline of the discovery of each element is also given in reference 1. Individual articles in the *Journal of Chemical Education* from time to time give excellent histories of individual elements.

D. W. Ball, in the *Journal of Chemical Education*, 1985, **62**, 787, and J. G. Stark and H. G. Wallace, in *Education in Chemistry*, 1970, 152, explain how the names of the elements were chosen.

Chemical properties

Descriptions

The brief descriptions of the elements, their names, and their reactivity towards air, water, acids, and alkalis have been taken from references 7–9. The *Encyclopedia of the chemical elements*, edited by C. A. Hampel (Reinhold Book Corporation, New York, 1968) gives a comprehensive, if somewhat dated account, of the uses of each element.

Radii

To convert radii, which are given in picometres (pm), to metres divide by 10^{12}. To convert to nanometres (nm) divide by 1000; and to convert to angstroms (Å) divide by 100.

The radius of an atom depends upon several factors: oxidation state, degree of ionization, and coordination number (for metals this is generally 12). When it is part of a molecule two radii are defined: the covalent radius, which refers to the role it plays in forming bonds, and the van der Waals radius, which refers to the radius it presents to the world beyond the molecule. Many textbooks quote some of these radii, but the best sources appear to be references 2 and 4.

Electronegativity

This quantity is well understood but ill-defined. It refers to the potential of an atom to attract electrons to itself. The higher the electronegativity the stronger is this ability. Fluorine is the most electronegative of all elements. The units of electronegativity are rarely quoted.

Effective nuclear charge, Z_e

Like electronegativity, this quantity is easier to understand than calculate. Although Z_e can be defined for any electron within an atom, only the Z_e for the valence shell electrons is of interest to the chemist. Z_e is the charge due to the protons of the nucleus less a screening factor due to the other electrons of the atom. There are several slightly different ways of calculating this screening, and consequently there are several values for Z_e. Those quoted were computed by J. C. Slater (*Physical Reviews*, 1930, **36**), E. Clementi and D. L. Raimondi, (*Journal of Chemical Physics*, 1963, **38**, 2686), E. Clementi, D. L. Raimondi, and W. P. Reinhardt (*Journal of Chemical Physics*, 1967, **47**, 1300), and C. Froese-Fischer (*Atomic Data*, 1972, **4**, 301, and *Atomic Data and Nuclear Data Tables*, 1973, **12**, 87).

Standard reduction potentials

In these diagrams the element is arranged with the highest oxidation state on the left. Potentials are given in volts. The higher the value of $E^\ominus$, the stronger is the oxidant as an oxidizing agent; the lower $E^\ominus$, the stronger is the reductant as a reducing agent. These diagrams are taken from *Standard potentials in aqueous solutions*, edited by A. J. Bard, R. Parsons, and J. Jordan. (Marcel Dekker (for IUPAC), New York, 1985). Other compilations of $E^\ominus$ data are to be found in references 1–6.

Oxidation states

Although it is merely a formalism, the concept of oxidation number is much used in describing the changes that happen to an element in its chemical reactions. Consequently it is a useful way of classifying and explaining the compounds of that element. From the multitude of compounds known for elements in their various oxidation states I have chosen where possible to give the oxides, hydroxides or acids, hydrides, fluorides, and chlorides (after which 'etc.' means the corresponding bromides and iodides), and the species present in aqueous solutions of simple salts of the element (denoted 'aq'). Salts, complexes, and organometallic compounds are also given if these are special; otherwise I have merely indicated that such substances exist, and references 7–9 should be consulted for further details.

Covalent bonds

To convert bond lengths, r, given in picometres, to metres, divide by 10^{12}; to convert them to nanometres (nm) divide by 1000. To convert them to angstroms (Å) divide by 100. To convert bond energies, i.e. bond enthalpies, E, given in kJ mol^{-1} to kcal mol^{-1}, divide by 4.184.

The values are taken from various sources, notably bond lengths from references 1 and 4, and bond enthalpies from references 1, 3, and 4, and from *Bond energies, ionization potentials and electron affinities*, by V. I. Vedeneyev, L. V. Gurvich, V. N. Kondrat'yev, V. A. Mededev, and Ye. L. Frankevich (Edward Arnold, London, 1966). *SI chemical data* by G. H. Aylward and T. J. V. Findley (Wiley, Sydney, 1971) also gives r and E for many bonds.

Physical properties

Melting points and boiling points

These are given in Kelvin (K). They can be converted to degrees Celsius (C) by subtracting 273.15. The values quoted are based on reference 1, but are also given with slight variations in all major data books (references 2–9).

Enthalpy of fusion, ΔH_{fusion}

These, given in kJ mol^{-1}, can be converted to kcal mol^{-1} by dividing by 4.184.

The values are taken mainly from R. Loebel's compilation in reference 1 (where they are given in c.g.s. units), supplemented by references 4 and 6, where they are in SI units.

Thermodynamic properties

To convert kJ mol^{-1} to kcal mol^{-1} divide by 4.184. To convert entropies in J K^{-1} mol^{-1} to eu (entropy units, i.e. cal K^{-1} mol^{-1}) divide by 4.184. To convert specific heats, C_p, to cal g^{-1} K^{-1} divide first by 4.184 and then by the relative atomic mass of the element concerned.

The thermodynamic properties are taken from *The NBS tables of chemical thermodynamic properties*, by D. D. Wagman, W. H. Evans, and V. B. Parker, which was published jointly by the American Chemical Society and the American Institute of Physics for the National Bureau of Standards, Washington DC, in 1982. Although thermodynamic data for the elements are to be found in references 1–5, the NBS compilation is preferred and is in SI units.

Density

Since the basic SI unit of weight is the kilogramme (kg) and of length the metre (m), the preferred unit of density is kg m^{-3}. The more common unit however is g cm^{-3} and to convert from kg m^{-3} to g cm^{-3} divide by 1000.

Many sources list densities at only one temperature, such as reference 2 (293 K), others at various temperatures, e.g. reference 6, which reports them in SI units. The densities used here are based mainly on reference 1 for the solid elements. In the same work can be found the densities of the liquid elements at their melting points, compiled by G. Lang.

Thermal conductivity

The SI unit for this property is watts per metre per Kelvin (W m^{-1} K^{-1}). To convert to W cm^{-1} K^{-1} divide by 100.

The values are taken from C. Y. Ho, R. W. Powell, and P. E. Liley (*Journal of Physical Chemistry Reference Data*, 1974, **3**, suppl. 1), with the values for carbon

being taken from reference 1 for the directions perpendicular and parallel to the graphite axis. Reference 2 also reports the thermal conductivity of the elements.

Electrical resistivity

The SI units are ohm metres (Ω m) and the electrical resistivity of metals are of the order of 10^{-8} Ω m. To convert to the more common units of μΩ cm multiply the Ω m values by 10^8.

The data are taken from reference 1. Reference 6, pp. 102–3, also gives the electrical resistivity in Ω m at temperatures of 78, 273, 373, 573, and 1473 K. Values in non-SI units are given in reference 3, pp. 580–684.

Mass magnetic susceptibility, χ

This is obtained from the volume magnetic susceptibility, κ, which is unitless, by dividing by the density (kg m^{-3} in SI units). To convert mass magnetic susceptibilities from the SI units of kg^{-1} m^3 to c.g.s. units of g^{-1} cm^3, multiply by $1000/4\pi$, i.e. 79.6. To convert to molar magnetic susceptibility multiply first by 79.6 then by the relative atomic mass of the element. The use of SI units for magnetic properties is discussed by T. I. Quickenden and R. C. Marshall, in the *Journal of Chemical Education*, 1972, **49**, 114.

The magnetic susceptibility data are taken from *Constantes sélectionnées. Diamagnétisme et paramagnétisme*, by G. Foëx (Masson et Cie, Paris, 1957), and *Modern magnetism* (4th edition), by L. F. Bates (Cambridge University Press, Cambridge, 1963). Values for certain of the lanthanides were taken from J. M. Lock (*Proceedings of the Physical Society*, 1957, **B70**, 476 and 566). Reference 1 reports molar magnetic susceptibilities in c.g.s. units.

Coefficient of linear thermal expansion, α

This is the same in SI and c.g.s. units, i.e. K^{-1}. It is often reported as $10^6\alpha$. The data are taken from reference 3 (pp. 580–684).

Molar volume (or atomic volume)

This represents a slight problem for SI whose unit of volume is m^3. Traditionally atomic volume is obtained by dividing the relative atomic mass of an element in g by the density in g cm^{-3}, and is consequently traditionally expressed in cm^3. To express the molar volume in m^3 divide the values given by 10^6. Expressing it in this way, however, implies a density measured in g m^{-3}, which is not a recognized way of expressing this quantity.

Since atomic volume for an element depends upon the density, it also depends upon the phase, the allotrope, and the temperature. The values reported here are based where possible on the solid state at room temperature (298 K or as indicated), and are taken mainly from C. N. Singman (*Journal of Chemical Education*, 1984, **61**, 137).

Lattice structure

To convert the cell parameters to angstroms (Å) divide by 100. To convert them to nm divide by 1000.

The abbreviation b.c.c. means body-centred cubic, f.c.c. means face-centred cubic, and h.c.p. means hexagonal close packed. The data are taken from *Landolt–Bornstein*, New Series, Group III, vol. 6, edited by K. H. Hellwege and A. M. Hellwege. Crystal structure data are also to be found in references 3, 5, and 9; and *The Structures of the Elements* by J. Donohue (John Wiley & Sons, 1974).

X-ray diffraction mass absorption coefficients

These are taken from *International tables for X-ray crystallography*, Vol. III,

Physical and Chemical Tables, section 3.2, published for the International Union of Crystallographers by The Kynock Press (Birmingham UK, 1962).

Nuclear properties

Thermal neutron capture cross-section

The barn is defined as 10^{-24} cm^2, which is 10^{-28} m^2. The values are taken from reference 1.

Key isotopes

This lists all stable isotopes and the longest lived radioactive nuclides, plus those used in research. The per cent natural abundance is given as 'trace' for certain short-lived nuclides that are part of a natural decay series. The half-life, $T_{1/2}$, is expressed in seconds (s), minutes (m), hours (h), days (d), or years (y). The decay mode is shown as β^- for negative emission, β^+ for positron emission, α for alpha decay, EC for electron capture, IT for isomeric transition, and SF for spontaneous fission. Some nuclei decay by two routes. The energy (in mega electronvolts, MeV) of the radiation is given in parentheses. γ indicates the emission of gamma radiation.

The nuclide data is taken from reference 1. Similar tables are given in references 2 and 3. The most comprehensive listing is to be found in the *Table of isotopes* (7th edition) by C. M. Lederer and V. S. Shirley (John Wiley & Sons, New York, 1978). This work lists the full data on 2600 known isotopes, and its compilation was supported by the US NBS Office of Standard Reference Data.

Nuclear spin (I) is reported in units of $h/2\pi$. The nuclear magnetic moment (μ) is reported in nuclear magnetrons with diamagnetic correction. Both I and μ are taken from *Nuclear spins and moments* by G. H. Fuller (*Journal of Physical Chemistry Reference Data*, 1976, **5**, 835).

Use is made of isotopes in various areas of research, denoted by 'NMR', 'tracer', or 'med' short for medical. Nuclear magnetic resonance spectroscopy (NMR) details are given directly below the Key Isotopes table. The radioactive isotopes used as tracers and in medicine are taken from the *Merck index* (10th edition), edited by M. Windholz (Merck & Co., Inc., Rahway (NY), 1983); *Isotopes, products and services catalog* (Oak Ridge National Laboratory, P.O. Box X, Oak Ridge, Tennessee 37831, USA); and *Biochemicals* (The Radiochemical Centre, P.O. Box 16, Amersham, Bucks, HP7 9LL, UK). *Radionuclide tracers* by M. F. L'Annunziata (Academic Press, New York, London, 1987), also contains appendices of available isotopes and their radiation characteristics.

Nuclear magnetic resonance

Where data for two nuclei are given, the one enclosed in square brackets is less frequently used for NMR studies. Nuclear spins are given in the Key Isotopes table.

Relative sensitivity is at constant field for equal numbers of nuclei. Absolute sensitivity is the relative sensitivity multiplied by the natural abundance (given in the Key Isotopes table as per cent, but here used as a fraction of unity). Also known as receptivity, it is commonly quoted relative to $^{13}C = 1.00$. The ratio of ^1H to ^{13}C receptivities is 1 to 5680.

The magnetogyric ratio is given in rad T^{-1} s^{-1}, but is often quoted as γ values which are in units of rad T^{-1} s^{-1} multiplied by 10^7. Thus for ^1H $\gamma = 26.75$, as opposed to 26.75×10^{-7} rad T^{-1} s^{-1}. γ is a constant of proportionality between frequency and field strength, and so the units are frequency (rad s^{-1})/field (T), i.e. rad T^{-1} s^{-1}.

Quadrupole moments are given in units of m^2. They can be converted to cm^2 by multiplying by 10^4, or to barns by multiplying by 10^{28}.

Frequency is quoted relative to the ^1H signal of $Si(CH_3)_4$, which is exactly 100 MHz in a magnetic field of 2.3488 T. For NMR spectrometers with ^1H at 60, 90, 200, 250, 360, or 400 MHz the frequency and field vary in direct proportion; in other words multiply the frequencies given in the tables by 0.6, 0.9, 2, 2.5, 3.6, or 4 respectively.

The data for the NMR tables were compiled from the *Handbook of high resolution multinuclear NMR*, by C. Brevard and P. Granger (John Wiley & Sons, New York, 1981), *NMR and the periodic table*, by R. K. Harris and B. E. Mann (Academic Press, London, 1978), *Multinuclear NMR* by J. Mason (Plenum Press, New York and London, 1987) and from Bruker Scientific Instruments publications.

Electron shell properties

Ground state electron configuration and term symbol

These are given in most inorganic textbooks, such as references 7 and 8. The data given here were taken from the *Handbook of atomic data*, by S. Fraga, J. Karwowski, and K. M. S. Saxena (Elsevier, Amsterdam, 1976).

Electron affinity

To convert the values given in kJ mol^{-1} to electron volts (eV) divide by 96.486; to convert to MJ mol^{-1} divide by 1000.

Electron affinity is conventionally reported as positive if the addition of an electron to an atom releases energy, as it almost invariably does for the step $M \rightarrow M^-$, and negative if the process is energy absorbing, as it is for the addition of a second electron, i.e. $M^- \rightarrow M^{2-}$. This energy convention is the opposite of that used in reporting ionization energies and most other energy changes.

Although many chemistry textbooks report some electron affinities, and the major compilations, references 1 and 2, give extensive lists (in eV), the data reported here come from H. Hotop and W. C. Lineberger (*Journal of Physical Chemistry Reference Data*, 1975, **4**, 539), also reported in eV. An earlier compilation by R. J. Zollweg (*Journal of Chemical Physics*, 1969, **50**) also gives some values for the heavier elements. Their variation from element to element is discussed by E. C. M. Chen and W. E. Wentworth (*Journal of Chemical Education*, 1975, **52**, 486).

Ionization energies (ionization potentials)

To convert from kJ mol^{-1} to electron volts (eV) divide the values given by 96.486; to convert to MJ mol^{-1} divide by 1000.

Ionization energies are known to a high degree of accuracy for removal of the first, second, third, fourth, and fifth electrons for most elements, and for subsequent electron removal from the lighter elements. Values given in parentheses are less reliable. For some elements ionization energies beyond the tenth are available.

Many textbooks and compilations list ionization energies, some in SI units such as reference 4. The NBS has at different times published ionization data, beginning with E. C. Moore's *Atomic energy levels* Volume III (NBS Circular 467, 1958). Reference 1 quotes *Analyses of optical spectra*, NSRDS-NBS 34 (Office of Standard Reference Data, NBS, Washington DC, 1970). The values for the lanthanides and actinides were taken from W. C. Martin, L. Hagan, J. Reader, and J. Sugar (*Journal of Physical Chemistry Reference Data*, 1974, **3**, 771).

Principal lines in atomic spectrum

The stronger lines in the spectrum are listed with the strongest shown in bold. Lines arising from the neutral atom are indicated by Species I, and those arising from the singly charged ion M^+ are Species II. The most sensitive line in the neutral atom spectrum is denoted by U1, the next most sensitive by U2, etc. The corresponding order of sensitivity for the M^+ ion spectrum is shown as V1, V2, etc. Application in atomic absorption spectrometry is indicated by AA and in atomic emission spectrometry by AE.

For some transuranium elements with many intense lines only the first seven lines identified as arising from the neutral atom are given.

Data are derived from *Tables of spectral line intensities* (2nd edition) by W. F. Meggers, C. H. Corliss, and B. F. Scribner (NBS Monograph 145, Part I, 1975); *Line spectra of the elements* by J. Reader and C. H. Corliss in reference 1; *Atomic energy levels and Grotrian diagrams* Vols I, II, and III, by S. Bashkin and J. O. Stoner (Elsevier, Amsterdam, 1976); *CRC handbook of spectroscopy*, Vol. 1, by J. W. Robinson (CRC Press, Boca Raton, FL, 1974); *Pye Unicam atomic absorption data book* (3rd edition) by P. J. Whiteside (Pye Unicam, Cambridge, UK, 1979); and the *Handbook of the physicochemical properties of the elements*, edited by G. V. Samsonov, pp. 30–67 (IFI-Plenum, New York, 1968).

Environmental and biological properties

Abundances

For atmosphere, Earth's crust, and seawater the units are parts per million (p.p.m.), defined in the case of the atmosphere as cubic centimetres per cubic metre, in the case of the Earth's crust as grammes per metric tonne (1000 kg), which is the same as milligrammes per kg, and for seawater as milligrammes per litre.

The atmospheric abundances are taken from reference 6; the crustal abundances from *Geochemistry* by W. S. Fyfe (Oxford University Press, Oxford, 1974), as reproduced in reference 7; the concentrations in seawater come from references 3 and 6 and the *McGraw-Hill encyclopedia of science and technology*, p. 627 (McGraw-Hill, New York, 1970).

Biological role

Information about essential elements is from the *Handbook of vitamins, minerals and hormones* (2nd edition) by R. J. Kutsky (Van Nostrand Reinhold, New York, 1981). Toxicological information about the elements depends on individual compounds, as well as the elements themselves. Useful sources of information are *Metal toxicology in mammals* (2 volumes) by T. D. Luckey and B. Venugopal (Plenum Press, New York, 1977), and the *Handbook of toxicity of inorganic compounds*, edited by H. G. Seiler and H. Sigel (Marcel Dekker, New York, 1988), which covers a majority of elements in alphabetical order.

General references

1. Weast, R. C. (ed.) *CRC handbook of chemistry and physics*, (63rd edn), CRC Press, Boca Raton, Fl 1982.
2. Dean, J. A. (ed.) *Lange's handbook of chemistry*, (13th edn). McGraw-Hill, New York, 1985.
3. Moses, A. J. *The practising scientist's handbook*. Van Nostrand Reinhold, New York, 1978.
4. Ball, M. C. and Norbury, A. H. *Physical data for inorganic chemists*. Longman, London, 1974.

5. Samsonov, G. V. (ed.), *Handbook of the physicochemical properties of the elements.* IFI-Plenum, New York, 1968.
6. Kaye, G. W. C. and Laby, T. H. *Tables of physical and chemical constants* (14th edn), Longman, London, 1973.
7. Greenwood, N. N. and Earnshaw, A. *Chemistry of the elements.* Pergamon Press, Oxford, 1984.
8. Cotton, F. A. and Wilkinson, G. *Advanced inorganic chemistry* (4th edn), John Wiley & Sons, New York, 1980.
9. Bailar, J. C., Emeleus, H. J., Nyholm, R., and Trotman-Dickenson, A. F. (eds). *Comprehensive inorganic chemistry*, (5 vols) Pergamon Press, Oxford, 1973.

THE ELEMENTS

Ac

Atomic number: **89**

Relative atomic mass (^{12}C=12.0000): **(227)**

Chemical properties

Occurs as part of ^{235}U decay series but made by neutron bombardment of ^{226}Ra. Soft, silvery-white metal which glows in the dark. Reacts with water to evolve H_2.

Radii/pm: Ac^{3+} 118; atomic 187.8

Electronegativity: 1.1 (Pauling); 1.00 (Allred-Rochow)

Effective nuclear charge: 1.80 (Slater)

Standard reduction potentials $E^{\ominus}$/V

	III		0
Acidic solution	Ac^{3+}	$\xrightarrow{-2.13}$	Ac
Basic solution	$Ac(OH)_3$	$\xrightarrow{-2.6}$	Ac

Oxidation states

Ac^0 (d^1 s^2)
Ac^{III} ([Rn]) Ac_2O_3, $Ac(OH)_3$ insoluble
AcH_2 and AcH_3 are probably Ac^{III} compounds

Physical properties

Melting point/K: 1320±50

Boiling point/K: 3470±300

ΔH_{fusion}/kJ mol^{-1}: 14.2

ΔH_{vap}/kJ mol^{-1}: 293

Thermodynamic properties (298.15 K, 0.1 MPa)

State	$\Delta_f H^{\ominus}$/kJ mol^{-1}	$\Delta_f G^{\ominus}$/kJ mol^{-1}	$S^{\ominus}$/J K^{-1} mol^{-1}	C_p/J K^{-1} mol^{-1}
Solid	0	0	56.5	27.2
Gas	406	366	188.1	20.84

Density/kg m^{-3}: 10 060 [293 K]

Thermal conductivity/W m^{-1} K^{-1}: 12 [300 K]

Electrical resistivity/Ω m: n.a.

Mass magnetic susceptibility/kg^{-1} m^3: n.a.

Molar volume/cm^3: 22.6

Coefficient of linear thermal expansion/K^{-1}: 14.9×10^{-6}

Lattice structure (cell dimensions/pm), space group

f.c.c. (a=531.1), Fm3m

X-ray diffraction: mass absorption coefficients (μ/ρ)/cm^2 g^{-1}:

CuK_α n.a. MoK_α n.a.

Actinium

Discovered in 1899 by A. Debierne at Paris, France

[Greek, *aktinos* = ray]

Nuclear properties

Atomic number: 89

Thermal neutron capture cross-section/barns: 810 ± 20 (^{227}Ac)

Number of isotopes (including nuclear isomers): 11

Isotope mass range: 221 → 231

Key isotopes

Nuclide	Atomic mass	Natural abundance (%)	Half life $T_{1/2}$	Decay mode and energy (MeV)	Nuclear spin I	Nuclear magnetic moment μ	Uses
^{225}Ac		0	10.0d	$\alpha(5.829)$			tracer
^{227}Ac	227.0278	trace	21.6y	$\beta(0.043\ 99\%$; α; γ	3/2	+1.1	NMR
^{228}Ac		trace	6.13h	$\beta^-(2.14)$; γ			

NMR

	^{227}Ac
Relative sensitivity ($^1H = 1.00$)	—
Absolute sensitivity ($^1H = 1.00$)	—
Receptivity ($^{13}C = 1.00$)	—
Magnetogyric ratio/rad T^{-1} s^{-1}	3.5×10^7
Quadrupole moment/m^2	1.7×10^{-28}
Frequency ($^1H = 100$ MHz; 2.3488 T)/MHz	13.1

Electron shell properties

Ground state electron configuration: [Rn]$6d^17s^2$

Term symbol: $^2D_{3/2}$

Electron affinity ($M \rightarrow M^-$)/kJ mol^{-1}: n.a.

Ionization energies/kJ mol^{-1}

1. $M \rightarrow M^+$	499	6. $M^{5+} \rightarrow M^{6+}$	(7 300)
2. $M^+ \rightarrow M^{2+}$	1170	7. $M^{6+} \rightarrow M^{7+}$	(9 200)
3. $M^{2+} \rightarrow M^{3+}$	1900	8. $M^{7+} \rightarrow M^{8+}$	(10 500)
4. $M^{3+} \rightarrow M^{4+}$	(4700)	9. $M^{8+} \rightarrow M^{9+}$	(11 900)
5. $M^{4+} \rightarrow M^{5+}$	(6000)	10. $M^{9+} \rightarrow M^{10+}$	(15 800)

Principal lines in atomic spectrum

Wavelength/nm	Species	Sensitivity	Application
386.312	II		
408.844	II		
416.840	II		
417.998	I		
729.040*	I		

* Lowest energy transition from ground state to nearest empty orbital

Abundance: Only traces in uranium ores

Biological role: None; toxic due to radioactivity

Al

Atomic number: 13

Relative atomic mass (^{12}C=12.0000): **26.981 54**

Chemical properties

Hard, strong, silvery-white metal, protected by oxide film from reacting with air and water. Soluble in hot concentrated HCl and NaOH solution. Hundreds of uses as metal and alloys.

Radii/pm: Al^{3+} 57; covalent 125; atomic 143.1: van der Waals 205

Electronegativity: 1.61 (Pauling); 1.47 (Allred-Rochow)

Effective nuclear charge: 3.50 (Slater); 4.07 (Clementi); 3.64 (Froese–Fischer)

Standard reduction potentials $E^{\ominus}$/V

	III		0
Acid solution	Al^{3+}	−1.676	Al
	AlF_6^{3-}	−2.067	Al
Alkaline solution	$Al(OH)_3$	−2.300	Al
	$Al(OH)_4^-$	−2.310	Al

Covalent bonds	r/pm	E/kJ mol^{-1}
Al–H	c. 170	285
Al–C	224	225
Al–O	162	585
Al–F	163	665
Al–Cl	206	498
Al–Al	286	c. 200

Oxidation states

Al^0 (s^2p^1)	
Al^I (s^2)	AlCl in gas phase
Al^{III} ([Ne])	Al_2O_3 (amphoteric)
	AlO(OH), $Al(OH)_3$
	$Al(H_2O)_6^{3+}$ (aq)
	Al^{3+} salts, AlH_3
	$LiAlH_4$, AlF_3,
	Na_3AlF_6, Al_2Cl_6

Physical properties

Melting point/K: 933.52

Boiling point/K: 2740

ΔH_{fusion}/kJ mol^{-1}: 10.67

ΔH_{vap}/kJ mol^{-1}: 290.8

Thermodynamic properties (298.15 K, 0.1 MPa)

State	$\Delta_f H^{\ominus}$/kJ mol^{-1}	$\Delta_f G^{\ominus}$/kJ mol^{-1}	$S^{\ominus}$/J K^{-1} mol^{-1}	C_p/J K^{-1} mol^{-1}
Solid	0	0	28.33	24.35
Gas	326.4	285.7	165.54	21.38

Density/kg m^{-3}: 2698 [293 K]; 2390 [liquid at m.p.]

Thermal conductivity/W m^{-1} K^{-1}: 237 [300 K]

Electrical resistivity/Ω m: 2.6548×10^{-8} [293 K]

Mass magnetic susceptibility/kg^{-1} m^3: $+7.7 \times 10^{-9}$ (s)

Molar volume/cm^3: 10.00

Coefficient of linear thermal expansion/K^{-1}: 23.03×10^{-6}

Lattice structure (cell dimensions/pm), space group

f.c.c. (a=404.959), Fm3m

X-ray diffraction: mass absorption coefficients (μ/ρ)/cm^2 g^{-1}: CuK_α 48.6 MoK_α 5.16

Aluminium (Aluminum)

Discovered in 1825 by Hans Christian Oersted at Copenhagen, Denmark

[Latin, *alumen* = alum]

Nuclear properties

Atomic number: 13

Thermal neutron capture cross-section/barns: 0.232

Number of isotopes (including nuclear isomers): 8

Isotope mass range: 24→30

Key isotopes

Nuclide	Atomic mass	Natural abundance (%)	Half life $T_{1/2}$	Decay mode and energy (MeV)	Nuclear spin I	Nuclear magnetic moment μ	Uses
^{26}Al		0	7.4×10^5y	β^+(4.003); EC; γ	5		tracer
^{27}Al	26.981 54	100	stable		5/2	+3.6413	NMR

NMR	^{27}Al
Relative sensitivity ($^1H = 1.00$)	0.21
Absolute sensitivity ($^1H = 1.00$)	0.21
Receptivity ($^{13}C = 1.00$)	1.17×10^3
Magnetogyric ratio/rad T^{-1} s^{-1}	6.9704×10^7
Quadrupole moment/m^2	0.4193×10^{-28}
Frequency ($^1H = 100$ MHz; 2.3488 T)/MHz	26.057

Reference: $Al(H_2O)_6^{3+}$

Electron shell properties

Ground state electron configuration: [Ne] $3s^2 3p^1$

Term symbol: $^2P_{1/2}$

Electron affinity ($M \rightarrow M^-$)/kJ mol^{-1}: 44

Ionization energies/kJ mol^{-1}

1. $M \rightarrow M^+$	577.4	6. $M^{5+} \rightarrow M^{6+}$	18 376
2. $M^+ \rightarrow M^{2+}$	1816.6	7. $M^{6+} \rightarrow M^{7+}$	23 293
3. $M^{2+} \rightarrow M^{3+}$	2744.6	8. $M^{7+} \rightarrow M^{8+}$	27 457
4. $M^{3+} \rightarrow M^{4+}$	11 575	9. $M^{8+} \rightarrow M^{9+}$	31 857
5. $M^{4+} \rightarrow M^{5+}$	14 839	10. $M^{9+} \rightarrow M^{10+}$	38 459

Principal lines in atomic spectrum

Wavelength/nm	Species	Sensitivity	Application
308.215	I	U4	AA AE
309.271	I	U3	AA AE
309.284	I		
349.401*	I	U2	AA AE
396.152	I	U1	AA AE

* Lowest energy transition from ground state to nearest empty orbital

Abundance: Earth's crust 83 000 p.p.m.; seawater 0.01 p.p.m.

Biological role: None; accumulates in body from daily intake (*c.* 20 mg per day); implicated in Alzheimer's disease (senile dementia)

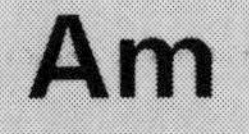

Atomic number: 95
Relative atomic mass (^{12}C=12.0000): (243)

Chemical properties

Radioactive, silvery metal produced in 100 g quantities as ^{243}Am from neutron bombardment of ^{239}Pu. Attacked by air, steam, and acids, but not alkalis.

Radii/pm: Am^{6+} 80; Am^{5+} 86; Am^{4+} 92; Am^{3+} 107; atomic 184
Electronegativity: 1.3 (Pauling); n.a. (Allred-Rochow)
Effective nuclear charge: 4.65 (Slater)

Standard reduction potentials $E^{\ominus}$/V

	VI		V		IV		III		0
Acid solution	AmO_2^{2+}	1.59	AmO^+	0.82	Am^{4+}	2.62	Am^{3+}	−2.07	Am
Alkaline solution	$AmO_2(OH)_2$	0.9	$AmO_2(OH)$	0.7	AmO_2	0.22	$Am(OH)_3$	−2.53	Am

Acid solution: AmO^+ → Am^{3+} 1.72; AmO_2^{2+} → Am^{4+} 1.20; Am^{4+} → Am −0.90

Oxidation states

Am^{II} (f^7)	AmO, $AmCl_2$, etc.
Am^{III} (f^6)	Am_2O_3, AmF_3, $AmCl_3$, etc. $[AmCl_6]^{3-}$, Am^{3+}(aq)
Am^{IV} (f^5)	AmO_2, AmF_4
Am^{V} (f^4)	AmO_2^+ (aq) } unstable due to reduction by radioactive
Am^{VI} (f^3)	AmO_2^{2+} (aq) } decay products

Physical properties

Melting point/K: 1267
Boiling point/K: 2880
ΔH_{fusion}/kJ mol^{-1}: 14.4
ΔH_{vap}/kJ mol^{-1}: 238.5

Thermodynamic properties (298.15 K, 0.1 MPa)

State	$\Delta_f H^{\ominus}$/kJ mol^{-1}	$\Delta_f G^{\ominus}$/kJ mol^{-1}	$S^{\ominus}$/J K^{-1} mol^{-1}	C_p/J K^{-1} mol^{-1}
Solid	0	0	n.a.	n.a.
Gas	n.a.	n.a.	n.a.	n.a.

Density/kg m^{-3}: 13 670 [293 K]
Thermal conductivity/W m^{-1} K^{-1}: 10 est. [300 K]
Electrical resistivity/Ω m: 68×10^{-8}
Mass magnetic susceptibility/kg^{-1} m^3: $+5 \times 10^{-8}$ (s)
Molar volume/cm^3: 17.78
Coefficient of linear thermal expansion/K^{-1}: n.a.

Lattice structure (cell dimensions/pm), space group

α-Am h.c.p. (a=346.80; c=1124.0), $P6_3/mmc$
β-Am f.c.c. (a=489.4), Fm3m
$T(\alpha \rightarrow \beta)$=1347 K

X-ray diffraction: mass absorption coefficients (μ/ρ)/cm^2 g^{-1}: n.a.

Americium

Discovered in 1944 by G. T. Seaborg, R. A. James, L. O. Morgan and A. Ghiorso at Chicago, USA

[English, *America*]

Nuclear properties

Atomic number: 95

Thermal neutron capture cross-section/barns: 180 ± 20 (^{243}Am)

Number of isotopes (including nuclear isomers): 13

Isotope mass range: $237 \rightarrow 247$

Key isotopes

Nuclide	Atomic mass	Natural abundance (%)	Half life $T_{1/2}$	Decay mode and energy (MeV)	Nuclear spin I	Nuclear magnetic moment μ	Uses
^{241}Am	241.0567	0	458y	$\alpha(5.640)$; γ	5/2	+1.59	tracer; medical
^{243}Am	243.0614	0	7.4×10^3y	$\alpha(5.439)$; γ	5/2	+1.4	NMR

NMR

	^{243}Am
Relative sensitivity ($^1H = 1.00$)	—
Absolute sensitivity ($^1H = 1.00$)	—
Receptivity ($^{13}C = 1.00$)	—
Magnetogyric ratio/rad T^{-1} s^{-1}	1.54×10^7
Quadrupole moment/m^2	4.9×10^{-28}
Frequency ($^1H = 100$ MHz; 2.3488 T)/MHz	5.76

Electron shell properties

Ground state electron configuration: $[Rn]5f^77s^2$

Term symbol: $^8S_{7/2}$

Electron affinity ($M \rightarrow M^-$)/kJ mol^{-1}: n.a.

Ionization energies/kJ mol^{-1}

1. $M \rightarrow M^+$	578.2	6. $M^{5+} \rightarrow M^{6+}$	
2. $M^+ \rightarrow M^{2+}$		7. $M^{6+} \rightarrow M^{7+}$	
3. $M^{2+} \rightarrow M^{3+}$		8. $M^{7+} \rightarrow M^{8+}$	
4. $M^{3+} \rightarrow M^{4+}$		9. $M^{8+} \rightarrow M^{9+}$	
5. $M^{4+} \rightarrow M^{5+}$		10. $M^{9+} \rightarrow M^{10+}$	

Principal lines in atomic spectrum

Wavelength/nm	Species	Sensitivity	Application
351.013	I		
356.919	I		
367.312	I		
428.926	I		
605.464	I		

Abundance: 0

Biological role: Toxic due to radioactivity

Sb

Atomic number: 51

Relative atomic mass ($^{12}C = 12.0000$): **121.75**

Chemical properties

Metalloid element, various allotropes, of which metal is bright, silvery, hard, and brittle. Occurs in nature as Sb_2S_3. Stable in dry air and not attacked by dilute acids or alkalis. Used to harden other metals as alloys.

Radii/pm: Sb^{5+} 62; Sb^{3+} 89; covalent 141; atomic 182; van der Waals 220; Sb^{2-} 245

Electronegativity: 2.05 (Pauling); 1.82 (Allred-Rochow)

Effective nuclear charge: 6.30 (Slater); 9.99 (Clementi); 12.37 (Froese–Fischer)

Standard reduction potentials $E^{\ominus}$/V

	V		IV		III		0		−III
Acid solution	Sb_2O_5	0.605			SbO^+	0.204	Sb		
Neutral solution	Sb_2O_5	1.055	Sb_2O_4	0.342	Sb_4O_6	0.150	Sb	−0.510	SbH_3
	(Sb_2O_5 → Sb_4O_6: 0.699)								
Alkaline solution	$Sb(OH)_6^-$	−0.465			$Sb(OH)_4^-$	−0.639	Sb	−1.338	SbH_3

Covalent bonds r/pm E/kJ mol^{-1}

	r/pm	E/kJ mol^{-1}
Sb–H	170.7	257
Sb–C	220	215
Sb–O	200	314
Sb–F	203	389
Sb–Cl	233	313
Sb–Sb	290	299

Oxidation states

Sb^{-III}	SbH_3
Sb^{III}	Sb_4O_6, SbO_3^{3-} (aq)
	SbF_3, $SbCl_3$, etc.
	$[SbF_5]^{2-}$, Sb_2S_3
Sb^{V}	Sb_4O_{10}, $[Sb(OH)_6]^-$ (aq)
	SbF_5, $SbCl_5$, $[SbCl_6]^-$
	$[SbBr_6]^-$

Physical properties

Melting point/K: 903.89

Boiling point/K: 1908

ΔH_{fusion}/kJ mol^{-1}: 20.9 ΔH_{vap}/kJ mol^{-1}: 165.8

Thermodynamic properties (298.15 K, 0.1 MPa)

State	$\Delta_f H^{\ominus}$/kJ mol^{-1}	$\Delta_f G^{\ominus}$/kJ mol^{-1}	$S^{\ominus}$/J K^{-1} mol^{-1}	C_p/J K^{-1} mol^{-1}
Solid	0	0	45.69	25.23
Gas	262.3	222.1	180.27	20.79

Density/kg m^{-3}: 6691 [293 K]; 6483 [liquid at m.p.]

Thermal conductivity/W m^{-1} K^{-1}: 243 [300 K]

Electrical resistivity/Ω m: 39.0×10^{-8} [273 K]

Mass magnetic susceptibility/kg^{-1} m^3: -1.0×10^{-8} (s)

Molar volume/cm^3: 18.20

Coefficient of linear thermal expansion/K^{-1}: 8.5×10^{-6}

Lattice structure (cell dimensions/pm), space group

Grey rhombohedral ($a = 430.84$; $c = 1124.7$), $R\bar{3}m$
(Grey) cubic ($a = 298.6$), Pm3m
Metal h.c.p. ($a = 336.9$, $c = 533$), $P6_3/mmc$

X-ray diffraction: mass absorption coefficients (μ/ρ)/cm^2 g^{-1}: CuK_α 270 MoK_α 33.1

Antimony

Probably known to the ancients and certainly to the alchemists

[Greek, *anti*+*monos* = not alone; Latin, *stibium*]

Nuclear properties

Atomic number: 51

Thermal neutron capture cross-section/barns: 5 ± 1

Number of isotopes (including nuclear isomers): 29

Isotope mass range: $112 \rightarrow 133$

Key isotopes

Nuclide	Atomic mass	Natural abundance (%)	Half life $T_{1/2}$	Decay mode and energy (MeV)	Nuclear spin I	Nuclear magnetic moment μ	Uses
^{121}Sb	120.9038	57.3	stable		5/2	+3.3592	NMR
^{122}Sb		0	2.80d	$\beta^-(1.972)$; β^+; EC; γ	2	−1.90	tracer
^{123}Sb	122.9041	42.7	stable		7/2	+2.5466	NMR
^{124}Sb		0	60.4d	$\beta^-(2.916)$; γ	3	±1.3	tracer
^{125}Sb		0	2.71y	$\beta^-(0.764)$; γ	7/2	±2.61	tracer

NMR	^{121}Sb	[^{123}Sb]
Relative sensitivity ($^1H = 1.00$)	0.16	4.57×10^{-2}
Absolute sensitivity ($^1H = 1.00$)	9.16×10^{-2}	1.95×10^{-2}
Receptivity ($^{13}C = 1.00$)	520	111
Magnetogyric ratio/rad T^{-1} s^{-1}	6.4016×10^7	3.4668×10^7
Quadrupole moment/m^2	-0.53×10^{-28}	-0.68×10^{-28}
Frequency ($^1H = 100$ MHz; 2.3488 T)/MHz	23.930	12.959

Reference: $Et_4N^+SbCl_6^-$

Electron shell properties

Ground state electron configuration: [Kr] $4d^{10}5s^25p^3$

Term symbol: $^4S_{3/2}$

Electron affinity ($M \rightarrow M^-$)/kJ mol^{-1}: 101

Ionization energies/kJ mol^{-1}

1.	$M \rightarrow M^+$	833.7	6.	$M^{5+} \rightarrow M^{6+}$	10400
2.	$M^+ \rightarrow M^{2+}$	1794	7.	$M^{6+} \rightarrow M^{7+}$	(12700)
3.	$M^{2+} \rightarrow M^{3+}$	2443	8.	$M^{7+} \rightarrow M^{8+}$	(15200)
4.	$M^{3+} \rightarrow M^{4+}$	4260	9.	$M^{8+} \rightarrow M^{9+}$	(17800)
5.	$M^{4+} \rightarrow M^{5+}$	5400	10.	$M^{9+} \rightarrow M^{10+}$	(20400)

Principal lines in atomic spectrum

Wavelength/nm	Species	Sensitivity	Application
206.833	I	U1	AA AE
217.581	I	U2	AA AE
231.147*	I		AA AE
252.852	I		AE
259.805	I		AE

* Lowest energy transition from ground state to nearest empty orbital

Abundance: Earth's crust 0.2 p.p.m.; seawater 0.0005 p.p.m.

Biological role: Toxic; stimulatory

Ar

Atomic number: 18

Relative atomic mass ($^{12}C = 12.0000$): **39.948**

Chemical properties

Colourless, odourless gas comprising 1 per cent of the atmosphere. Obtained from liquid air. Used as inert atmosphere in lamps and high temperature metallurgy.

Radii/pm: atomic 174; van der Waals 191

Electronegativity: n.a. (Pauling); n.a. (Allred-Rochow)

Effective nuclear charge: 6.75 (Slater); 6.76 (Clementi); 7.52 (Froese–Fischer)

Oxidation states

Ar^0	$Ar_8(H_2O)_{46}$ and $Ar(quinol)_3$. These are not true compounds but clathrates in which argon atoms are trapped inside lattice of other molecules

Physical properties

Melting point/K: 83.78

Boiling point/K: 87.29

ΔH_{fusion}/kJ mol^{-1}: 1.21

ΔH_{vap}/kJ mol^{-1}: 6.53

Thermodynamic properties (298.15 K, 0.1 MPa)

State	$\Delta_f H^\ominus$/kJ mol^{-1}	$\Delta_f G^\ominus$/kJ mol^{-1}	$S^\ominus$/J K^{-1} mol^{-1}	C_p/J K^{-1} mol^{-1}
Gas	0	0	154.843	20.786

Density/kg m^{-3}: 1656 [40 K]; 13806 [liquid b.p.], 1.784 [273 K]

Thermal conductivity/W m^{-1} K^{-1}: 0.0177 [300 K]$_g$

Mass magnetic susceptibility/kg^{-1} m^3: -6.16×10^{-9} (g)

Molar volume/cm^3: 24.12 [40 K]

Lattice structure (cell dimensions/pm), space group

f.c.c. (40 K) ($a = 531.088$), Fm3m

X-ray diffraction: mass absorption coefficients (μ/ρ)/cm^2 g^{-1}:

CuK_α 123 MoK_α 13.5

Argon

Discovered in 1894 by Lord Rayleigh and Sir William Ramsay, UK

[Greek, *argos* = inactive]

Nuclear properties

Atomic number: 18

Thermal neutron capture cross-section/barns: 0.650 ± 0.030

Number of isotopes (including nuclear isomers): 8

Isotope mass range: 35→42

Key isotopes

Nuclide	Atomic mass	Natural abundance (%)	Half life $T_{1/2}$	Decay mode and energy (MeV)	Nuclear spin I	Nuclear magnetic moment μ	Uses
^{36}Ar	35.967 55	0.337	stable		0		
^{37}Ar	36.9667	0	35d	EC (0.814); no γ	3/2	+0.95	
^{38}Ar	37.962 72	0.063	stable		0		
^{39}Ar	38.964	0	269y	β^-(3.44); no γ	7/2	−1.3	
^{40}Ar	39.9624	99.600	stable		0		

Electron shell properties

Ground state electron configuration: [Ne] $3s^23p^6$

Term symbol: $^1S_0$

Electron affinity $(M \rightarrow M^-)$/kJ mol^{-1}: −35 (calc.)

Ionization energies/kJ mol^{-1}

1. $M \rightarrow M^+$	1520.4	6. $M^{5+} \rightarrow M^{6+}$	8811
2. $M^+ \rightarrow M^{2+}$	2665.2	7. $M^{6+} \rightarrow M^{7+}$	12 021
3. $M^{2+} \rightarrow M^{3+}$	3928	8. $M^{7+} \rightarrow M^{8+}$	13 844
4. $M^{3+} \rightarrow M^{4+}$	5770	9. $M^{8+} \rightarrow M^{9+}$	40 759
5. $M^{4+} \rightarrow M^{5+}$	7238	10. $M^{9+} \rightarrow M^{10+}$	46 186

Principal lines in atomic spectrum

Wavelength/nm	Species	Sensitivity	Application
106.660* (vac.)	I		
696.543	I	U3	AE
706.722	I	U3	AE
750.387	I	U4	AE
811.531	I	U2	AE

* Lowest energy transition from ground state to nearest empty orbital

Abundance: Atmosphere 9300 p.p.m. (volume); earth's crust 0.04 p.p.m.; sea-water 0.6 p.p.m.

Biological role: None

As

Atomic number: 33

Relative atomic mass (^{12}C=12.0000): **74.9216**

Chemical properties

Metalloid, several allotropes. Grey arsenic is metallic—soft and brittle, tarnishes, burns in O_2, resists attack by water, acids, and alkalis. Attacked by hot acids and molten NaOH. Uses: alloys, semiconductors.

Radii/pm: As^{5+} 46; As^{3+} 69; covalent 121; atomic 125; van der Waals 200

Electronegativity: 2.18 (Pauling); 2.20 (Allred-Rochow)

Effective nuclear charge: 6.30 (Slater); 7.45 (Clementi); 8.98 (Froese–Fischer)

Standard reduction potentials $E^{\ominus}$/V

	V		III		0		−III
Acid solution	H_3AsO_4	0.560	$HAsO_2$	0.240	As	−0.225	AsH_3
Alkaline solution	AsO_4^{3-}	−0.67	AsO_2^-	−0.68	As	1.37	AsH_3

Covalent bonds	r/pm	E/kJ mol^{-1}
As–H	151.9	c. 245
As–C	198	200
As–O	178	477
As–F	171	464
As–Cl	216	293
As–As	244	348

Oxidation states

As^{-III}	AsH_3
As^{III}	As_4O_6, H_3AsO_3, $H_2AsO_3^-$ (aq), AsF_3, $AsCl_3$, etc.
As^{V}	As_4O_{10}, H_3AsO_4, $H_2AsO_4^-$, etc. (aq), $NaAsO_3$, AsF_5

Physical properties

Melting point/K: 1090 (α) under pressure

Boiling point/K: 889 (sublimes)

ΔH_{fusion}/kJ mol^{-1}: 27.7

ΔH_{vap}/kJ mol^{-1}: 31.9

Thermodynamic properties (298.15 K, 0.1 MPa)

State	$\Delta_f H^{\ominus}$/kJ mol^{-1}	$\Delta_f G^{\ominus}$/kJ mol^{-1}	$S^{\ominus}$/J K^{-1} mol^{-1}	C_p/J K^{-1} mol^{-1}
Solid (α)	0	0	35.1	24.64
Gas	302.5	261.0	174.21	20.786

Density/kg m^{-3}: 5780 (α); 4700 (β) [293 K]

Thermal conductivity/W m^{-1} K^{-1}: 50.0 (α) [300 K]

Electrical resistivity/Ω m: 26×10^{-8} [273 K]

Mass magnetic susceptibility/kg^{-1} m^3: -9.17×10^{-10} (α); -3.97×10^{-9} (β)

Molar volume/cm^3: 12.95 (α); 15.9 (β)

Coefficient of linear thermal expansion/K^{-1}: 4.7×10^{-6}

Lattice structure (cell dimensions/pm), space group

α-As rhombohedral (a=413.18; α=54°10′), R$\bar{3}$m, metallic form
β-As hexagonal (a=376.0, c=10.548), yellow
grey amorphous

T(α→β)=501 K
T(β→grey)=room temperature

X-ray diffraction: mass absorption coefficients (μ/ρ)/cm^2 g^{-1}: CuK_α 83.4 MoK_α 69.7

Arsenic

Probably first isolated by Albertus Magnus (1193–1280)

[.Greek, *arsenikon* = yellow orpiment]

Nuclear properties

Atomic number: 33

Thermal neutron capture cross-section/barns: 4.30 ± 0.10

Number of isotopes (including nuclear isomers): 14

Isotope mass range: 69→81

Key isotopes

Nuclide	Atomic mass	Natural abundance (%)	Half life $T_{1/2}$	Decay mode and energy (MeV)	Nuclear spin I	Nuclear magnetic moment μ	Uses
^{73}As			80.3d	EC(0.37); γ			tracer
^{74}As	73.924		17.9d	β^-(1.36); β^+; EC; γ			tracer
^{75}As	74.9216	100	stable		3/2	+1.439	NMR
^{76}As	75.922		26.5h	β^-(2.97); γ	2	−0.905	tracer

NMR

	^{75}As
Relative sensitivity ($^1H = 1.00$)	2.51×10^{-2}
Absolute sensitivity ($^1H = 1.00$)	2.51×10^{-2}
Receptivity ($^{13}C = 1.00$)	143
Magnetogyric ratio/rad T^{-1} s^{-1}	4.5804×10^7
Quadrupole moment/m^2	0.3×10^{-28}
Frequency ($^1H = 100$ MHz; 2.3488 T)/MHz	17.126

Reference: $KAsF_6$

Electron shell properties

Ground state electron configuration: [Ar] $3d^{10}4s^24p^3$

Term symbol: $^4S_{3/2}$

Electron affinity ($M \rightarrow M^-$)/kJ mol^{-1}: 77

Ionization energies/kJ mol^{-1}

1. $M \rightarrow M^+$	947.0	6. $M^{5+} \rightarrow M^{6+}$	12 305
2. $M^+ \rightarrow M^{2+}$	1798	7. $M^{6+} \rightarrow M^{7+}$	(15 400)
3. $M^{2+} \rightarrow M^{3+}$	2735	8. $M^{7+} \rightarrow M^{8+}$	(18 900)
4. $M^{3+} \rightarrow M^{4+}$	4837	9. $M^{8+} \rightarrow M^{9+}$	(22 600)
5. $M^{4+} \rightarrow M^{5+}$	6042	10. $M^{9+} \rightarrow M^{10+}$	(26 400)

Principal lines in atomic spectrum

Wavelength/nm	Species	Sensitivity	Application
193.696	I		AA
197.197*	I		AA
228.812	I	U2	AE
234.984	I	U3	AE
245.653	I	U4	AE

* Lowest energy transition from ground state to nearest empty orbital

Abundance: Earth's crust 1.8 p.p.m.; seawater 0.003 p.p.m.

Biological role: May be essential, but toxic in small doses; stimulatory; suspected carcinogen

At

Atomic number: 85

Relative atomic mass ($^{12}C = 12.0000$): (210)

Chemical properties

Radioactive element obtainable in various ways, e.g. neutron bombardment of ^{200}Bi produces ^{211}At, but not in weighable amounts.

Radii/pm: At^{5+} 57; At^{-} 227

Electronegativity: 2.2 (Pauling); 1.96 (Allred-Rochow)

Effective nuclear charge: 7.60 (Slater); 15.16 (Clementi); 19.61 (Froese–Fischer)

Standard reduction potentials $E^{\ominus}$/V

	V		I		0		−I
Acid solution	$HAtO_3$	1.4	HAtO	0.7	At_2	0.2	At^-
Alkaline solution	AtO_3^-	0.5	AtO^-	0.0	At_2	0.2	At^-

Covalent bonds r/pm E/kJ mol^{-1}

	r/pm	E/kJ mol^{-1}
At–At	*c.* 290 (estimated)	110

Oxidation states

At^{-I} ([Rn])	At^- (aq)
At^{I} (s^2p^4)	$AtBr_2^-$
At^{III} (s^2p^2)	AtO_3^- (aq)

Physical properties

Melting point/K: 575 (est.)

Boiling point/K: 610 (est.)

ΔH_{fusion}/kJ mol^{-1}: 23.8

ΔH_{vap}/kJ mol^{-1}: n.a.

Thermodynamic properties (298.15 K, 0.1 MPa)

State	$\Delta_f H^{\ominus}$/kJ mol^{-1}	$\Delta_f G^{\ominus}$/kJ mol^{-1}	$S^{\ominus}$/J K^{-1} mol^{-1}	C_p/J K^{-1} mol^{-1}
Solid (α)	0	0	n.a.	n.a.
Gas	n.a.	n.a.	n.a.	n.a.

Density/kg m^{-3}: n.a.

Thermal conductivity/W m^{-1} K^{-1}: 1.7 [300 K]

Mass magnetic susceptibility/kg^{-1} m^3: n.a.

Molar volume/cm^3: n.a.

Lattice structure (cell dimensions/pm), space group

n.a.

X-ray diffraction: mass absorption coefficients (μ/ρ)/cm^2 g^{-1}: n.a.

Astatine

Produced by D. R. Corson, K. R. Mackenzie, and E. Segré in 1940 at the University of California, USA

[Greek, *astatos* = unstable]

Nuclear properties

Atomic number: 85

Thermal neutron capture cross-section/barns: n.a.

Number of isotopes (including nuclear isomers): 21

Isotope mass range: 200→219

Key isotopes

Nuclide	Atomic mass	Natural abundance (%)	Half life $T_{1/2}$	Decay mode and energy (MeV)	Nuclear spin I	Nuclear magnetic moment μ	Uses
^{210}At		0	8.3h	EC(3.87 calc.); α(5.63) < 1%			
^{211}At	210.9875	0	7.21h	EC(0.97 calc.) 59%; α(5.981) 41%	9/2		

Electron shell properties

Ground state electron configuration: [Xe] $4f^{14}5d^{10}6s^26p^5$

Term symbol: $^3P_{3/2}$

Electron affinity ($M \rightarrow M^-$)/kJ mol^{-1}: 270

Ionization energies/kJ mol^{-1}

1. $M \rightarrow M^+$	930		6. $M^{5+} \rightarrow M^{6+}$	(7 500)
2. $M^+ \rightarrow M^{2+}$	1600		7. $M^{6+} \rightarrow M^{7+}$	(8 800)
3. $M^{2+} \rightarrow M^{3+}$	(2900)		8. $M^{7+} \rightarrow M^{8+}$	(13 300)
4. $M^{3+} \rightarrow M^{4+}$	(4000)		9. $M^{8+} \rightarrow M^{9+}$	(15 400)
5. $M^{4+} \rightarrow M^{5+}$	(4900)		10. $M^{9+} \rightarrow M^{10+}$	(17 700)

Principal lines in atomic spectrum

Wavelength/nm	Species	Sensitivity	Application
216.225	I		
224.401	I		

Abundance: Trace amount in some minerals

Biological role: None; toxic due to radioactivity

Ba

Atomic number: 56

Relative atomic mass ($^{12}C = 12.0000$): **137.327**

Chemical properties

Relatively soft, silvery-white metal. Obtained from BaO on heating with aluminium. Attacked by air and water. Principal ore is barytes, $BaSO_4$.

Radii/pm: Ba^{2+} 143; atomic 217.3; covalent 198

Electronegativity: 0.89 (Pauling); 0.97 (Allred-Rochow)

Effective nuclear charge: 2.85 (Slater); 7.58 (Clementi); 10.27 (Froese–Fischer)

Standard reduction potentials $E^{\ominus}$/V

IV		II		0		−II
			−1.110 (Ba²⁺ → BaH₂)			
BaO_2 (hyd.)	2.365 (→ Ba^{2+})	Ba^{2+}	−2.92	Ba	−0.685	BaH_2
	1.626 (→ BaO)	BaO (hyd.)	−2.166	Ba		
			−0.741 (BaO → BaH₂)			

Oxidation states

Ba II ([Xe])	BaH_2, BaO, $Ba(OH)_2$ basic, BaO_2 (peroxide), Ba^{2+}(aq), BaF_2, $BaCl_2$, etc., $BaCO_3$, $BaSO_4$ insoluble, many other salts

Physical properties

Melting point/K: 1002

Boiling point/K: 1910

ΔH_{fusion}/kJ mol^{-1}: 7.66

ΔH_{vap}/kJ mol^{-1}: 150.9

Thermodynamic properties (298.15 K, 0.1 MPa)

State	$\Delta_f H^{\ominus}$/kJ mol^{-1}	$\Delta_f G^{\ominus}$/kJ mol^{-1}	$S^{\ominus}$/J K^{-1} mol^{-1}	C_p/J K^{-1} mol^{-1}
Solid	0	0	62.8	28.07
Gas	180	146	170.243	20.786

Density/kg m^{-3}: 3594 [293 K]; 3325 [liquid at m.p.]

Thermal conductivity/W m^{-1} K^{-1}: 18.4 [300 K]

Electrical resistivity/Ω m: 50×10^{-8} [273 K]

Mass magnetic susceptibility/kg^{-1} m^3: $+1.9 \times 10^{-9}$ (s)

Molar volume/cm^3: 38.21

Coefficient of linear thermal expansion/K^{-1}: $18.1–21.0 \times 10^{-6}$

Lattice structure (cell dimensions/pm), space group

b.c.c. ($a = 502.5$), Im3m

High pressure form: ($a = 390.1$, $c = 615.5$), $P6_3mmc$

X-ray diffraction: mass absorption coefficients (μ/ρ)/cm^2 g^{-1}: CuK_α 330 MoK_α 43.5

Barium

Isolated in 1808 by Sir Humphrey Davy at London, UK

[Greek, *barys* = heavy]

Nuclear properties

Atomic number: 56

Thermal neutron capture cross-section/barns: 1.2±0.1

Number of isotopes (including nuclear isomers): 25

Isotope mass range: 123→143

Key isotopes

Nuclide	Atomic mass	Natural abundance (%)	Half life $T_{1/2}$	Decay mode and energy (MeV)	Nuclear spin I	Nuclear magnetic moment μ	Uses
^{130}Ba	129.9062	0.106	stable				
^{132}Ba	131.9057	0.101	stable				
^{133}Ba		0	7.2y	EC(0.488); γ			tracer
^{134}Ba	133.9043	2.417	stable		0		
^{135}Ba	134.9056	6.592	stable		3/2	+0.8365	NMR
^{135m}Ba		0	28.7h	IT(0.268); γ			tracer
^{136}Ba	135.9044	7.854	stable		0		
^{137}Ba		11.32	stable		3/2	+0.9357	NMR
^{138}Ba	137.9050	71.70	stable		0		
^{140}Ba		0	12.80d	β^-(1.05); γ			tracer

NMR	[^{135}Ba]	^{137}Ba
Relative sensitivity (^1H=1.00)	4.90×10^{-4}	7.76×10^{-4}
Absolute sensitivity (^1H=1.00)	3.22×10^{-4}	7.76×10^{-4}
Receptivity (^{13}C=1.00)	1.83	4.41
Magnetogyric ratio/rad T^{-1} s^{-1}	2.6575×10^7	2.9728×10^7
Quadrupole moment/m^2	0.18×10^{-28}	0.28
Frequency (^1H=100 MHz; 2.3488 T)/MHz	9.934	11.113

Reference: $BaCl_2$ (aq)

Electron shell properties

Ground state electron configuration: [Xe]$6s^2$

Term symbol: $^1S_0$

Electron affinity ($M \rightarrow M^-$)/kJ mol^{-1}: −46

Ionization energies/kJ mol^{-1}

1. $M \rightarrow M^+$	502.8	6. $M^{5+} \rightarrow M^{6+}$	(7700)
2. $M^+ \rightarrow M^{2+}$	965.1	7. $M^{6+} \rightarrow M^{7+}$	(9000)
3. $M^{2+} \rightarrow M^{3+}$	(3600)	8. $M^{7+} \rightarrow M^{8+}$	(10200)
4. $M^{3+} \rightarrow M^{4+}$	(4700)	9. $M^{8+} \rightarrow M^{9+}$	(13500)
5. $M^{4+} \rightarrow M^{5+}$	(6000)	10. $M^{9+} \rightarrow M^{10+}$	(15100)

Principal lines in atomic spectrum

Wavelength/nm	Species	Sensitivity	Application
350.111	I		AA
455.403	II	V1	AA AE
493.409	II	V2	AA AE
553.548	I	U1	AA AE
577.767	I	U2	AA AE
791.134*	I		

* Lowest energy transition from ground state to nearest empty orbital

Abundance: Earth's crust 390 p.p.m.; seawater 0.03 p.p.m.

Biological role: Toxic; stimulatory

Bk

Atomic number: 97
Relative atomic mass (^{12}C=12.0000): (247)

Chemical properties

Radioactive silvery metal, made in mg quantities as ^{294}Bk by neutron bombardment of ^{239}Pu. Attacked by oxygen, steam, and acids, but not by alkalis

Radii/pm: Bk^{2+} 118; Bk^{3+} 98; Bk^{4+} 87
Electronegativity: 1.3 (Pauling); n.a. (Allred-Rochow)
Effective nuclear charge: 1.65 (Slater)

Standard reduction potentials $E^{\ominus}$/V

	III		IV		0
Acid solution	Bk^{4+}	1.67	Bk^{3+}	−2.01	Bk

(Bk^{4+} → Bk: −1.05)

Oxidation states

Bk (f^9)	BkO
Bk^{III} (f^8)	Bk_2O_3, BkF_3, $BkCl_3$ etc., $[BkCl_6]^{3-}$, Bk^{3+} (aq)
Bk^{IV} (f^7)	BkO_2, BkF_4

Physical properties

Melting point/K: n.a.
Boiling point/K: n.a.
ΔH_{fusion}/kJ mol^{-1}: n.a.
ΔH_{vap}/kJ mol^{-1}: n.a.

Thermodynamic properties (298.15 K, 0.1 MPa)

State	$\Delta_f H^{\ominus}$/kJ mol^{-1}	$\Delta_f G^{\ominus}$/kJ mol^{-1}	$S^{\ominus}$/J K^{-1} mol^{-1}	C_p/J K^{-1} mol^{-1}
Solid	0	0	n.a.	n.a.
Gas	n.a.	n.a.	n.a.	n.a.

Density/kg m^{-3}: 14 790 [293 K]
Thermal conductivity/W m^{-1} K^{-1}: 10 (est.) [300 K]
Electrical resistivity/Ω m: n.a.
Mass magnetic susceptibility/kg^{-1} m^3: n.a.
Molar volume/cm^3: 16.70
Coefficient of linear thermal expansion/K^{-1}: n.a.

Lattice structure (cell dimensions/pm), space group

n.a.

X-ray diffraction: mass absorption coefficients (μ/ρ)/cm^2 g^{-1}:
n.a.

Berkelium

Produced in December 1949 by S. G. Thompson, A. Ghiorso, and G. T. Seaborg at Berkeley, California, USA

[English, *Berkeley*]

Nuclear properties

Atomic number: 97

Thermal neutron capture cross-section/barns: 1000 ± 500 (^{249}Bk)

Number of isotopes (including nuclear isomers): 8

Isotope mass range: 243→250

Key isotopes

Nuclide	Atomic mass	Natural abundance (%)	Half life $T_{1/2}$	Decay mode and energy (MeV)	Nuclear spin I	Nuclear magnetic moment μ	Uses
^{247}Bk	247.0702	0	1.4 × 10^{3}y	α(5.86); γ			
^{249}Bk		0	314d	β^-	7/2		

Electron shell properties

Ground state electron configuration: [Rn]5f^{9}7s^2

Term symbol: $^6H_{15/2}$

Electron affinity (M→M$^-$)/kJ mol^{-1}: n.a.

Ionization energies/kJ mol^{-1}

1. M → M^+	601	6. M^{5+} → M^{6+}
2. M^+ → M^{2+}		7. M^{6+} → M^{7+}
3. M^{2+} → M^{3+}		8. M^{7+} → M^{8+}
4. M^{3+} → M^{4+}		9. M^{8+} → M^{9+}
5. M^{4+} → M^{5+}		10. M^{9+} → M^{10+}

Selected lines in atomic spectrum

Wavelength/nm	Species	Sensitivity	Application
329.972	I		
325.219	I		
328.875	I		
329.935	I		
333.526	I		
340.828	I		
342.695	I		

Abundance: Nil

Biological role: Toxic due to radioactivity

Be

Atomic number: 4
Relative atomic mass ($^{12}C = 12.0000$): **9.012 18**

Chemical properties

Silvery-white, lustrous, relatively soft metal, obtained e.g. by the electrolysis of fused $BeCl_2$. Unaffected by air or water. Used in alloys with copper and nickel. Occurs as beryl $Be_3Al_2Si_6O_{18}$.

Radii/pm: Be^{2+} 34; atomic 113.3; covalent 89
Electronegativity: 1.57 (Pauling); 1.47 (Allred-Rochow)
Effective nuclear charge: 1.95 (Slater); 1.91 (Clementi); 2.27 (Froese–Fischer)

Standard reduction potentials $E^{\ominus}/V$

II		0
Be^{2+}	$\xrightarrow{-1.97}$	Be

Covalent bonds r/pm E/kJ mol^{-1}

Bond	r/pm	E/kJ mol^{-1}
Be–H	163	226
Be–O	133	523
Be–F	143	615
Be–Cl	177	293
Be–C	193	
Be–Be	222.6	

Oxidation states

Be^{II}	BeO, $Be(OH)_2$, BeH_2, $[Be(H_2O)_4]^{2+}$aq, BeF_2, $BeCl_2$ etc., $BeCO_3$, salts

Physical properties

Melting point/K: 1551±5
Boiling point/K: 3243 (under pressure)
ΔH_{fusion}/kJ mol^{-1}: 9.80
ΔH_{vap}/kJ mol^{-1}: 308.8

Thermodynamic properties (298.15 K, 0.1 MPa)

State	$\Delta_f H^{\ominus}$/kJ mol^{-1}	$\Delta_f G^{\ominus}$/kJ mol^{-1}	$S^{\ominus}$/J K^{-1} mol^{-1}	C_p/J K^{-1} mol^{-1}
Solid	0	0	9.50	16.44
Gas	324.6	286.6	136.269	20.786

Density/kg m^{-3}: 1847.7 [293 K]
Thermal conductivity/W m^{-1} K^{-1}: 200 [300 K]
Electrical resistivity/Ω m: 4.0×10^{-8} [293 K]
Mass magnetic susceptibility/kg^{-1} m^3: -1.3×10^{-8} (s)
Molar volume/cm^3: 4.88
Coefficient of linear thermal expansion/K^{-1}: 11.5×10^{-6}

Lattice structure (cell dimensions/pm), space group

α-Be h.c.p. (a=228.55, c=358.32), P6_3/mmc
β-Be b.c.c. (a=255.15), Im3m

$T(\alpha \rightarrow \beta) = 1523$ K

X-ray diffraction: mass absorption coefficients (μ/ρ)/cm^2 g^{-1}:
CuK_α 1.50 MoK_α 0.298

Contents

Discovered in 1797 by N.-L. Vauquelin. Isolated in 1828 by F. Wöhler at Berlin, Germany and independently by A. A. B. Bussy at Paris, France

[Greek, *beryllos* = beryl]

Beryllium

Nuclear properties

Atomic number: 4

Thermal neutron capture cross-section/barns: 0.0092 ± 0.0005

Number of isotopes (including nuclear isomers): 6

Isotope mass range: $6 \rightarrow 11$

Key isotopes

Nuclide	Atomic mass	Natural abundance (%)	Half life $T_{1/2}$	Decay mode and energy (MeV)	Nuclear spin I	Nuclear magnetic moment μ	Uses
^{7}Be	7.0169	0	53.37d	EC(0.862); γ			tracer
^{9}Be	9.012 18	100	stable		3/2	−1.177 45	NMR
^{10}Be	10.0135	0	2.5×10^{6}y	β^-(0.555); no γ			

NMR

	^{9}Be
Relative sensitivity ($^{1}H = 1.00$)	1.39×10^{-2}
Absolute sensitivity ($^{1}H = 1.00$)	1.39×10^{-2}
Receptivity ($^{13}C = 1.00$)	78.8
Magnetogyric ratio/rad T^{-1} s^{-1}	3.7589×10^{7}
Quadrupole moment/m^2	5.2×10^{-30}
Frequency ($^{1}H = 100$ MHz; 2.3488 T)/MHz	14.053

Reference: $Be(NO_3)_2$ (aq)

Electron shell properties

Ground state electron configuration: $[He]2s^2$

Term symbol: $^{1}S_0$

Electron affinity ($M \rightarrow M^-$)/kJ mol^{-1}: −18

Ionization energies/kJ mol^{-1}

1. $M \rightarrow M^+$	899.4
2. $M^+ \rightarrow M^{2+}$	1 757.1
3. $M^{2+} \rightarrow M^{3+}$	14 848
4. $M^{3+} \rightarrow M^{4+}$	21 006

Principal lines in atomic spectrum

Wavelength/nm	Species	Sensitivity	Application
234.861	I	U1	AA AE
332.134	I	U2	AE
332.109	I	U3	AE
332.101	I	U4	AE
467.342	II		

Abundance: Earth's crust 2 p.p.m.; seawater 5×10^{-5} p.p.m.

Biological role: Toxic; carcinogenic

Bi

Atomic number: 83
Relative atomic mass ($^{12}C = 12.0000$): 208.9804

Chemical properties

Brittle metal, silvery lustre with pink tinge. Stable to oxygen and water. Dissolves in concentrated nitric acid. Basic oxide. Used in alloys and medicines.

Radii/pm: Bi^{5+} 74; Bi^{3+} 96; covalent 152; atomic 155; van der Waals 240

Electronegativity: 2.02 (Pauling); 1.67 (Allred-Rochow)

Effective nuclear charge: 6.30 (Slater); 13.34 (Clementi); 16.90 (Froese–Fischer)

Standard reduction potentials $E^{\ominus}$/V

$$\overset{\text{V}}{Bi^{5+}} \xrightarrow{\text{ca. 2}} \overset{\text{III}}{Bi^{3+}} \xrightarrow{0.317} \overset{0}{Bi} \xrightarrow{-0.97} \overset{-\text{III}}{BiH_3}$$

Covalent bonds

	r/pm	E/kJ mol^{-1}
Bi–H	n.a.	194
Bi–C	230	143
Bi–O	232	339
Bi–F	235	314
Bi–Cl	248	285
Bi–Bi	309	200

Oxidation states

Bi^{-III}	BiH_3
Bi^{I}	Bi^+, cation clusters Bi_3^+, Bi_5^{3+}, Bi_9^{5+} etc.
Bi^{III}	Bi_2O_3, $Bi(OH)_3$, Bi^{3+}(aq), BiOCl, BiF_3, $BiCl_3$ etc. $[BiBr_6]^{3-}$, salts
Bi^{V}	Bi_2O_5 unstable, $[Bi(OH)_6]^-$(aq), $NaBiO_3$, BiF_5, $KBiF_6$

Physical properties

Melting point/K: 544.5
Boiling point/K: 1833±5
ΔH_{fusion}/kJ mol^{-1}: 10.48
ΔH_{vap}/kJ mol^{-1}: 179.1

Thermodynamic properties (298.15 K, 0.1 MPa)

State	$\Delta_f H^{\ominus}$/kJ mol^{-1}	$\Delta_f G^{\ominus}$/kJ mol^{-1}	$S^{\ominus}$/J K^{-1} mol^{-1}	C_p/J K^{-1} mol^{-1}
Solid	0	0	56.74	25.52
Gas	207.1	168.2	187.005	20.786

Density/kg m^{-3}: 9747 [293 K]; 10050 [liquid at m.p.]
Thermal conductivity/W m^{-1} K^{-1}: 7.87 [300 K]
Electrical resistivity/Ω m: 106.8×10^{-8} [273 K]
Mass magnetic susceptibility/kg^{-1} m^3: -1.684×10^{-8} (s)
Molar volume/cm^3: 21.44
Coefficient of linear thermal expansion/K^{-1}: 13.4×10^{-6}

Lattice structure (cell dimensions/pm), space group

Rhombohedral (a=454.950, c=1186.225), R$\bar{3}$m

X-ray diffraction: mass absorption coefficients (μ/ρ)/cm^2 g^{-1}:
CuK$_\alpha$ 240 MoK$_\alpha$ 120

Known in the fifteenth century, discoverer unknown

[German, *bisemutum*]

Bismuth

Nuclear properties

Atomic number: 83

Thermal neutron capture cross-section/barns: 0.034

Number of isotopes (including nuclear isomers): 19

Isotope mass range: 199→215

Key isotopes

Nuclide	Atomic mass	Natural abundance (%)	Half life $T_{1/2}$	Decay mode and energy (MeV)	Nuclear spin I	Nuclear magnetic moment μ	Uses
^{206}Bi		0	6.3d	EC(0.98); γ	6	+4.56	tracer
^{207}Bi		0	30.2y	EC(2.40); γ			
^{209}Bi	208.9804	100	stable		9/2	+4.080	NMR
^{210m}Bi		0	3×10^6y	α(4.96); β^-; γ			
^{210}Bi		trace	5.01d	β^-(1.16); α; no γ	1	−0.0442	tracer
^{212}Bi		trace	60.6m	β^-(2.25); α(6.206); γ			
^{214}Bi		trace	19.7m	β^-(3.28); α(5.616); γ			

NMR	^{209}Bi
Relative sensitivity ($^1H=1.00$)	0.13
Absolute sensitivity ($^1H=1.00$)	0.13
Receptivity ($^{13}C=1.00$)	777
Magnetogyric ratio/rad T^{-1} s^{-1}	4.2986×10^7
Quadrupole moment/m^2	-0.4×10^{-28}
Frequency ($^1H=100$ MHz; 2.3488 T)/MHz	16.069
Reference: $KBiF_6$	

Electron shell properties

Ground state electron configuration: [Xe]$4f^{14}5d^{10}6s^26p^3$

Term symbol: $^4S_{3/2}$

Electron affinity ($M\rightarrow M^-$)/kJ mol^{-1}: 101

Ionization energies/kJ mol^{-1}

1. $M\rightarrow M^+$	703.2	6. $M^{5+}\rightarrow M^{6+}$	8 520
2. $M^+\rightarrow M^{2+}$	1610	7. $M^{6+}\rightarrow M^{7+}$	(10 300)
3. $M^{2+}\rightarrow M^{3+}$	2466	8. $M^{7+}\rightarrow M^{8+}$	(12 300)
4. $M^{3+}\rightarrow M^{4+}$	4372	9. $M^{8+}\rightarrow M^{9+}$	(14 300)
5. $M^{4+}\rightarrow M^{5+}$	5400	10. $M^{9+}\rightarrow M^{10+}$	(16 300)

Principal lines in atomic spectrum

Wavelength/nm	Species	Sensitivity	Application
202.121	I		AA
206.170	I		AA AE
222.825	I		AA
289.798	I	U2	AA AE
306.772	I	U1	AA AE

Abundance: Earth's crust 0.008 p.p.m.; seawater 0.0002 p.p.m.

Biological role: None; non-toxic

B

Atomic number: 5
Relative atomic mass (^{12}C=12.0000): **10.81**

Chemical properties

Occurs as large deposits of borax in US and Turkey. Amorphous boron is dark powder unreactive to oxygen, water, acids, and alkalis. Forms metal borides with most metals. Used in glass, bleaches and fireproofing.

Radii/pm: B^{3+} 23; covalent 88; atomic 83; van der Waals 208

Electronegativity: 2.04 (Pauling); 2.01 (Allred-Rochow)

Effective nuclear charge: 2.60 (Slater); 2.42 (Clementi); 2.27 (Froese–Fischer)

Standard reduction potentials $E^{\ominus}$/V

III		0
$B(OH)_3$	−0.890	B
BF_4^-	−1.284	B

Covalent bonds	r/pm	E/kJ mol^{-1}
B–H	119	381
B–H–B	132	439
B–C	156	372
B–O	136	523
B–F	129	644
B–Cl	174	444
B–B	175	335

Oxidation states

B^{III}	B_2O_3, H_3BO_3 (=$B(OH)_3$), borates e.g. borax $Na_2[B_4O_5(OH)_4].8H_2O$, B_2H_6, B_4H_{10} etc., $NaBH_4$, BF_3, BCl_3 etc.

Physical properties

Melting point/K: 2573
Boiling point/K: 3931
ΔH_{fusion}/kJ mol^{-1}: 22.2
ΔH_{vap}/kJ mol^{-1}: 504.5

Thermodynamic properties (298.15 K, 0.1 MPa)

State	$\Delta_f H^{\ominus}$/kJ mol^{-1}	$\Delta_f G^{\ominus}$/kJ mol^{-1}	$S^{\ominus}$/J K^{-1} mol^{-1}	C_p/J K^{-1} mol^{-1}
Solid (α)	0	0	5.86	11.09
Gas	562.7	518.8	153.45	20.799

Density/kg m^{-3}: 2340 (β-rhomb.) [293 K]
Thermal conductivity/W m^{-1} K^{-1}: 27.0 [300 K]
Electrical resistivity/Ω m: 18 000 [273 K]
Mass magnetic susceptibility/kg^{-1} m^3: -7.8×10^{-9} (s)
Molar volume/cm^3: 4.62
Coefficient of linear thermal expansion/K^{-1}: 5×10^{-6}

Lattice structure (cell dimensions/pm), space group

Tetragonal (a=874.0; c=506), $P4_2/nnm$
α-B rhombohedral (a=506.7, α=58°4′), $R\bar{3}m$
β-B rhombohedral (a=1014.5, α=65°12′), $R\bar{3}m$, R32, R3m
Rhombic (a=1015, b=895, c=1790)
Monoclinic (a=1013, b=893, c=1786, α≃90°, β≃90°, γ≃90°) or triclinic
Hexagonal (a=1198, c=954)

X-ray diffraction: mass absorption coefficients (μ/ρ)/cm^2 g^{-1}:
CuK_α 2.39 MoK_α 0.392

Discovered in 1808 by L. J. Lussac and L. J. Thenard in Paris, France and Sir Humphrey Davy in London, UK

[Arabic, *buraq*]

Boron

Nuclear properties

Atomic number: 5

Thermal neutron capture cross-section/barns: 3837 ^{10}B; 0.005 ^{11}B

Number of isotopes (including nuclear isomers): 6

Isotope mass range: 8→13

Key isotopes

Nuclide	Atomic mass	Natural abundance (%)	Half life $T_{1/2}$	Decay mode and energy (MeV)	Nuclear spin I	Nuclear magnetic moment μ	Uses
^{10}B	10.0129	20.0	stable		3	+1.8006	NMR
^{11}B	11.009 31	80.0	stable		3/2	+2.6885	NMR

NMR

	[^{10}B]	^{11}B
Relative sensitivity (^1H=1.00)	1.99×10^{-2}	0.17
Absolute sensitivity (^1H=1.00)	3.90×10^{-3}	0.13
Receptivity (^{13}C=1.00)	22.1	754
Magnetogyric ratio/rad T^{-1} s^{-1}	2.8740×10^7	8.5794×10^7
Quadrupole moment/m^2	7.4×10^{-30}	3.55×10^{-30}
Frequency (^1H=100 MHz; 2.3488 T)/MHz	10.746	32.084

Reference: Et_2O/BF_3

Electron shell properties

Ground state electron configuration: [He]$2s^2 2p^1$

Term symbol: $^2P_{1/2}$

Electron affinity ($M \rightarrow M^-$)/kJ mol^{-1}: 23

Ionization energies/kJ mol^{-1}

1. $M \rightarrow M^+$ 800.6
2. $M^+ \rightarrow M^{2+}$ 2 427
3. $M^{2+} \rightarrow M^{3+}$ 3 660
4. $M^{3+} \rightarrow M^{4+}$ 25 025
5. $M^{4+} \rightarrow M^{5+}$ 32 822

Principal lines in atomic spectrum

Wavelength/nm	Species	Sensitivity	Application
249.678*	I	U2	AA AE
249.773	I	U1	AA AE
345.129	II	V2	AE
1116.004	I		

* Lowest energy transition from ground state to nearest empty orbital

Abundance: Earth's crust 9 p.p.m.; seawater 4.8 p.p.m.

Biological role: Essential for plants: toxic in excess

Br

Atomic number: 35
Relative atomic mass (^{12}C=12.0000): **79.904**

Chemical properties

Deep red, dense, sharp smelling liquid, Br_2, obtained from natural brines or seawater by treatment with Cl_2. Compounds used in fuel additives, pesticides, flame-retardants, and photography.

Radii/pm: covalent 114.2; Br^- 196; van der Waals 195
Electronegativity: 2.96 (Pauling); 2.74 (Allred-Rochow)
Effective nuclear charge: 7.60 (Slater); 9.03 (Clementi); 10.89 (Froese–Fischer)

Standard reduction potentials E°/V

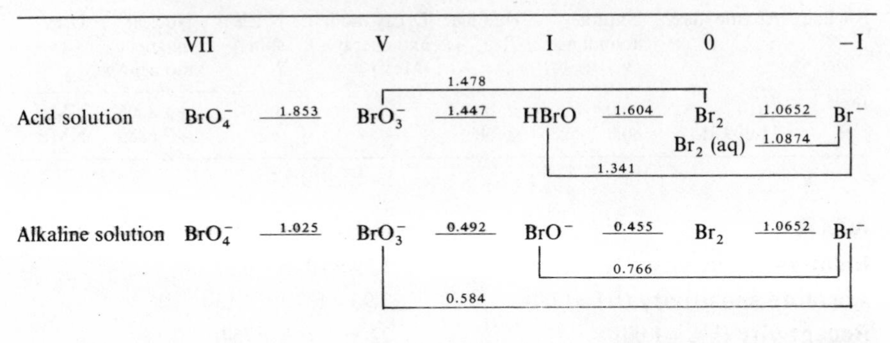

Covalent bonds

	r/pm	E/kJ mol^{-1}
Br–H	140.8	366
Br–C	194	285
Br–O	160	234
Br–F	176	285
Br–Br	229	193
Br–B	187	410
Br–Si	215	310
Br–P	218	264

Oxidation states

Br^{-I} ([Kr])	Br^-(aq), HBr, KBr etc.
Br^{I} (s^2p^4)	Br_2O, $BrCl_2^-$
Br^{III} (s^2p^2)	BrF_3, BrF_4^-
Br^{IV} (s^2p^1)	BrO_2
Br^{V} (s^2)	BrO_3^-(aq), BrF_5, BrF_6^-
Br^{VII} (d^{10})	$KBrO_4$, BrF_6^+

Physical properties

Melting point/K: 265.9
Boiling point/K: 331.93
ΔH_{fusion}/kJ mol^{-1}: 10.8
ΔH_{vap}/kJ mol^{-1}: 30.5

Thermodynamic properties (298.15 K, 0.1 MPa)

State	$\Delta_f H^\circ$/kJ mol^{-1}	$\Delta_f G^\circ$/kJ mol^{-1}	S°/J K^{-1} mol^{-1}	C_p/J K^{-1} mol^{-1}
Liquid	0	0	152.231	75.689
Gas (atom)	111.884	82.396	175.022	20.786

Density/kg m^{-3}: 4050 [123 K]; 3122.6 [293 K]; 7.59 [gas]
Thermal conductivity/W m^{-1} K^{-1}: 0.122 [300 K]$_l$
Mass magnetic susceptibility/kg^{-1} m^3: -4.44×10^{-9} (l)
Molar volume/cm^3: 19.73 [123 K]

Lattice structure (cell dimensions/pm), space group

Orthorhombic (120 K) (a=673.7; b=454.8; c=876.1), Cmca

X-ray diffraction: mass absorption coefficients (μ/ρ)/cm^2 g^{-1}:
CuK$_\alpha$ 99.6 MoK$_\alpha$ 79.8

Discovered in 1826 by A. J. Balard at Montpellier, France and C. Löwig in Germany

[Greek, *bromos* = stench]

Bromine

Nuclear properties

Atomic number: 35

Thermal neutron capture cross-section/barns: 6.8 ± 0.1

Number of isotopes (including nuclear isomers): 19

Isotope mass range: 74→90

Key isotopes

Nuclide	Atomic mass	Natural abundance (%)	Half life $T_{1/2}$	Decay mode and energy (MeV)	Nuclear spin I	Nuclear magnetic moment μ	Uses
^{77}Br		0	57h	β^+, EC(1.365); γ	3/2		tracer
^{79}Br	79.9183	50.69	stable		3/2	+2.1055	NMR
^{81}Br	80.9163	49.31	stable		3/2	+2.2696	NMR
^{82}Br	81.917	0	35.5h	β^-(3.092); γ	5	±1.626	tracer

NMR

	[^{79}Br]	^{81}Br
Relative sensitivity (^1H = 1.00)	7.86×10^{-2}	9.85×10^{-2}
Absolute sensitivity (^1H = 1.00)	3.97×10^{-2}	4.87×10^{-2}
Receptivity (^{13}C = 1.00)	226	277
Magnetogyric ratio/rad $T^{-1} s^{-1}$	6.7023×10^7	7.2246×10^7
Quadrupole moment/m^2	0.33×10^{-28}	0.28×10^{-28}
Frequency (^1H = 100 MHz; 2.3488 T)/MHz	25.053	27.006

Reference: NaBr(aq)

Electron shell properties

Ground state electron configuration: $[Ar]3d^{10}4s^24p^5$

Term symbol: $^2P_{3/2}$

Electron affinity ($M \rightarrow M^-$)/kJ mol^{-1}: 324.5

Ionization energies/kJ mol^{-1}

1. $M \rightarrow M^+$	1139.9	6. $M^{5+} \rightarrow M^{6+}$	8 550
2. $M^+ \rightarrow M^{2+}$	2104	7. $M^{6+} \rightarrow M^{7+}$	9 940
3. $M^{2+} \rightarrow M^{3+}$	3500	8. $M^{7+} \rightarrow M^{8+}$	18 600
4. $M^{3+} \rightarrow M^{4+}$	4560	9. $M^{8+} \rightarrow M^{9+}$	(23 900)
5. $M^{4+} \rightarrow M^{5+}$	5760	10. $M^{9+} \rightarrow M^{10+}$	(28 100)

Principal lines in atomic spectrum

Wavelength/nm	Species	Sensitivity	Application
154.065* (vac)	I		AE
470.486	II	V1	AE
478.550	II	V2	AE
481.671	II	V3	AE
635.073	I		

* Lowest energy transition from ground state to nearest empty orbital

Abundance: Earth's crust 2.5 p.p.m.; seawater 65 p.p.m.

Biological role: None; Br^- slightly toxic; Br_2 very toxic

Cd

Atomic number: 48

Relative atomic mass (^{12}C=12.0000): 112.411

Chemical properties

Silvery metal obtained as by-product from zone refining. Tarnishes in air, soluble in acids but not alkalis. Some used in rechargeable batteries and alloys.

Radii/pm: Cd^{2+} 103; Cd^{+} 114; covalent 141; atomic 148.9

Electronegativity: 1.69 (Pauling); 1.46 (Allred-Rochow)

Effective nuclear charge: 4.35 (Slater); 8.19 (Clementi); 11.58 (Froese–Fischer)

Standard reduction potentials $E^{\ominus}$/V

	II		0
Acid solution	Cd^{2+}	$\xrightarrow{-0.4025}$	Cd
Alkaline solution	$Cd(OH)_2$	$\xrightarrow{-0.824}$	Cd
	$[Cd(NH_3)_4]^{2+}$	$\xrightarrow{-0.622}$	Cd
	$[Cd(CN)_4]^{2-}$	$\xrightarrow{-1.09}$	Cd

Oxidation states

Cd^{I} ($d^{10}s^1$)	rare $Cd_2[AlCl_4]_2$
Cd^{II} (d^{10})	CdO (basic), CdS, $Cd(OH)_2$, CdF_2, $CdCl_2$ etc., many salts, $[Cd(H_2O)_6]^{2+}$ (aq), many complexes, e.g. $[Cd(SCN)_4]^{2-}$

Physical properties

Melting point/K: 594.1

Boiling point/K: 1038

ΔH_{fusion}/kJ mol^{-1}: 6.11

ΔH_{vap}/kJ mol^{-1}: 100.0

Thermodynamic properties (298.15 K, 0.1 MPa)

State	$\Delta_f H^{\ominus}$/kJ mol^{-1}	$\Delta_f G^{\ominus}$/kJ mol^{-1}	$S^{\ominus}$/J K^{-1} mol^{-1}	C_p/J K^{-1} mol^{-1}
Solid	0	0	51.76	25.98
Gas	112.01	77.41	167.746	20.786

Density/kg m^{-3}: 8650 [293 K]; 7996 [liquid at m.p.]

Thermal conductivity/W m^{-1} K^{-1}: 96.8 [300 K]

Electrical resistivity/Ω m: 6.83×10^{-8} [273 K]

Mass magnetic susceptibility/kg^{-1} m^3: -2.21×10^{-9} (s)

Molar volume/cm^3: 13.00

Coefficient of linear thermal expansion/K^{-1}: 29.8×10^{-6}

Lattice structure (cell dimensions/pm), space group

h.c.p. (a=297.94; c=561.86), $P6_3/mmc$

X-ray diffraction: mass absorption coefficients (μ/ρ)/cm^2 g^{-1}:
CuK_α 231 MoK_α 27.5

Discovered in 1817 by F. Stromeyer at Göttingen, Germany

[Latin, *cadmia* = calomine]

Cadmium

Nuclear properties

Atomic number: 48

Thermal neutron capture cross-section/barns: 2450 ± 20

Number of isotopes (including nuclear isomers): 22

Isotope mass range: 103→119

Key isotopes

Nuclide	Atomic mass	Natural abundance (%)	Half life $T_{1/2}$	Decay mode and energy (MeV)	Nuclear spin I	Nuclear magnetic moment μ	Uses
^{106}Cd	105.907	1.25	stable				
^{108}Cd	107.9040	0.89	stable		0		
^{109}Cd		0	450d	EC(0.16); γ	5/2	−0.8270	tracer
^{110}Cd	109.9030	12.51	stable		0		
^{111}Cd	110.9042	12.81	stable		1/2	−0.5943	NMR
^{112}Cd	111.9028	24.13	stable		0		
^{113}Cd	112.9046	12.22	stable		1/2	−0.6217	NMR
^{114}Cd	113.9036	28.72	stable		0		
^{115m}Cd		0	43d	β^-(1.62); γ	11/2	−1.0400	tracer
^{115}Cd		0	53.5h	β^-(1.45); γ	1/2	−0.648	tracer
^{116}Cd	115.9050	7.47	stable		0		

NMR

	[^{111}Cd]	^{113}Cd
Relative sensitivity (^1H = 1.00)	9.54×10^{-3}	1.09×10^{-2}
Absolute sensitivity (^1H = 1.00)	1.21×10^{-3}	1.33×10^{-3}
Receptivity (^{13}C = 1.00)	6.93	7.6
Magnetogyric ratio/rad T^{-1} s^{-1}	-5.6714×10^7	-5.9328×10^7
Frequency (^1H = 100 MHz; 2.3488 T)/MHz	21.205	22.182

References: $Cd(ClO_4)_2$ (aq) and $Cd(CH_3)_2$

Electron shell properties

Ground state electron configuration: [Kr]$4d^{10}5s^2$

Term symbol: $^1S_0$

Electron affinity ($M \rightarrow M^-$)/kJ mol^{-1}: −26

Ionization energies/kJ mol^{-1}

1. $M \rightarrow M^+$	867.6	6. $M^{5+} \rightarrow M^{6+}$	(9 100)
2. $M^+ \rightarrow M^{2+}$	1631	7. $M^{6+} \rightarrow M^{7+}$	(11 100)
3. $M^{2+} \rightarrow M^{3+}$	3616	8. $M^{7+} \rightarrow M^{8+}$	(14 100)
4. $M^{3+} \rightarrow M^{4+}$	(5300)	9. $M^{8+} \rightarrow M^{9+}$	(16 400)
5. $M^{4+} \rightarrow M^{5+}$	(7000)	10. $M^{9+} \rightarrow M^{10+}$	(18 800)

Principal lines in atomic spectrum

Wavelength/nm	Species	Sensitivity	Application
214.441	II	V1	AE
226.502	II	V2	AE
228.802	I	U1	AA, AE
326.106*	I		AA, AE
643.847	I		AE

* Lowest energy transition from ground state to nearest empty orbital

Abundance: Earth's crust 0.16 p.p.m.; seawater 0.0001 p.p.m.

Biological role: Toxic; stimulatory; carcinogenic; teratogenic

Cs

Atomic number: 55
Relative atomic mass ($^{12}C=12.0000$): **132.9054**

Chemical properties

Silvery-white, very soft metal; reacts rapidly with oxygen and explosively with water. Found as pollucite $Cs_4Al_4Si_9O_{26}.H_2O$ but most obtained as by-product of lithium production. Little used.

Radii/pm: Cs^+ 165; atomic 265.4; covalent 235; van der Waals 262

Electronegativity: 0.79 (Pauling); 0.86 (Allred-Rochow)

Effective nuclear charge: 2.20 (Slater); 6.36 (Clementi); 8.56 (Froese–Fischer)

Standard reduction potentials $E^\ominus$/V

I		0
Cs^+	−2.923	Cs

Oxidation states

Cs^{-I} (s^2)	caesium metal in liquid ammonia
Cs^I ([Xe])	Cs_2O, Cs_2O_2, CsO_2, CsOH, CsH, CsF, CsCl etc., $[Cs(H_2O)_x]^+$(aq), Cs_2CO_3, many salts and salt hydrates, some complexes with crown ethers etc.

Physical properties

Melting point/K: 301.55
Boiling point/K: 951.6
ΔH_{fusion}/kJ mol^{-1}: 2.09
ΔH_{vap}/kJ mol^{-1}: 66.5

Thermodynamic properties (298.15 K, 0.1 MPa)

State	$\Delta_f H^\ominus$/kJ mol^{-1}	$\Delta_f G^\ominus$/kJ mol^{-1}	$S^\ominus$/J K^{-1} mol^{-1}	C_p/J K^{-1} mol^{-1}
Solid	0	0	85.23	32.17
Gas	76.065	49.121	175.595	20.786

Density/kg m^{-3}: 1873 [293 K]; 1843 [liquid at m.p.]
Thermal conductivity/W m^{-1} K^{-1}: 35.9 [300 K]
Electrical resistivity/Ω m: 20.0×10^{-8} [293 K]
Mass magnetic susceptibility/kg^{-1} m^3: $+2.8 \times 10^{-9}$ (s)
Molar volume/cm^3: 70.96
Coefficient of linear thermal expansion/K^{-1}: 97×10^{-6}

Lattice structure (cell dimensions/pm), space group

b.c.c. (78 K) ($a=614$), Im3m

High pressure forms: ($a=598.4$), Fm3m
($a=580.0$), Fm3m

X-ray diffraction: mass absorption coefficients (μ/ρ)/cm^2 g^{-1}:
CuK_α 318 MoK_α 41.3

Discovered by R. Bunsen and G. R. Kirchoff in 1860 at Heidelberg, Germany

[Latin, *caesius* = sky blue]

Caesium (Cesium)

Nuclear properties

Atomic number: 55

Thermal neutron capture cross-section/barns: 30.0 ± 1.5

Number of isotopes (including nuclear isomers): 22

Isotope mass range: 123 → 144

Key isotopes

Nuclide	Atomic mass	Natural abundance (%)	Half life $T_{1/2}$	Decay mode and energy (MeV)	Nuclear spin I	Nuclear magnetic moment μ	Uses
^{133}Cs	132.9051	100	stable		7/2	+2.5779	NMR
^{134}Cs		0	2.046y	β^-(2.062); γ	4	+2.989	tracer
^{135}Cs		0	3×10^6y	β^-(0.210); no γ	7/2	+2.7280	
^{137}Cs		0	30.23y	β^-(1.176); γ	7/2	+2.8372	tracer, medical

NMR

	^{133}Cs
Relative sensitivity (^{1}H = 1.00)	4.74×10^{-2}
Absolute sensitivity (^{1}H = 1.00)	4.74×10^{-2}
Receptivity (^{13}C = 1.00)	269
Magnetogyric ratio/rad T^{-1} s^{-1}	3.5087×10^7
Quadrupole moment/m^2	-3×10^{-31}
Frequency (^{1}H = 100 MHz; 2.3488 T)/MHz	13.117

Reference: 0.5M CsBr(aq)

Electron shell properties

Ground state electron configuration: [Xe]$6s^1$

Term symbol: $^2S_{1/2}$

Electron affinity ($M \rightarrow M^-$)/kJ mol^{-1}: 45.5

Ionization energies/kJ mol^{-1}

1. $M \rightarrow M^+$	375.7	6. $M^{5+} \rightarrow M^{6+}$	(7100)
2. $M^+ \rightarrow M^{2+}$	2420	7. $M^{6+} \rightarrow M^{7+}$	(8300)
3. $M^{2+} \rightarrow M^{3+}$	(3400	8. $M^{7+} \rightarrow M^{8+}$	(11300)
4. $M^{3+} \rightarrow M^{4+}$	(4400)	9. $M^{8+} \rightarrow M^{9+}$	(12700)
5. $M^{4+} \rightarrow M^{5+}$	(6000)	10. $M^{9+} \rightarrow M^{10+}$	(23700)

Principal lines in atomic spectrum

Wavelength/nm	Species	Sensitivity	Application
455.528	I	U3	AA AE
459.317	I	U4	AA AE
460.376	II		
852.110	I	U1	AA AE
894.350*	I	U2	AA AE

* Lowest energy transition from ground state to nearest empty orbital

Abundance: Earth's crust 2.6 p.p.m.; seawater 0.0005 p.p.m.

Biological role: None; non-toxic

Ca

Atomic number: 20

Relative atomic mass ($^{12}C=12.0000$): **40.078**

Chemical properties

Silvery white, relatively soft metal obtained from fused calcium chloride by electrolysis. Protected by oxide/nitride film and can be worked as a metal. Attacked by oxygen and water. Used in alloys.

Radii/pm: Ca^{2+} 106; covalent 174; atom (α form) 197.3

Electronegativity: 1.00 (Pauling); 1.04. (Allred-Rochow)

Effective nuclear charge: 2.85 (Slater); 4.40 (Clementi); 5.69 (Froese–Fischer)

Standard reduction potentials $E^\ominus$/V

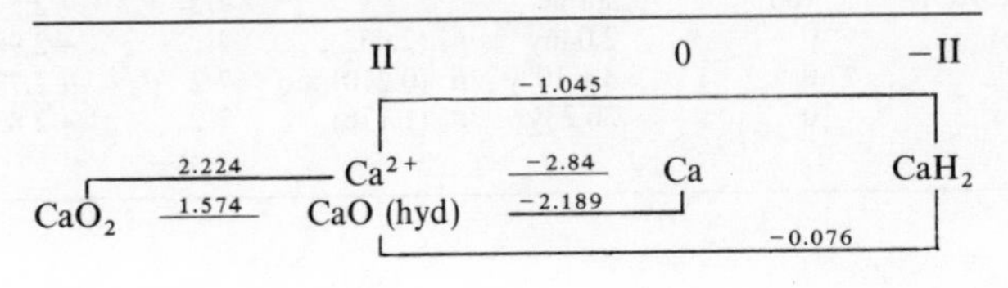

Oxidation states

Ca^{II} ([Ar])	CaO, CaO_2 (peroxide), $Ca(OH)_2$, CaH_2, CaF_2, $CaCl_2$ etc., Ca^{2+}(aq), $CaCO_3$, $CaSO_4.2H_2O$ (gypsum) $CaSO_4.\frac{1}{2}H_2O$ (plaster of Paris) CaC_2 (calcium carbide), many salts, few complexes

Physical properties

Melting point/K: 1112

Boiling point/K: 1757

ΔH_{fusion}/kJ mol^{-1}: 9.33

ΔH_{vap}/kJ mol^{-1}: 150.6

Thermodynamic properties (298.15 K, 0.1 MPa)

State	$\Delta_f H^\ominus$/kJ mol^{-1}	$\Delta_f G^\ominus$/kJ mol^{-1}	$S^\ominus$/J K^{-1} mol^{-1}	C_p/J K^{-1} mol^{-1}
Solid (α)	0	0	41.42	25.31
Gas	178.2	144.3	154.884	20.786

Density/kg m^{-3}: 1550 [293 K]; 1365 [liquid at m.p.]

Thermal conductivity/W m^{-1} K^{-1}: 200 [300 K]

Electrical resistivity/Ω m: 3.43×10^{-8} [293 K]

Mass magnetic susceptibility/kg^{-1} m^3: $+1.4\times10^{-8}$ (s)

Molar volume/cm^3: 25.86

Coefficient of linear thermal expansion/K^{-1}: 22×10^{-6}

Lattice structure (cell dimensions/pm), space group

α-Ca f.c.c. ($a=558.84$), Fm3m
β-Ca f.c.c. ($a=448.0$), Im3m
γ-Ca h.c.p. ($a=397$, $c=649$), $P6_3/mmc$, [may contain H]

$T(\alpha\rightarrow\beta)=573$ K
$T(\beta\rightarrow\gamma)=723$ K

X-ray diffraction: mass absorption coefficients (μ/ρ)/cm^2 g^{-1}:
CuK_α 162 MoK_α 18.3

Isolated in 1808 by Sir Humphry Davy in London, UK

[Latin, *calx* = lime]

Calcium

Nuclear properties

Atomic number: 20

Thermal neutron capture cross-section/barns: 0.44 ± 0.02

Number of isotopes (including nuclear isomers): 14

Isotope mass range: $37 \rightarrow 50$

Key isotopes

Nuclide	Atomic mass	Natural abundance (%)	Half life $T_{1/2}$	Decay mode and energy (MeV)	Nuclear spin I	Nuclear magnetic moment μ	Uses
^{40}Ca	39.962 59	96.941	stable		0		
^{42}Ca	41.958 63	0.647	stable				
^{43}Ca	42.958 78	0.135	stable		7/2	−1.3172	NMR
^{44}Ca	43.955 49	2.086	stable				
^{45}Ca	44.956	0	165d	β^-(0.252); no γ			tracer
^{46}Ca	45.9537	0.004	stable				
^{47}Ca	46.954	0	4.53d	β^-(1.979); γ			tracer
^{48}Ca	47.9524	0.187	stable				

NMR

	^{43}Ca
Relative sensitivity (^{1}H = 1.00)	6.40×10^{-3}
Absolute sensitivity (^{1}H = 1.00)	9.28×10^{-6}
Receptivity (^{13}C = 1.00)	0.0527
Magnetogyric ratio/rad T^{-1} s^{-1}	-1.8001×10^7
Quadrupole moment/m^2	-0.05×10^{-28}
Frequency (^{1}H = 100 MHz; 2.3488 T)/MHz	6.728

Reference: $CaCl_2$(aq)

Electron shell properties

Ground state electron configuration: [Ar]$4s^2$

Term symbol: $^1S_0$

Electron affinity ($M \rightarrow M^-$)/kJ mol^{-1}: −186

Ionization energies/kJ mol^{-1}

1.	$M \rightarrow M^+$	589.7	6.	$M^{5+} \rightarrow M^{6+}$	10 496
2.	$M^+ \rightarrow M^{2+}$	1145	7.	$M^{6+} \rightarrow M^{7+}$	12 320
3.	$M^{2+} \rightarrow M^{3+}$	4910	8.	$M^{7+} \rightarrow M^{8+}$	14 207
4.	$M^{3+} \rightarrow M^{4+}$	6474	9.	$M^{8+} \rightarrow M^{9+}$	18 191
5.	$M^{4+} \rightarrow M^{5+}$	8144	10.	$M^{9+} \rightarrow M^{10+}$	20 385

Principal lines in atomic spectrum

Wavelength/nm	Species	Sensitivity	Application
239.856	I		AA
393.366	II	V1	AE
422.673	I	U1	AA AE
443.496	I	U3	AE
445.478	I	U2	AE
657.278*	I		

* Lowest energy transition from ground state to nearest empty orbital

Abundance: Earth's crust 46 600 p.p.m. (limestone and dolomite deposits); seawater 400 p.p.m.

Biological role: Essential constituent of cells, fluids, and bones

Cf

Atomic number: 98
Relative atomic mass (^{12}C=12.0000): (251)

Chemical properties

Radioactive, silvery metal obtained in mg quantities as ^{249}Cf and ^{252}Cf from neutron bombardment of ^{239}Pu. Attacked by oxygen, steam, and acids, but not alkalis. ^{252}Cf used in cancer therapy.

Radii/pm: Cf^{4+} 86; Cf^{3+} 98; Cf^{2+} 117
Electronegativity: 1.3 (Pauling); n.a. (Allred-Rochow)
Effective nuclear charge: 1.65 (Slater)

Standard reduction potentials $E^{\ominus}$/V

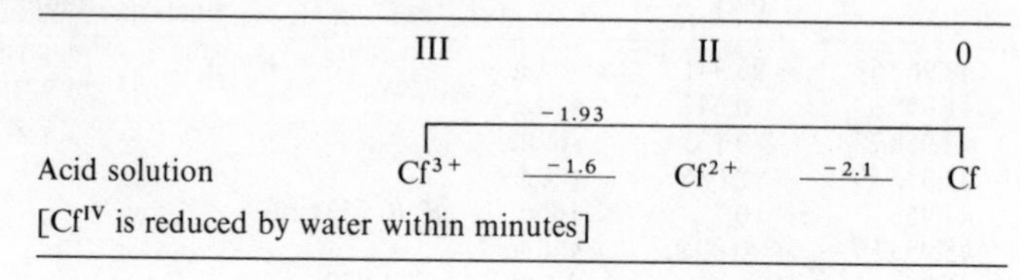

III
II
0
−1.93
Acid solution
Cf^{3+}
−1.6
Cf^{2+}
−2.1
Cf

[Cf^{IV} is reduced by water within minutes]

Oxidation states

Cf^{II} (f^{10})	CfO ?, $CfBr_2$, CfI_2
Cf^{III} (f^9)	Cf_2O_3, CfF_3, $CfCl_3$ etc. Cf^{3+}(aq), $[Cf(C_5H_5)_3]$
Cf^{IV} (f^8)	CfO_2, CfF_4

Physical properties

Melting point/K: n.a.
Boiling point/K: n.a.
ΔH_{fusion}/kJ mol^{-1}: n.a.
ΔH_{vap}/kJ mol^{-1}: n.a.

Thermodynamic properties (298.15 K, 0.1 MPa)

State	$\Delta_f H^{\ominus}$/kJ mol^{-1}	$\Delta_f G^{\ominus}$/kJ mol^{-1}	$S^{\ominus}$/J K^{-1} mol^{-1}	C_p/J K^{-1} mol^{-1}
Solid	0	0	n.a.	25.98
Gas	n.a.	n.a.	n.a.	n.a.

Density/kg m^{-3}: n.a.
Thermal conductivity/W m^{-1} K^{-1}: 10 (est.) [300 K]
Electrical resistivity/Ω m: n.a.
Mass magnetic susceptibility/kg^{-1} m^3: n.a.
Molar volume/cm^3: n.a.
Coefficient of linear thermal expansion/K^{-1}: n.a.

Lattice structure (cell dimensions/pm), space group

Cubic

X-ray diffraction: mass absorption coefficients (μ/ρ)/cm^2 g^{-1}:
n.a.

Produced in 1950 by S. G. Thompson, K. Street Jr, A. Ghiorso, and G. T. Seaborg at Berkeley, California, USA

[English, *California*]

Californium

Nuclear properties

Atomic number: 98

Thermal neutron capture cross-section/barns: 2100 ± 1000 (^{251}Cf)

Number of isotopes (including nuclear isomers): 12

Isotope mass range: 242→254

Key isotopes

Nuclide	Atomic mass	Natural abundance (%)	Half life $T_{1/2}$	Decay mode and energy (MeV)	Nuclear spin I	Nuclear magnetic moment μ	Uses
^{249}Cf	249.0748	0	360y	α(6.295)			
^{251}Cf		0	900y	α(5.94)			
^{252}Cf		0	2.65y	α(6.217); SF			tracer, medical

Electron shell properties

Ground state electron configuration: $[Rn]5f^{10}7s^2$

Term symbol: $^5I_8$

Electron affinity ($M \rightarrow M^-$)/kJ mol^{-1}: n.a.

Ionization energies/kJ mol^{-1}

1. $M \rightarrow M^+$	608	6. $M^{5+} \rightarrow M^{6+}$	
2. $M^+ \rightarrow M^{2+}$		7. $M^{6+} \rightarrow M^{7+}$	
3. $M^{2+} \rightarrow M^{3+}$		8. $M^{7+} \rightarrow M^{8+}$	
4. $M^{3+} \rightarrow M^{4+}$		9. $M^{8+} \rightarrow M^{9+}$	
5. $M^{4+} \rightarrow M^{5+}$		10. $M^{9+} \rightarrow M^{10+}$	

Selected lines in atomic spectrum

Wavelength/nm	Species	Sensitivity	Application
339.222	I		
353.149	I		
354.098	I		
359.877	I		
360.532	I		
361.749	I		
366.270	I		

Abundance: Nil

Biological role: Toxic due to radioactivity

C

Atomic number: 6

Relative atomic mass ($^{12}C=12.0000$): **12.011**

Chemical properties

Found naturally as element (graphite and diamond) but mainly as hydrocarbons (methane gas, oil, and coal) and carbonates (limestone and dolomite). Used as coke (steel making), carbon black (printing), and activated charcoal (sugar refining, etc.).

Radii/pm: covalent bond 77; double 67; triple 60; C^{4-} 260; van der Waals 185

Electronegativity: 2.55 (Pauling); 2.50 (Allred-Rochow)

Effective nuclear charge: 3.25 (Slater); 3.14 (Clementi); 2.87 (Froese–Fischer)

Standard reduction potentials $E^{\ominus}$/V

	IV		II		0		−IV		
Acid solution	CO_2	−0.106	CO	0.517	C	0.132	CH_4		
	CO_2	−0.20	HCO_2H	0.034	HCHO	0.232	CH_3OH	0.59	CH_4
Alkaline solution	CO_2	−1.01	HCO_2^-	−1.07	HCHO	0.59	CH_3OH	−0.2	CH_4

Covalent bonds*

	r/pm	E/kJ mol^{-1}
C–H	109.3	339
C–C	154	348
C=C	134	614
C≡C	120	839
C=N	130	615
C≡N	116	891
C=O	123	745
C≡O	112.8	1074

*See also other elements for bonds to carbon

Oxidation states

This concept is rarely used in discussing carbon and its compounds because of their subtleties of bonding. However, for simple compounds with a single carbon we can use it:

C^{-IV} CH_4
C^{II} CO
C^{IV} CO_2, CO_3^{2-}, CF_4, etc.

Physical properties

Melting point/K: *c.* 3820 (diamond)

Boiling point/K: 5100 (sublimes)

ΔH_{fusion}/kJ mol^{-1}: 105.0

ΔH_{vap}/kJ mol^{-1}: 710.9

Thermodynamic properties (298.15 K, 0.1 MPa)

State	$\Delta_f H^{\ominus}$/kJ mol^{-1}	$\Delta_f G^{\ominus}$/kJ mol^{-1}	$S^{\ominus}$/J K^{-1} mol^{-1}	C_p/J K^{-1} mol^{-1}
Solid (graphite)	0	0	5.740	8.527
Solid (diamond)	1.895	2.900	2.377	6.113
Gas	716.682	671.257	158.096	20.838

Density/kg m^{-3}: 3513 (diam.); 2260 (graph.) [293 K]

Thermal conductivity/W m^{-1} K^{-1}: 990–2320 (diam.); $5.7^{\perp}$, $1960^{\parallel}$ (graph.) [298 K]

Electrical resistivity/Ω m: 10^{11} (diam.); 1.375×10^{-5} (graph.) [293 K]

Mass magnetic susceptibility/kg^{-1} m^3: -6.3×10^{-9} (graph.); -6.2×10^{-9} (diam.)

Molar volume/cm^3: 3.42 (diam.)

Coefficient of linear thermal expansion/K^{-1}: 1.19×10^{-6} (diam.)

Lattice structure (cell dimensions/pm), space group

Cubic diamond ($a=356.703$), Fd3m
Hexagonal graphite ($a=246.12$, $c=670.78$), $P6_3mc$
Rhombohedral graphite ($a=364.2$, $\alpha=39°30'$), $R\bar{3}m$
Hexagonal diamond ($a=252$, $c=412$), $P6_3/mmc$
Hexagonal ($a=894.8$, $c=1408$)

X-ray diffraction: mass absorption coefficients (μ/ρ)/cm^2 g^{-1}:
CuK_α 4.60 MoK_α 0.625

Occurs naturally as graphite (and diamond); known to prehistoric humans

[Latin, *carbo*=charcoal]

Carbon

Nuclear properties

Atomic number: 6
Thermal neutron capture cross-section/barns: 0.0034 ± 0.0002
Number of isotopes (including nuclear isomers): 7
Isotope mass range: $9 \rightarrow 16$

Key isotopes

Nuclide	Atomic mass	Natural abundance (%)	Half life $T_{1/2}$	Decay mode and energy (MeV)	Nuclear spin I	Nuclear magnetic moment μ	Uses
^{12}C	12.0000	98.90	stable		0		
^{13}C	13.003 35	1.10	stable		1/2	+0.7024	NMR
^{14}C	14.0032	trace	5730y	β^-(0.156); no γ	0		tracer

NMR

	^{13}C
Relative sensitivity (^1H=1.00)	1.59×10^{-2}
Absolute sensitivity (^1H=1.00)	1.76×10^{-4}
Receptivity (^{13}C=1.00)	1.00 (by definition)
Magnetogyric ratio/rad T^{-1} s^{-1}	6.7263×10^7
Frequency (^1H=100 MHz; 2.3488 T)/MHz	25.144

Reference: $Si(CH_3)_4$

Electron shell properties

Ground state electron configuration: $[He]2s^22p^2$
Term symbol: $^3P_0$
Electron affinity ($M \rightarrow M^-$)/kJ mol^{-1}: 122.5

Ionization energies/kJ mol^{-1}

1. $M \rightarrow M^+$ 1086.2
2. $M^+ \rightarrow M^{2+}$ 2352
3. $M^{2+} \rightarrow M^{3+}$ 4620
4. $M^{3+} \rightarrow M^{4+}$ 6222
5. $M^{4+} \rightarrow M^{5+}$ 37827
6. $M^{5+} \rightarrow M^{6+}$ 47270

Principal lines in atomic spectrum

Wavelength/nm	Species	Sensitivity	Application
247.856	I	U2	AE
283.671	II	V4	AE
283.760	II	V5	AE
296.485*	I		
426.700	II	V3	AE
426.726	II	V2	AE

* Lowest energy transition from ground state to nearest empty orbital

Abundance: Atmosphere CO_2 335 p.p.m., CH_4 1.7 p.p.m.; CO up to 0.02 p.p.m. (all by volume); Earth's crust 180 p.p.m.; seawater 28 p.p.m.

Biological role: Basis of all life (part of DNA molecule)

Ce

Atomic number: 58

Relative atomic mass (^{12}C=12.0000): 140.115

Chemical properties

Reactive, grey metal; most abundant of the lanthanide metals. Obtained from bastnaesite and monazite. Tarnishes in air, burns when heated, reacts rapidly with water, dissolves in acids. Used in glass, flints, ceramics.

Radii/pm: Ce^{3+} 107; Ce^{4+} 94; atomic 182.5; covalent 165

Electronegativity: 1.12 (Pauling); 1.06 (Allred-Rochow)

Effective nuclear charge: 2.85 (Slater); 10.80 (Clementi); 10.57 (Froese–Fischer)

Standard reduction potentials $E^{\ominus}$/V

	IV		III		0
Acid solution	Ce^{4+}	1.72	Ce^{3+}	−2.34	Ce
Alkaline solution	CeO_2	−0.7	$Ce(OH)_3$	−2.78	Ce

Oxidation states

Ce^{III} ($4f^1$)	Ce_2O_3, $Ce(OH)_3$, CeF_3, $CeCl_3$ etc., Ce^{3+} salts, $[Ce(H_2O)_6]^{3+}$ (aq), complexes
Ce^{IV} ([Xe])	CeO_2, CeF_4, $CaCl_6^{2-}$, $[Ce(NO_3)_6]^{2-}$ (aq)

Physical properties

Melting point/K: 1072

Boiling point/K: 3699

ΔH_{fusion}/kJ mol^{-1}: 8.87

ΔH_{vap}/kJ mol^{-1}: 398

Thermodynamic properties (298.15 K, 0.1 MPa)

State	$\Delta_f H^{\ominus}$/kJ mol^{-1}	$\Delta_f G^{\ominus}$/kJ mol^{-1}	$S^{\ominus}$/J K^{-1} mol^{-1}	C_p/J K^{-1} mol^{-1}
Solid	0	0	72.0	26.94
Gas	423	385	191.776	23.075

Density/kg m^{-3}: 8240 (α); 6749 (β); 6773 (γ); 6700 (δ) [298 K]

Thermal conductivity/W m^{-1} K^{-1}: 11.4 [300 K]

Electrical resistivity/Ω m: 73×10^{-8} [273 K]

Mass magnetic susceptibility/kg^{-1} m^3: $+2.17 \times 10^{-7}$ (s)

Molar volume/cm^3: 17.00

Coefficient of linear thermal expansion/K^{-1}: 8.5×10^{-6}

Lattice structure (cell dimensions/pm), space group

α-Ce f.c.c. (a=485), Fm3m
β-Ce hexagonal (a=367.3, c=1180.2), $P6_3/mmc$
γ-Ce f.c.c. (a=516.01), Fm3m
δ-Ce f.c.c. (a=412), Im3m
$T(\beta \rightarrow \gamma)$=441 K

X-ray diffraction: mass absorption coefficients (μ/ρ)/cm^2 g^{-1}:
CuK_α 352 MoK_α 48.2

Discovered in 1803 by J. J. Berzelius and W. Hisinger at Vestmanland, Sweden. First isolated by W. F. Hillebrand and T. H. Norton in 1875 at Washington, DC, USA
[Named after asteroid Ceres, discovered in 1801]

Cerium

Nuclear properties

Atomic number: 58
Thermal neutron capture cross-section/barns: 0.73±0.08
Number of isotopes (including nuclear isomers): 19
Isotope mass range: 132→148

Key isotopes

Nuclide	Atomic mass	Natural abundance (%)	Half life $T_{1/2}$	Decay mode and energy (MeV)	Nuclear spin I	Nuclear magnetic moment μ	Uses
^{136}Ce		0.19	stable				
^{138}Ce	137.9057	0.25	stable				
^{139}Ce		0	140d	EC(0.27); γ	3/2	±0.9	tracer
^{140}Ce	139.9053	88.48	stable				
^{141}Ce		0	32.5d	β^- (0.581); γ	7/2	±0.97	tracer
^{142}Ce	141.9090	11.08	stable				
^{143}Ce		0	33h		3/2	*c.* ±1	tracer
^{144}Ce		0	284.9h	β^- (0.31); γ			

Electron shell properties

Ground state electron configuration: [Xe]$4f^2 6s^2$
Term symbol: $^3H_4$
Electron affinity ($M \rightarrow M^-$)/kJ mol^{-1}: ≤50

Ionization energies/kJ mol^{-1}

1. $M \rightarrow M^+$	527.4	6. $M^{5+} \rightarrow M^{6+}$	(8 200)
2. $M^+ \rightarrow M^{2+}$	1047	7. $M^{6+} \rightarrow M^{7+}$	(9 700)
3. $M^{2+} \rightarrow M^{3+}$	1949	8. $M^{7+} \rightarrow M^{8+}$	(11 800)
4. $M^{3+} \rightarrow M^{4+}$	3547	9. $M^{8+} \rightarrow M^{9+}$	(13 200)
5. $M^{4+} \rightarrow M^{5+}$	(6800)	10. $M^{9+} \rightarrow M^{10+}$	(14 700)

Principal lines in atomic spectrum

Wavelength/nm	Species	Sensitivity	Application
399.924	II		AE
418.660	II		AE
522.349	I		AA
569.699	I		AA
739.777*	I		

* Lowest energy transition from ground state to nearest empty orbital

Abundance: Earth's crust 66 p.p.m.; seawater 0.0004 p.p.m.
Biological role: None; stimulatory

Cl

Atomic number: 17

Relative atomic mass (^{12}C=12.0000): 35.4527

Chemical properties

Yellow-green, dense, sharp smelling gas, Cl_2; produced on large scale by electrolysis of sodium chloride. Used as bleaching agent and to make organochlorine solvents and polymers (PVC)

Radii/pm: Cl^- 181; covalent 99; van der Waals 181

Electronegativity: 3.16 (Pauling); 2.83 (Allred-Rochow)

Effective nuclear charge: 6.10 (Slater); 6.12 (Clementi); 6.79 (Froese–Fischer)

Standard reduction potentials $E^{\ominus}$/V

VII V III I 0 −I

Acid solution

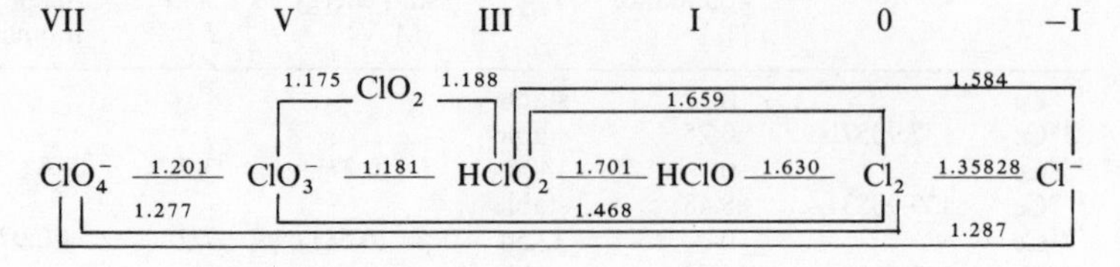

Alkaline solution

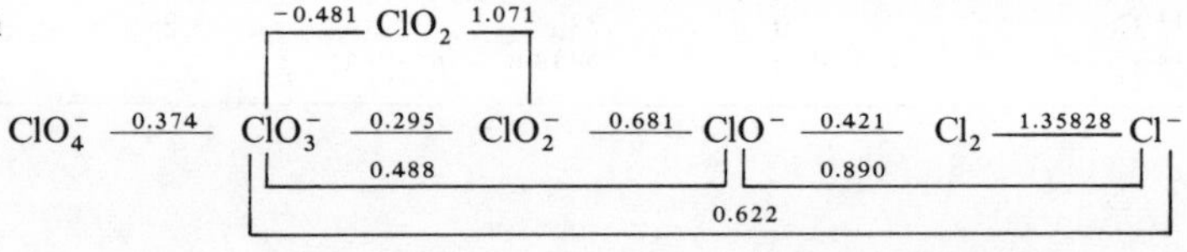

Covalent bonds r/pm E/kJ mol^{-1}

Bond	r/pm	E/kJ mol^{-1}
Cl–O	170	206
Cl–Cl	198.8	242
Cl–F	163	257

For other bonds see other elements

Oxidation states

Cl^{-I}	Cl^-(aq), HCl, NaCl etc
Cl^{I}	Cl_2O, HOCl, salts, ClO^-(aq), ClF
Cl^{III}	$NaClO_2$, ClF_3
Cl^{IV}	ClO_2
Cl^{V}	$HClO_3$, salts, ClO_3^- (aq), ClF_5, F_3ClO
Cl^{VI}	Cl_2O_6
Cl^{VII}	Cl_2O_7, $HClO_4$, salts, ClO_4^- (aq), $FClO_3$

Physical properties

Melting point/K: 172.17

Boiling point/K: 239.18 **Triple point**/K: 172.17

ΔH_{fusion}/kJ mol^{-1}: 6.41 ΔH_{vap}/kJ mol^{-1}: 20.4033

Thermodynamic properties (298.15 K, 0.1 MPa)

State	$\Delta_f H^{\ominus}$/kJ mol^{-1}	$\Delta_f G^{\ominus}$/kJ mol^{-1}	$S^{\ominus}$/J K^{-1} mol^{-1}	C_p/J K^{-1} mol^{-1}
Gas (Cl_2)	0	0	223.066	33.907
Gas (atoms)	121.679	105.680	165.198	21.840

Density/kg m^{-3}: 2030 [113 K]; 1507 [239 K]; 3.214 [273 K]

Thermal conductivity/W m^{-1} K^{-1}: 0.0089 [300 K]$_g$

Critical temperature/K: 416.9 **Critical pressure**/kPa: 7977

Mass magnetic susceptibility/kg^{-1} m^3: -7.2×10^{-9} (g)

Molar volume/cm^3: 17.46 [113 K]

Coefficient of linear thermal expansion/K^{-1}: n.a.

Lattice structure (cell dimensions/pm), space group

Tetragonal (a=856; c=612), P4/ncm
Orthorhombic (a=624, b=448, c=826), Cmca

T (tetragonal →orthorhombic)=100 K

X-ray diffraction: mass absorption coefficients (μ/ρ)/cm^2 g^{-1}:
CuK_α 106 MoK_α 11.4

Discovered in 1774 by C. W. Scheele at Uppsala, Sweden

[Greek, *chloros* = pale green]

Chlorine

Nuclear properties

Atomic number: 17

Thermal neutron capture cross-section/barns: 44±2 (^{35}Cl); 0.403±0.1 (^{37}Cl)

Number of isotopes (including nuclear isomers): 11

Isotope mass range: 32→40

Key isotopes

Nuclide	Atomic mass	Natural abundance (%)	Half life $T_{1/2}$	Decay mode and energy (MeV)	Nuclear spin I	Nuclear magnetic moment μ	Uses
^{35}Cl	34.968 85	75.77	stable		3/2	+0.821 81	NMR
^{36}Cl	35.9797	0	3.1×10^5y	β^-(0.712); β^+, EC; (2%); no γ	2	+1.2853	tracer
^{37}Cl	36.9658	24.23	stable		3/2	+0.684 07	NMR

NMR

	^{35}Cl	^{37}Cl
Relative sensitivity (^{1}H = 1.00)	4.70×10^{-3}	2.71×10^{-3}
Absolute sensitivity (^{1}H = 1.00)	3.55×10^{-3}	6.63×10^{-4}
Receptivity (^{13}C = 1.00)	20.2	3.8
Magnetogyric ratio/rad T^{-1} s^{-1}	2.6210×10^7	2.1718×10^7
Quadrupole moment/m^2	-8.0×10^{-30}	-6.32×10^{-30}
Frequency (^{1}H = 100 MHz; 2.3488 T)/MHz	9.798	8.156

Reference: NaCl (aq)

Electron shell properties

Ground state electron configuration: [Ne] $3s^23p^5$

Term symbol: $^2P_{3/2}$

Electron affinity ($M\rightarrow M^-$)/kJ mol^{-1}: 348.7

Ionization energies/kJ mol^{-1}

1. $M \rightarrow M^+$	1251.1	6. $M^{5+}\rightarrow M^{6+}$	9 362
2. $M^+\rightarrow M^{2+}$	2297	7. $M^{6+}\rightarrow M^{7+}$	11 020
3. $M^{2+}\rightarrow M^{3+}$	3826	8. $M^{7+}\rightarrow M^{8+}$	33 610
4. $M^{3+}\rightarrow M^{4+}$	5158	9. $M^{8+}\rightarrow M^{9+}$	38 600
5. $M^{4+}\rightarrow M^{5+}$	6540	10. $M^{9+}\rightarrow M^{10+}$	43 960

Principal lines in atomic spectrum

Wavelength/nm	Species	Sensitivity	Application
138.99* (vac)	I		
479.455	II	V2	AE
481.006	II	V3	AE
481.947	II	V4	AE
837.594	I		

* Lowest energy transition from ground state to nearest empty orbital

Abundance: Earth's crust 126 p.p.m.; seawater 19 000 p.p.m.

Biological role: Essential as Cl^-; toxic as Cl_2 gas

Cr

Atomic number: 24
Relative atomic mass (^{12}C=12.0000): 51.9961

Chemical properties

Hard blue-white metal; main ore chromite $FeCr_2O_4$. Soluble in HCl and H_2SO_4 but not HNO_3, H_3PO_4 or $HClO_4$ due to formation of protective layer. Resists oxidation in air. Main use in alloys and chrome plating.

Radii/pm: Cr^{2+} 84; Cr^{3+} 64; Cr^{4+} 56; atomic 124.9

Electronegativity: 1.66 (Pauling); 1.56 (Allred-Rochow)

Effective nuclear charge: 3.45 (Slater); 5.13 (Clementi); 6.92 (Froese–Fischer)

Standard reduction potentials $E^\ominus$/V

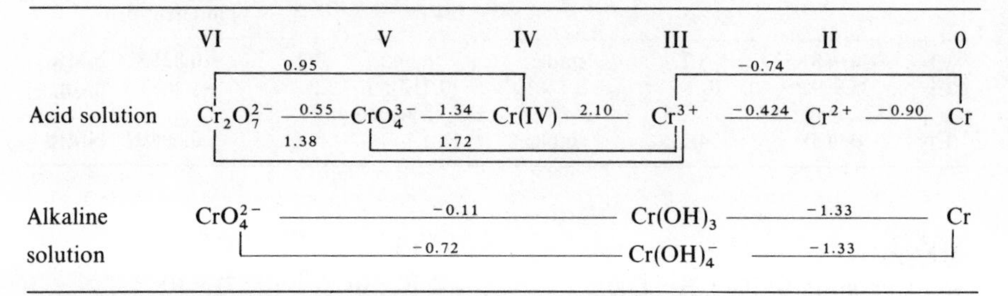

Oxidation states

Cr^{-II} (d^8)	$Na_2[Cr(CO)_5]$
Cr^{-I} (d^7)	$Na_2[Cr_2(CO)_{10}]$
Cr^0 (d^6)	$Cr(CO)_6$
Cr^I (d^5)	$[Cr(bipyridyl)_3]$
Cr^{II} (d^4)	CrO, CrF_2, $CrCl_2$ etc., CrS
Cr^{III} (d^3)	Cr_2O_3, CrF_3, $CrCl_3$ etc. $[Cr(H_2O)_6]^{3+}$ (aq), $Cr(OH)_3$, salts, complexes
Cr^{IV} (d^2)	CrO_2, CrF_4
Cr^V (d^1)	CrF_5
Cr^{VI} (d^0)	CrO_3, $Na_2Cr_2O_7$, CrO_4^{2-}, $CrOF_4$

Physical properties

Melting point/K: 2130±20

Boiling point/K: 2945

ΔH_{fusion}/kJ mol^{-1}: 15.3

ΔH_{vap}/kJ mol^{-1}: 341.8

Thermodynamic properties (298.15 K, 0.1 MPa)

State	$\Delta_f H^\ominus$/kJ mol^{-1}	$\Delta_f G^\ominus$/kJ mol^{-1}	$S^\ominus$/J K^{-1} mol^{-1}	C_p/J K^{-1} mol^{-1}
Solid	0	0	23.47	23.35
Gas	396.6	351.8	174.50	20.79

Density/kg m^{-3}: 7190 [293 K]; 6460 [liquid at m.p.]

Thermal conductivity/W m^{-1} K^{-1}: 93.7 [300 K]

Electrical resistivity/Ω m: 12.7×10^{-8} [273 K]

Mass magnetic susceptibility/kg^{-1} m^3: $+4.45\times10^{-8}$ (s)

Molar volume/cm^3: 7.23

Coefficient of linear thermal expansion/K^{-1}: 6.2×10^{-6}

Lattice structure (cell dimensions/pm), space group

b.c.c. (a=288.46), Im3m

X-ray diffraction: mass absorption coefficients (μ/ρ)/cm^2 g^{-1}:
CuK_α 260 MoK_α 31.1

Discovered and isolated in 1780 by N.-L. Vanquelin at Paris, France

[Greek, *chroma* = colour]

Chromium

Nuclear properties

Atomic number: 24

Thermal neutron capture cross-section/barns: 3.1 ± 0.2

Number of isotopes (including nuclear isomers): 9

Isotope mass range: 48 → 56

Key isotopes

Nuclide	Atomic mass	Natural abundance (%)	Half life $T_{1/2}$	Decay mode and energy (MeV)	Nuclear spin I	Nuclear magnetic moment μ	Uses
^{50}Cr	49.9461	4.35	stable				
^{51}Cr	50.945	0	27.8d	EC (0.752); γ	7/2	±0.934	tracer, medical
^{52}Cr	51.9405	83.79	stable				
^{53}Cr	52.9407	9.50	stable		3/2	−0.4735	NMR
^{54}Cr	53.9389	2.36	stable				

NMR

	^{53}Cr
Relative sensitivity (^1H = 1.00)	9.03×10^{-4}
Absolute sensitivity (^1H = 1.00)	8.62×10^{-3}
Receptivity (^{13}C = 1.00)	0.49
Magnetogyric ratio/rad T^{-1} s^{-1}	-1.5120×10^7
Quadrupole moment/m^2	$\pm 0.3 \times 10^{-28}$
Frequency (^1H = 100 MHz; 2.3488 T)/MHz	5.652

Reference: CrO_4^{2-} (aq)

Electron shell properties

Ground state electron configuration: $[Ar]3d^54s^1$

Term symbol: $^7S_3$

Electron affinity ($M \rightarrow M^-$)/kJ mol^{-1}: 94

Ionization energies/kJ mol^{-1}

1. $M \rightarrow M^+$	652.7	6. $M^{5+} \rightarrow M^{6+}$	8 738
2. $M^+ \rightarrow M^{2+}$	1592	7. $M^{6+} \rightarrow M^{7+}$	15 550
3. $M^{2+} \rightarrow M^{3+}$	2987	8. $M^{7+} \rightarrow M^{8+}$	17 830
4. $M^{3+} \rightarrow M^{4+}$	4740	9. $M^{8+} \rightarrow M^{9+}$	20 220
5. $M^{4+} \rightarrow M^{5+}$	6690	10. $M^{9+} \rightarrow M^{10+}$	23 580

Principal lines in atomic spectrum

Wavelength/nm	Species	Sensitivity	Application
357.869	I		AA
359.349	I		AA
425.435	I	U1	AA, AE
427.480	I	U2	AA, AE
428.972*	I	U3	AA, AE
520.844	I	U4	AA, AE

* Lowest energy transition from ground state to nearest empty orbital

Abundance: Earth's crust 122 p.p.m.; seawater 5×10^{-5} p.p.m.

Biological role: Essential trace element; chromates toxic; stimulatory; carcinogenic

Co

Atomic number: 27
Relative atomic mass (^{12}C=12.0000): 58.93320

Chemical properties

Lustrous, silvery-blue, hard metal; ferromagnetic and used to make magnetic alloys. Also used in ceramics, paints and as catalysts. Stable in air, slowly attacked by dilute acids. ^{60}Co useful radio-isotope.

Radii/pm: Co^{2+} 82; Co^{3+} 64; atomic 125.3; covalent 116

Electronegativity: 1.88 (Pauling); 1.70 (Allred-Rochow)

Effective nuclear charge: 3.90 (Slater); 5.58 (Clementi); 7.63 (Froese–Fischer)

Standard reduction potentials $E^{\ominus}$/V

	IV		III		II		0
Acid solution	CoO_2	1.416	Co^{3+}	1.92	Co^{2+}	−0.277	Co
Alkaline solution	CoO_2	0.7	$Co(OH)_3$	0.17	$Co(OH)_2$	−0.733	Co

Oxidation states

Co^{-I} (d^{10})	rare $[Co(CO)_4]^-$
Co^0 (d^9)	rare $[Co_2(CO)_8]$
Co^I (d^8)	rare $[Co(NCCH_2)_5]^+$
Co^{II} (d^7)	CoO, Co_3O_4 (=$Co^{II}Co_2^{III}O_4$), $Co(OH)_2$, CoF_2, $CoCl_2$ etc., $[Co(H_2O)_6]^{2+}$ (aq), many salts and complexes
Co^{III} (d^6)	$Co(OH)_3$, CoF_3, $[Co(H_2O)_6]^{3+}$(aq), $[Co(NH_3)_6]^{3+}$, many complexes
Co^{IV} (d^5)	CoO_2?, CoS_2, $[CoF_6]^{2-}$
Co^V (d^4)	K_3CoO_4

Physical properties

Melting point/K: 1768
Boiling point/K: 3143
ΔH_{fusion}/kJ mol^{-1}: 15.2
ΔH_{vap}/kJ mol^{-1}: 382.4

Thermodynamic properties (298.15 K, 0.1 MPa)

State	$\Delta_f H^{\ominus}$/kJ mol^{-1}	$\Delta_f G^{\ominus}$/kJ mol^{-1}	$S^{\ominus}$/J K^{-1} mol^{-1}	C_p/J K^{-1} mol^{-1}
Solid	0	0	30.04	24.81
Gas	424.7	380.3	179.515	23.020

Density/kg m^{-3}: 8900 [293 K]; 7670 [liquid at m.p.]
Thermal conductivity/W m^{-1} K^{-1}: 100 [300 K]
Electrical resistivity/Ω m: 6.24×10^{-8} [293 K]
Mass magnetic susceptibility/kg^{-1} m^3: Ferromagnetic
Molar volume/cm^3: 6.62
Coefficient of linear thermal expansion/K^{-1}: 13.36×10^{-6}

Lattice structure (cell dimensions/pm), space group

α-Co f.c.c. (a=354.41), Fm3m
ε-Co h.c.p. (a=250.7; c=406.9), $P6_3/mmc$
T (α→ε)=690 K

X-ray diffraction: mass absorption coefficients (μ/ρ)/cm^2 g^{-1}:
CuK_α 313 MoK_α 42.5

Discovered in 1735 by G. Brandt at Stockholm, Sweden

[German, *kobald* = goblin]

Cobalt

Nuclear properties

Atomic number: 27

Thermal neutron capture cross-section/barns: 37.5±0.2

Number of isotopes (including nuclear isomers): 14

Isotope mass range: 54→63

Key isotopes

Nuclide	Atomic mass	Natural abundance (%)	Half life $T_{1/2}$	Decay mode and energy (MeV)	Nuclear spin I	Nuclear magnetic moment μ	Uses
^{56}Co	55.940	0	77d	β^+ (4.57); EC; γ	4	±3.83	tracer
^{57}Co	56.936	0	270d	EC (0.837); γ	7/2	+4.72	tracer, medical
^{58}Co	57.936	0	71.3d	β^+ (2.309); EC; γ	2	+4.04	tracer, medical
^{59}Co	58.9332	100	stable		7/2	+4.616	NMR
^{60}Co	59.934	0	5.26y	β^- (2.819); γ	5	+3.79	tracer, medical

NMR

	^{59}Co
Relative sensitivity (^1H=1.00)	0.28
Absolute sensitivity (^1H=1.00)	0.28
Receptivity (^{13}C=1.00)	1570
Magnetogyric ratio/rad T^{-1} s^{-1}	6.3472×10^7
Quadrupole moment/m^2	0.40×10^{-28}
Frequency (^1H=100 MHz; 2.3488 T)/MHz	23.614
Reference: $K_3Co(CN)_6$	

Electron shell properties

Ground state electron configuration: [Ar] $3d^7 4s^2$

Term symbol: $^4F_{9/2}$

Electron affinity ($M \rightarrow M^-$)/kJ mol^{-1}: 102

Ionization energies/kJ mol^{-1}

1. $M \rightarrow M^+$	760.0	6. $M^{5+} \rightarrow M^{6+}$	9840
2. $M^+ \rightarrow M^{2+}$	1646	7. $M^{6+} \rightarrow M^{7+}$	12400
3. $M^{2+} \rightarrow M^{3+}$	3232	8. $M^{7+} \rightarrow M^{8+}$	15100
4. $M^{3+} \rightarrow M^{4+}$	4950	9. $M^{8+} \rightarrow M^{9+}$	17900
5. $M^{4+} \rightarrow M^{5+}$	7670	10. $M^{9+} \rightarrow M^{10+}$	26600

Principal lines in atomic spectrum

Wavelength/nm	Species	Sensitivity	Application
240.725	I		AA
242.493	I		AA
345.350	I	U1	AE
346.580	I	U2	AA, AE
352.981	I	U3	AE
423.400*	I		

* Lowest energy transition from ground state to nearest empty orbital

Abundance: Earth's crust 29 p.p.m.; seawater 0.0005 p.p.m.

Biological role: Essential trace element (vitamin B_{12}); carcinogenic

Cu

Atomic number: 29
Relative atomic mass (^{12}C=12.0000): 63.546

Chemical properties

Reddish metal, malleable and ductile, with high electrical and thermal conductivities. Main ore is chalcopyrite, $CuFeS_2$. Resistant to air and water but slowly weathers to green patina of carbonate. Historically important alloys.

Radii/pm: Cu^+ 96; Cu^{2+} 72; atomic 127.8; covalent 117

Electronegativity: 1.90 (Pauling); 1.75 (Allred-Rochow)

Effective nuclear charge: 4.20 (Slater); 5.84 (Clementi); 8.07 (Froese–Fischer)

Standard reduction potentials $E^\ominus$/V

(II → 0: 0.340)

II		I		0
Cu^{2+}	0.159	Cu^+	0.520	Cu
$Cu(NH_3)_4^{2+}$	0.10	$Cu(NH_3)_4^+$	−0.100	Cu
$Cu(CN)_2$	1.12	$Cu(CN)_2^-$	−0.44	Cu

Oxidation states

Cu^{-I} ($d^{10}s^2$)	—
Cu^0 ($d^{10}s^1$)	rare, $[Cu(CO)_3]$ at 10 K
Cu^I (d^{10})	Cu_2O, CuCl, $K[Cu(CN)_2]$
Cu^{II} (d^9)	CuO, $CuCl_2$, Cu^{2+} (aq), Cu^{2+} salts
Cu^{III} (d^8)	rare $K_3[CuF_6]$
Cu^{IV} (d^7)	rare $Cs_2[CuF_6]$

Physical properties

Melting point/K: 1356.6
Boiling point/K: 2840
ΔH_{fusion}/kJ mol^{-1}: 13.0
ΔH_{vap}/kJ mol^{-1}: 306.7

Thermodynamic properties (298.15 K, 0.1 MPa)

State	$\Delta_f H^\ominus$/kJ mol^{-1}	$\Delta_f G^\ominus$/kJ mol^{-1}	$S^\ominus$/J K^{-1} mol^{-1}	C_p/J K^{-1} mol^{-1}
Solid	0	0	33.150	24.435
Gas	338.32	298.58	166.38	20.786

Density/kg m^{-3}: 8960 [293 K]; 7940 [liquid at m.p.]
Thermal conductivity/W m^{-1} K^{-1}: 401 [300 K]
Electrical resistivity/Ω m: 1.6730×10^{-8} [293 K]
Mass magnetic susceptibility/kg^{-1} m^3: -1.081×10^{-9} (s)
Molar volume/cm^3: 7.09
Coefficient of linear thermal expansion/K^{-1}: 16.5×10^{-6}

Lattice structure (cell dimensions/pm), space group

f.c.c. (a=361.47), Fm3m

X-ray diffraction: mass absorption coefficients (μ/ρ)/cm^2 g^{-1}:
CuK_α 52.9 MoK_α 50.9

Known to ancient civilizations

[Latin, *cuprum* = Cyprus]

Copper

Nuclear properties

Atomic number: 29

Thermal neutron capture cross-section/barns: 3.8 ± 0.1

Number of isotopes (including nuclear isomers): 11

Isotope mass range: $58 \rightarrow 68$

Key isotopes

Nuclide	Atomic mass	Natural abundance (%)	Half life $T_{1/2}$	Decay mode and energy (MeV)	Nuclear spin I	Nuclear magnetic moment μ	Uses
^{63}Cu	62.9298	69.17	stable		3/2	+2.2228	NMR
^{64}Cu	63.930	0	12.9h	β^- (0.573) 40%; β^+, EC; γ	1	−0.216	tracer, medical
^{65}Cu	64.9278	30.83	stable		3/2	+2.3812	NMR
^{67}Cu			61.88h	β^- (0.576); γ			tracer

NMR

	^{63}Cu	^{65}Cu
Relative sensitivity ($^1H = 1.00$)	9.31×10^{-2}	0.11
Absolute sensitivity ($^1H = 1.00$)	6.43×10^{-2}	3.52×10^{-2}
Receptivity ($^{13}C = 1.00$)	365	201
Magnetogyric ratio/rad T^{-1} s^{-1}	7.0965×10^7	7.6018×10^7
Quadrupole moment/m^2	-0.211×10^{-28}	-0.195×10^{-28}
Frequency ($^1H = 100$ MHz; 2.3488 T)/MHz	26.505	28.394

Reference: $Cu(MeCN)_4^+ BF_4^-$ in MeCN

Electron shell properties

Ground state electron configuration: $[Ar]3d^{10}4s^1$

Term symbol: $^2S_{1/2}$

Electron affinity ($M \rightarrow M^-$)/kJ mol^{-1}: 118.3

Ionization energies/kJ mol^{-1}

1. $M \rightarrow M^+$	745.4	6. $M^{5+} \rightarrow M^{6+}$	(9940)
2. $M^+ \rightarrow M^{2+}$	1958	7. $M^{6+} \rightarrow M^{7+}$	(13400)
3. $M^{2+} \rightarrow M^{3+}$	3554	8. $M^{7+} \rightarrow M^{8+}$	(16000)
4. $M^{3+} \rightarrow M^{4+}$	5326	9. $M^{8+} \rightarrow M^{9+}$	(19200)
5. $M^{4+} \rightarrow M^{5+}$	7709	10. $M^{9+} \rightarrow M^{10+}$	(22400)

Principal lines in atomic spectrum

Wavelength/nm	Species	Sensitivity	Application
216.509	I		AA
217.894	I		AA
324.754	I	U1	AA, AE
327.396*	I	U2	AA, AE
515.324	I	U4	AE
521.820	I	U3	AE

* Lowest energy transition from ground state to nearest empty orbital

Abundance: Earth's crust 68 p.p.m.; seawater 0.003 p.p.m.

Biological role: Essential; excess copper is toxic

Cm

Atomic number: 96

Relative atomic mass (^{12}C=12.0000): **(247)**

Chemical properties

Radioactive, silvery metal obtained in gramme quantities as ^{242}Cm and ^{244}Cm from neutron bombardment of ^{239}Pu. Attacked by oxygen, steam, and acids but not alkalis.

Radii/pm: Cm^{2+} 119; Cm^{3+} 99; Cm^{4+} 88

Electronegativity: 1.3 (Pauling); n.a. (Allred-Rochow)

Effective nuclear charge: 1.80 (Slater)

Standard reduction potentials $E^{\ominus}$/V

	IV		III		0
Acid solution	Cm^{4+}	3.2	Cm^{3+}	−2.06	Cm
Alkaline solution	CmO_2	0.7	$Cm(OH)_3$	−2.5	Cm

Oxidation states

Cm^{II} ($f^7 d^1$)	CmO
Cm^{III} (f^7)	Cm_2O_3, $Cm(OH)_3$, CmF_3, $CmCl_3$ etc., Cm^{3+} (aq)
Cm^{IV} (f^6)	CmO_2, CmF_4, Cm^{4+} (aq) very unstable

Physical properties

Melting point/K: 1610±40

Boiling point/K: n.a.

ΔH_{fusion}/kJ mol^{-1}: n.a.

ΔH_{vap}/kJ mol^{-1}: n.a.

Thermodynamic properties (298.15 K, 0.1 MPa)

State	$\Delta_f H^{\ominus}$/kJ mol^{-1}	$\Delta_f G^{\ominus}$/kJ mol^{-1}	$S^{\ominus}$/J K^{-1} mol^{-1}	C_p/J K^{-1} mol^{-1}
Solid	0	0	n.a.	n.a.
Gas	n.a.	n.a.	n.a.	n.a.

Density/kg m^{-3}: 13 300 [293 K]

Thermal conductivity/W m^{-1} K^{-1}: 10 (est.) [300 K]

Electrical resistivity/Ω m: n.a.

Mass magnetic susceptibility/kg^{-1} m^3: n.a.

Molar volume/cm^3: 18.6

Coefficient of linear thermal expansion/K^{-1}: n.a.

Lattice structure (cell dimensions/pm), space group

n.a.

X-ray diffraction: mass absorption coefficients (μ/ρ)/cm^2 g^{-1}: n.a.

Prepared by G. T. Seaborg, R. A. James, and A. Ghiorso in 1944 at Berkeley, California, USA

[Named after Pierre and Marie Curie]

Curium

Nuclear properties

Atomic number: 96

Thermal neutron capture cross-section/barns: 180 (^{247}Cm)

Number of isotopes (including nuclear isomers): 13

Isotope mass range: 238→250

Key isotopes

Nuclide	Atomic mass	Natural abundance (%)	Half life $T_{1/2}$	Decay mode and energy (MeV)	Nuclear spin I	Nuclear magnetic moment μ	Uses
^{242}Cm	242.0588	0	163d	α (6.12); γ	0		available in mg quantities
^{244}Cm	244.0629	0	17.6y	α (5.902); γ			available in mg quantities
^{247}Cm	247.0704	0	1.6×10^7y	α (*c*. 5.3); γ	9/2	±0.4	
^{248}Cm		0	4.7×10^5y	α (5.161); SF			

Electron shell properties

Ground state electron configuration: $[Rn]5f^76d^17s^2$

Term symbol: $^9D_2$

Electron affinity ($M \to M^-$)/kJ mol^{-1}: n.a.

Ionization energies/kJ mol^{-1}

1. $M \to M^+$	581	6. $M^{5+} \to M^{6+}$	
2. $M^+ \to M^{2+}$		7. $M^{6+} \to M^{7+}$	
3. $M^{2+} \to M^{3+}$		8. $M^{7+} \to M^{8+}$	
4. $M^{3+} \to M^{4+}$		9. $M^{8+} \to M^{9+}$	
5. $M^{4+} \to M^{5+}$		10. $M^{9+} \to M^{10+}$	

Selected lines in atomic spectrum

Wavelength/nm	Species	Sensitivity	Application
299.939	I		
310.969	I		
311.641	I		
313.716	I		
314.733	I		
315.510	I		
315.860	I		

Abundance: Nil

Biological role: None; toxic due to radioactivity

Dy

Atomic number: 66
Relative atomic mass ($^{12}C=12.0000$): 162.50

Chemical properties

Reactive, hard, silvery metal of the rare earth group. Obtained from monazite and bastnaesite ores. Oxidized by oxygen, reacts rapidly with water, dissolves in acids.

Radii/pm: Dy^{3+} 91; atomic 177.3; covalent 159
Electronegativity: 1.22 (Pauling); 1.10 (Allred-Rochow)
Effective nuclear charge: 2.85 (Slater); 8.34 (Clementi); 11.49 (Froese–Fischer)

Standard reduction potentials $E^{\ominus}$/V

	IV		III		II		0
Acid solution	Dy^{4+}	5.7	Dy^{3+}	−2.5	Dy^{2+}	−2.2	Dy
Alkaline solution	DyO_2	3.5	$Dy(OH)_3$		−2.80		Dy

(Acid solution: Dy^{3+} → Dy −2.29)

Oxidation states

Dy^{II} (f^{10})	$DyCl_2$, DyI_2
Dy^{III} (f^9)	Dy_2O_3, $Dy(OH)_3$, DyF_3, $DyCl_3$ etc., $[Dy(H_2O)_x]^{3+}$ (aq), Dy^{3+} salts, $DyCl_6^{3-}$
Dy^{IV} (f^8)	Cs_3DyF_7

Physical properties

Melting point/K: 1685
Boiling point/K: 2835
ΔH_{fusion}/kJ mol^{-1}: 17.2
ΔH_{vap}/kJ mol^{-1}: 293

Thermodynamic properties (298.15 K, 0.1 MPa)

State	$\Delta_f H^{\ominus}$/kJ mol^{-1}	$\Delta_f G^{\ominus}$/kJ mol^{-1}	$S^{\ominus}$/J K^{-1} mol^{-1}	C_p/J K^{-1} mol^{-1}
Solid	0	0	74.77	28.16
Gas	290.4	254.4	196.63	20.79

Density/kg m^{-3}: 8550 [293 K]
Thermal conductivity/W m^{-1} K^{-1}: 10.7 [300 K]
Electrical resistivity/Ω m: 57.0×10^{-8} [273 K]
Mass magnetic susceptibility/kg^{-1} m^3: $+8.00\times10^{-6}$ (s)
Molar volume/cm^3: 19.00
Coefficient of linear thermal expansion/K^{-1}: 10.0×10^{-6}

Lattice structure (cell dimensions/pm), space group

Rhombic ($a=359.5$; $b=618.3$; $c=567.7$), Cmcm
h.c.p. ($a=359.03$; $c=564.75$), $P6_3/mmc$
b.c.c. ($a=398$), Im3m

T (rhombic→h.c.p.) = 86 K
High pressure form: ($a=334$, $c=245$), $R\bar{3}m$

X-ray diffraction: mass absorption coefficients (μ/ρ)/cm^2 g^{-1}:
CuK_α 286 MoK_α 70.6

Discovered in 1886 by Paul-Émile Lecoq de Boisbaudran at Paris, France

[Greek, *dysprositos* = hard to obtain]

Dysprosium

Nuclear properties

Atomic number: 66

Thermal neutron capture cross-section/barns: 90 ± 20

Number of isotopes (including nuclear isomers): 21

Isotope mass range: 149 → 167

Key isotopes

Nuclide	Atomic mass	Natural abundance (%)	Half life $T_{1/2}$	Decay mode and energy (MeV)	Nuclear spin I	Nuclear magnetic moment μ	Uses
^{156}Dy	155.9238	0.06	stable				
^{158}Dy	157.9240	0.10	stable				
^{160}Dy	159.9248	2.34	stable				
^{161}Dy	160.9266	19.0	stable		5/2	−0.482	NMR
^{162}Dy	161.9265	25.5	stable				
^{163}Dy	162.9284	24.9	stable		5/2	+0.676	NMR
^{164}Dy	163.9288	28.1	stable				

NMR

	^{161}Dy	^{163}Dy
Relative sensitivity (^{1}H = 1.00)	4.17×10^{-4}	1.12×10^{-3}
Absolute sensitivity (^{1}H = 1.00)	7.87×10^{-5}	2.79×10^{-4}
Receptivity (^{13}C = 1.00)	0.509	1.79
Magnetogyric ratio/rad T^{-1} s^{-1}	-0.9206×10^{7}	1.2750×10^{7}
Quadrupole moment/m^2	1.4×10^{-28}	1.6×10^{-28}
Frequency (^{1}H = 100 MHz; 2.3488 T)/MHz	3.294	4.583

Electron shell properties

Ground state electron configuration: $[Xe]4f^{10}6s^2$

Term symbol: $^{5}I_{8}$

Electron affinity ($M \rightarrow M^-$)/kJ mol^{-1}: n.a.

Ionization energies/kJ mol^{-1}

1. $M \rightarrow M^+$	571.9	6. $M^{5+} \rightarrow M^{6+}$
2. $M^+ \rightarrow M^{2+}$	1126	7. $M^{6+} \rightarrow M^{7+}$
3. $M^{2+} \rightarrow M^{3+}$	2200	8. $M^{7+} \rightarrow M^{8+}$
4. $M^{3+} \rightarrow M^{4+}$	4001	9. $M^{8+} \rightarrow M^{9+}$
5. $M^{4+} \rightarrow M^{5+}$		10. $M^{9+} \rightarrow M^{10+}$

Principal lines in atomic spectrum

Wavelength/nm	Species	Sensitivity	Application
353.170	II		
400.045	II		AE
404.597	I		AA, AE
416.797	I		AA, AE
418.682	I		AA
421.172	I		AA, AE

Abundance: Earth's crust 4.5 p.p.m.; seawater 2×10^{-7} p.p.m.

Biological role: None; low toxicity; stimulatory

Es

Atomic number: 99

Relative atomic mass ($^{12}C=12.0000$): **(254)**

Chemical properties

Radioactive, silvery metal obtained in mg quantities as ^{254}Es from neutron bombardment of ^{239}Pu. Attacked by oxygen, steam, and acids, but not alkalis.

Radii/pm: Es^{2+} 116; Es^{3+} 98; Es^{4+} 85

Electronegativity: 1.3 (Pauling); n.a. (Allred-Rochow)

Effective nuclear charge: 1.65 (Slater)

Standard reduction potentials $E^{\ominus}$/V

	III		II		0
Acid solution	Es^{3+}	$\xrightarrow{-1.5}$	Es^{2+}	$\xrightarrow{-2.2}$	Es
		Es^{3+} to Es: −2.0			

Oxidation states

Es^{II} (f^{11})	transient state
Es^{III} (f^{10})	Es_2O_3, $EsCl_3$, $EsBr_3$, Es^{3+} (aq), EsOCl

Physical properties

Melting point/K: n.a.

Boiling point/K: n.a.

ΔH_{fusion}/kJ mol^{-1}: n.a.

ΔH_{vap}/kJ mol^{-1}: n.a.

Thermodynamic properties (298.15 K, 0.1 MPa)

State	$\Delta_f H^{\ominus}$/kJ mol^{-1}	$\Delta_f G^{\ominus}$/kJ mol^{-1}	$S^{\ominus}$/J K^{-1} mol^{-1}	C_p/J K^{-1} mol^{-1}
Solid	0	0	n.a.	n.a.
Gas	n.a.	n.a.	n.a.	n.a.

Density/kg m^{-3}: n.a.

Thermal conductivity/W m^{-1} K^{-1}: 10 (est.) [300 K]

Electrical resistivity/Ω m: n.a.

Mass magnetic susceptibility/kg^{-1} m^3: n.a.

Molar volume/cm^3: n.a.

Coefficient of linear thermal expansion/K^{-1}: n.a.

Lattice structure (cell dimensions/pm), space group

n.a.

X-ray diffraction: mass absorption coefficients ((μ/ρ)/cm^2 g^{-1}): n.a.

Discovered in the debris of the 1952 thermonuclear explosion in the Pacific by G. R. Choppin, S. G. Thompson, A. Ghiorso, and B. G. Harvey
[Named after Albert Einstein]

Einsteinium

Nuclear properties

Atomic number: 99
Thermal neutron capture cross-section/barns: <40 (^{254}Es)
Number of isotopes (including nuclear isomers): 12
Isotope mass range: 246→256

Key isotopes

Nuclide	Atomic mass	Natural abundance (%)	Half life $T_{1/2}$	Decay mode and energy (MeV)	Nuclear spin I	Nuclear magnetic moment μ	Uses
^{253}Es		0	20.7d	α			Available
^{254}Es	254.0881	0	207d	α (6.623); γ			

Electron shell properties

Ground state electron configuration: $[Rn]4f^{11}7s^2$
Term symbol: $^5I_{15/2}$
Electron affinity ($M \rightarrow M^-$)/kJ mol^{-1}: ≤50

Ionization energies/kJ mol^{-1}

1. $M \rightarrow M^+$	619	6. $M^{5+} \rightarrow M^{6+}$	
2. $M^+ \rightarrow M^{2+}$		7. $M^{6+} \rightarrow M^{7+}$	
3. $M^{2+} \rightarrow M^{3+}$		8. $M^{7+} \rightarrow M^{8+}$	
4. $M^{3+} \rightarrow M^{4+}$		9. $M^{8+} \rightarrow M^{9+}$	
5. $M^{4+} \rightarrow M^{5+}$		10. $M^{9+} \rightarrow M^{10+}$	

Selected lines in atomic spectrum

Wavelength/nm	Species	Sensitivity	Application
342.848	I		
349.811	I		
351.433	I		
352.349	I		
355.534	I		
516.174	I		
520.440	I		

Abundance: Nil
Biological role: None; toxic due to radioactivity

Er

Atomic number: 68
Relative atomic mass (^{12}C=12.0000): 167.26

Chemical properties

Silver-grey metal of the rare earth group. Obtained from bastnaesite, monazite, and euxenite. Slowly tarnishes in air, slowly reacts with water, dissolves in acids. Some used in infrared absorbing glass.

Radii/pm: Er^{3+} 89; atomic 175.7; covalent 157
Electronegativity: 1.24 (Pauling); 1.11 (Allred-Rochow)
Effective nuclear charge: 2.85 (Slater); 8.48 (Clementi); 11.70 (Froese–Fischer)

Standard reduction potentials $E^{\ominus}$/V

	III		0
Acid solution	Er^{3+}	$\xrightarrow{-2.32}$	Er
Alkaline solution	$Er(OH)_3$	$\xrightarrow{-2.84}$	Er

Oxidation states

Er^{III} (f^{11})	Er_2O_3, $Er(OH)_3$, ErF_3, $ErCl_3$ etc., $[Er(H_2O)_x]^{3+}$ (aq), Er^{3+} salts, $ErCl_6^{3-}$, complexes etc.

Physical properties

Melting point/K: 1802
Boiling point/K: 3136
ΔH_{fusion}/kJ mol^{-1}: 17.2
ΔH_{vap}/kJ mol^{-1}: 280

Thermodynamic properties (298.15 K, 0.1 MPa)

State	$\Delta_f H^{\ominus}$/kJ mol^{-1}	$\Delta_f G^{\ominus}$/kJ mol^{-1}	$S^{\ominus}$/J K^{-1} mol^{-1}	C_p/J K^{-1} mol^{-1}
Solid	0	0	73.18	28.12
Gas	317.1	280.7	195.59	20.79

Density/kg m^{-3}: 9066 [298 K]
Thermal conductivity/W m^{-1} K^{-1}: 14.3 [300 K]
Electrical resistivity/Ω m: 87×10^{-8} [298 K]
Mass magnetic susceptibility/kg^{-1} m^3: $+3.33 \times 10^{-6}$ (s)
Molar volume/cm^3: 18.44
Coefficient of linear thermal expansion/K^{-1}: 9.2×10^{-6}

Lattice structure (cell dimensions/pm), space group

α-Er h.c.p. (a=355.88; c=558.74), $P6_3/mmc$
β-Er b.c.c. (a=394), Im3m

T (α→β)=1640 K

X-ray diffraction: mass absorption coefficients (μ/ρ)/cm^2 g^{-1}:
CuK_α 134 MoK_α 77.3

Discovered in 1842 by C. G. Mosander at Stockholm, Sweden

[Named after Ytterby, Sweden]

Erbium

Nuclear properties

Atomic number: 68

Thermal neutron capture cross-section/barns: 0.16 ± 0.03

Number of isotopes (including nuclear isomers): 16

Isotope mass range: $158 \rightarrow 172$

Key isotopes

Nuclide	Atomic mass	Natural abundance (%)	Half life $T_{1/2}$	Decay mode and energy (MeV)	Nuclear spin I	Nuclear magnetic moment μ	Uses
^{162}Er	161.9288	0.14	stable				
^{164}Er	163.9293	1.56	stable				
^{166}Er	165.9304	33.4	stable				
^{167}Er	166.9320	22.9	stable		7/2	−0.564	NMR
^{168}Er	167.9324	27.1	stable				
^{169}Er		0	9.4d	β^- (0.34); γ	1/2	+0.513	tracer
^{170}Er	169.9355	14.9	stable				
^{171}Er		0	7.52h	β^- (1.490); γ	5/2	±0.70	tracer

NMR

	^{167}Er
Relative sensitivity (^{1}H = 1.00)	5.07×10^{-4}
Absolute sensitivity (^{1}H = 1.00)	1.16×10^{-4}
Receptivity (^{13}C = 1.00)	0.665
Magnetogyric ratio/rad T^{-1} s^{-1}	-0.7752×10^7
Quadrupole moment/m^2	2.83×10^{-28}
Frequency (^{1}H = 100 MHz; 2.3488 T)/MHz	2.890

Electron shell properties

Ground state electron configuration: $[Xe]4f^{12}6s^2$

Term symbol: $^3H_6$

Electron affinity ($M \rightarrow M^-$)/kJ mol^{-1}: ≤ 50

Ionization energies/kJ mol^{-1}

1. $M \rightarrow M^+$	588.7	6. $M^{5+} \rightarrow M^{6+}$	
2. $M^+ \rightarrow M^{2+}$	1151	7. $M^{6+} \rightarrow M^{7+}$	
3. $M^{2+} \rightarrow M^{3+}$	2194	8. $M^{7+} \rightarrow M^{8+}$	
4. $M^{3+} \rightarrow M^{4+}$	4115	9. $M^{8+} \rightarrow M^{9+}$	
5. $M^{4+} \rightarrow M^{5+}$		10. $M^{9+} \rightarrow M^{10+}$	

Principal lines in atomic spectrum

Wavelength/nm	Species	Sensitivity	Application
349.910	II		AE
369.265	II		AE
386.285	I		AA
390.632	II		AE
400.796	I		AA

Abundance: Earth's crust 3.5 p.p.m.; seawater 2×10^{-7} p.p.m.

Biological role: None; low toxicity; stimulatory

Eu

Atomic number: 63

Relative atomic mass (^{12}C=12.0000): 151.965

Chemical properties

Rare, and most reactive of the rare earth metals. Obtained from monazite and bastnaesite ores. Soft, silvery metal which reacts quickly with oxygen and water. Little used.

Radii/pm: Eu^{2+} 112; Eu^{3+} 98; atomic 204.2; covalent 185

Electronegativity: n.a. (Pauling); 1.01 (Allred-Rochow)

Effective nuclear charge: 2.85 (Slater); 8.11 (Clementi); 11.17 (Froese–Fischer)

Standard reduction potentials $E^{\ominus}$/V

	III		II		0
Acid solution	Eu^{3+}	−0.35	Eu^{2+}	−2.80	Eu
	Eu^{3+} → Eu: −1.99				
Alkaline solution	$Eu(OH)_3$		−2.51		Eu

Oxidation states

Eu^{II} (f^7)	EuO, EuS, EuF_2, $EuCl_2$ etc.
Eu^{III} (f^6)	Eu_2O_3, $Eu(OH)_3$, EuF_3, $EuCl_3$ etc., $[Eu(H_2O)_x]^{3+}$ (aq), Eu^{3+} salts, complexes

Physical properties

Melting point/K: 1095

Boiling point/K: 1870

ΔH_{fusion}/kJ mol^{-1}: 10.5

ΔH_{vap}/kJ mol^{-1}: 176

Thermodynamic properties (298.15 K, 0.1 MPa)

State	$\Delta_f H^{\ominus}$/kJ mol^{-1}	$\Delta_f G^{\ominus}$/kJ mol^{-1}	$S^{\ominus}$/J K^{-1} mol^{-1}	C_p/J K^{-1} mol^{-1}
Solid	0	0	77.78	27.66
Gas	175.3	142.2	188.795	20.786

Density/kg m^{-3}: 5243 [293 K]

Thermal conductivity/W m^{-1} K^{-1}: 13.9 [300 K]

Electrical resistivity/Ω m: 90.0×10^{-8} [298 K]

Mass magnetic susceptibility/kg^{-1} m^3: $+2.81 \times 10^{-6}$ (s)

Molar volume/cm^3: 28.98

Coefficient of linear thermal expansion/K^{-1}: 32×10^{-6}

Lattice structure (cell dimensions/pm), space group

b.c.c. (a=458.20), Im3m

X-ray diffraction: mass absorption coefficients (μ/ρ)/cm^2 g^{-1}:
CuK_α 425 MoK_α 61.5

Discovered in 1901 by E.-A. Demarçay at Paris, France

[Named after Europe]

Europium

Nuclear properties

Atomic number: 63

Thermal neutron capture cross-section/barns: 4100 ± 100

Number of isotopes (including nuclear isomers): 21

Isotope mass range: 144→160

Key isotopes

Nuclide	Atomic mass	Natural abundance (%)	Half life $T_{1/2}$	Decay mode and energy (MeV)	Nuclear spin I	Nuclear magnetic moment μ	Uses
^{151}Eu	150.9196	47.8	stable		5/2	+3.463	NMR
^{151}Eu		0	12.7y	EC, β^- (1.82), β^+; γ	3	−1.937	tracer
^{152m_1}Eu		0	9.2h	β^-; γ	0		tracer
^{153}Eu	152.9209	52.2	stable		5/2		NMR

NMR

	^{151}Eu	^{153}Eu
Relative sensitivity (^{1}H = 1.00)	0.18	1.52×10^{-2}
Absolute sensitivity (^{1}H = 1.00)	8.51×10^{-2}	7.98×10^{-3}
Receptivity (^{13}C = 1.00)	4.64×10^2	45.7
Magnetogyric ratio/rad T^{-1} s^{-1}	6.5477×10^7	2.9371×10^7
Quadrupole moment/m^2	1.16×10^{-28}	2.9×10^{-28}
Frequency (^{1}H = 100 MHz; 2.3488 T)/MHz	24.801	10.951

Electron shell properties

Ground state electron configuration: [Xe]$4f^7 6s^2$

Term symbol: $^8S_{7/2}$

Electron affinity ($M \rightarrow M^-$)/kJ mol^{-1}: ≤50

Ionization energies/kJ mol^{-1}

1. $M \rightarrow M^+$	546.7	6. $M^{5+} \rightarrow M^{6+}$	
2. $M^+ \rightarrow M^{2+}$	1085	7. $M^{6+} \rightarrow M^{7+}$	
3. $M^{2+} \rightarrow M^{3+}$	2404	8. $M^{7+} \rightarrow M^{8+}$	
4. $M^{3+} \rightarrow M^{4+}$	4110	9. $M^{8+} \rightarrow M^{9+}$	
5. $M^{4+} \rightarrow M^{5+}$		10. $M^{9+} \rightarrow M^{10+}$	

Principal lines in atomic spectrum

Wavelength/nm	Species	Sensitivity	Application
412.974	II		AE
420.505	II		AE
459.402	I		AA, AE
462.722	I		AA, AE
466.187	I		AA, AE

Abundance: Earth's crust 2.1 p.p.m.; seawater 4×10^{-8} p.p.m.

Biological role: None; low toxicity

Fm

Atomic number: 100
Relative atomic mass (^{12}C=12.0000): (257)

Chemical properties

Radioactive metal obtained in microgramme quantities as ^{253}Fm from neutron bombardment of ^{239}Pu.

Radii/pm: Fm^{2+} 115; Fm^{3+} 97; Fm^{4+} 84
Electronegativity: 1.3 (Pauling); n.a. (Allred-Rochow)
Effective nuclear charge: 1.65 (Slater)

Standard reduction potentials $E^{\ominus}$/V

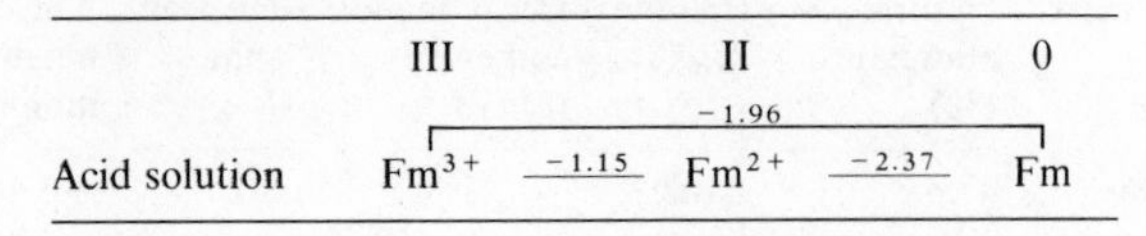

Oxidation states

Fm^{II} (f^{12})	?
Fm^{III} (f^{11})	$[Fm(H_2O)_x]^{3+}$ (aq)

Physical properties

Melting point/K: n.a.
Boiling point/K: n.a.
ΔH_{fusion}/kJ mol^{-1}: n.a.
ΔH_{vap}/kJ mol^{-1}: n.a.

Thermodynamic properties (298.15 K, 0.1 MPa)

State	$\Delta_f H^{\ominus}$/kJ mol^{-1}	$\Delta_f G^{\ominus}$/kJ mol^{-1}	$S^{\ominus}$/J K^{-1} mol^{-1}	C_p/J K^{-1} mol^{-1}
Solid	0	0	n.a.	n.a.
Gas	n.a.	n.a.	n.a.	n.a.

Density/kg m^{-3}: n.a.
Thermal conductivity/W m^{-1} K^{-1}: 10 (est.) [300 K]
Electrical resistivity/Ω m: n.a.
Mass magnetic susceptibility/kg^{-1} m^3: n.a.
Molar volume/cm^3: n.a.
Coefficient of linear thermal expansion/K^{-1}: n.a.

Lattice structure (cell dimensions/pm), space group

n.a.

X-ray diffraction: mass absorption coefficients (μ/ρ)/cm^2 g^{-1}:
n.a.

Discovered in the debris of the 1952 thermonuclear explosion in the Pacific by G. R. Choppin, S. G. Thompson, A. Ghioros, and B. G. Harvey
[Named after Enrico Fermi]

Fermium

Nuclear properties

Atomic number: 100
Thermal neutron capture cross-section/barns: 26±3 (^{255}Fm)
Number of isotopes (including nuclear isomers): 10
Isotope mass range: 248→257

Key isotopes

Nuclide	Atomic mass	Natural abundance (%)	Half life $T_{1/2}$	Decay mode and energy (MeV)	Nuclear spin I	Nuclear magnetic moment μ	Uses
^{254}Fm		0	3.24h	α			
^{255}Fm		0	20.0h	α			
^{257}Fm		0	80d	α (6.871); γ			

Electron shell properties

Ground state electron configuration: [Rn]5f^{12}7s^2
Term symbol: $^3H_6$
Electron affinity (M→M$^-$)/kJ mol^{-1}: n.a.

Ionization energies/kJ mol^{-1}

1. M →M^+	627	6. M^{5+}→M^{6+}	
2. M^+→M^{2+}		7. M^{6+}→M^{7+}	
3. M^{2+}→M^{3+}		8. M^{7+}→M^{8+}	
4. M^{3+}→M^{4+}		9. M^{8+}→M^{9+}	
5. M^{4+}→M^{5+}		10. M^{9+}→M^{10+}	

Principal lines in atomic spectrum

Wavelength/nm	Species	Sensitivity	Application
n.a.			

Abundance: Nil
Biological role: None; toxic due to radioactivity

F

Atomic number: 9

Relative atomic mass (^{12}C=12.0000): 19.998 4032

Chemical properties

Pale yellow gas, F_2, most reactive of all elements. Produced by electrolysis of molten KF.2HF. Used to make UF_6, SF_6, and fluorinating agents such as ClF_3. Organic compounds, polymers, and salts all used, notably CaF_2 as flux in metallurgy, and AlF_3 in aluminium production.

Radii/pm: F^- 133; covalent 64; atomic 71.7; van der Waals 135

Electronegativity: 3.98 (Pauling); 4.10 (Allred-Rochow)

Effective nuclear charge: 5.20 (Slater); 5.10 (Clementi); 4.61 (Froese–Fischer)

Standard reduction potentials $E^{\ominus}$/V

0		−I
F_2	2.866	F^- (aq)
F_2	2.979	HF_2^-
F_2	3.053	HF (aq)

Oxidation states

F^{-I} ([Ne])	F^- (aq), HF, KHF_2, CaF_2 many salts and derivatives of other elements

Covalent bonds r/pm E/kJ mol^{-1}

	r/pm	E/kJ mol^{-1}
F–F	141.7	159
F–O	147	190
F–N	137	272

For other bonds see other elements

Physical properties

Melting point/K: 53.53

Boiling point/K: 85.01

ΔH_{fusion}/kJ mol^{-1}: 1.02

ΔH_{vap}/kJ mol^{-1}: 3.26

Thermodynamic properties (298.15 K, 0.1 MPa)

State	$\Delta_f H^{\ominus}$/kJ mol^{-1}	$\Delta_f G^{\ominus}$/kJ mol^{-1}	$S^{\ominus}$/J K^{-1} mol^{-1}	C_p/J K^{-1} mol^{-1}
Gas (F_2)	0	0	202.78	31.30
Gas atoms	78.99	61.91	158.754	22.744

Density/kg m^{-3}: n.a. [solid]; 1516 [liquid, 85 K]; 1.696 [gas, 273 K]

Thermal conductivity/W m^{-1} K^{-1}: 0.0279 [300 K]

Mass magnetic susceptibility/kg^{-1} m^3: n.a.

Molar volume/cm^3: 18.05 [85 K]

Lattice structure (cell dimensions/pm), space group

α-F_2 rhombic (cell dimensions n.a.), C2/m
β-F_2 cubic (a=667), Pm3n

$T(\alpha\rightarrow\beta)$=45.6 K

X-ray diffraction: mass absorption coefficients (μ/ρ)/cm^2 g^{-1}:
CuK_α 16.4 MoK_α 1.80

First isolated in 1886 by H. Moissan at Paris, France

[Latin, *fluere* = to flow]

Fluorine

Nuclear properties

Atomic number: 9

Thermal neutron capture cross-section/barns: 0.098 ± 0.007

Number of isotopes (including nuclear isomers): 6

Isotope mass range: 17→22

Key isotopes

Nuclide	Atomic mass	Natural abundance (%)	Half life $T_{1/2}$	Decay mode and energy (MeV)	Nuclear spin I	Nuclear magnetic moment μ	Uses
^{18}F		0	109.7m	β^+ (1.65), EC (3%), no γ			tracer, medical
^{19}F	18.998 40	100	stable		1/2	+2.6283	NMR

NMR

	^{19}F
Relative sensitivity (^1H = 1.00)	0.83
Absolute sensitivity (^1H = 1.00)	0.83
Receptivity (^{13}C = 1.00)	4730
Magnetogyric ratio/rad T^{-1} s^{-1}	25.1665×10^7
Frequency (^1H = 100 MHz; 2.3488 T)/MHz	94.077

Reference: $CFCl_3$

Electron shell properties

Ground state electron configuration: [He]$2s^22p^5$

Term symbol: $^2P_{3/2}$

Electron affinity ($M \rightarrow M^-$)/kJ mol^{-1}: 322

Ionization energies/kJ mol^{-1}

1. $M \rightarrow M^+$	1 681	6. $M^{5+} \rightarrow M^{6+}$	15 164
2. $M^+ \rightarrow M^{2+}$	3 374	7. $M^{6+} \rightarrow M^{7+}$	17 867
3. $M^{2+} \rightarrow M^{3+}$	6 050	8. $M^{7+} \rightarrow M^{8+}$	92 036
4. $M^{3+} \rightarrow M^{4+}$	8 408	9. $M^{8+} \rightarrow M^{9+}$	106 432
5. $M^{4+} \rightarrow M^{5+}$	11 023		

Principal lines in atomic spectrum

Wavelength/nm	Species	Sensitivity	Application
95.482* (vac)	I		
685.603	I	U3	AE
690.248	I	U2	AE
703.747	I		
712.787			

* Lowest energy transition from ground state to nearest empty orbital

Abundance: Earth's crust 544 p.p.m.; seawater 1.3 p.p.m.

Biological role: Essential; excess fluoride toxic; HF and F_2 very toxic

Fr

Atomic number: 87

Relative atomic mass (^{12}C=12.0000): (223)

Chemical properties

Intensely radioactive, short-lived metal element obtained in minute quantities from actinium obtained from the neutron bombardment of radium.

Radii/pm: Fr^+ 180; atomic *c.* 270

Electronegativity: 0.7 (Pauling); 0.86 (Allred-Rochow)

Effective nuclear charge: 2.20 (Slater)

Standard reduction potentials $E^{\ominus}$/V

I		0
Fr^+	*c.* −2.9	Fr

Oxidation states

Fr^I ([Rn])	little studied, Fr^+ (aq), $FrClO_4$ insoluble

Physical properties

Melting point/K: 300

Boiling point/K: 950

ΔH_{fusion}/kJ mol^{-1}: n.a.

ΔH_{vap}/kJ mol^{-1}: n.a.

Thermodynamic properties (298.15 K, 0.1 MPa)

State	$\Delta_f H^{\ominus}$/kJ mol^{-1}	$\Delta_f G^{\ominus}$/kJ mol^{-1}	$S^{\ominus}$/J K^{-1} mol^{-1}	C_p/J K^{-1} mol^{-1}
Solid	0	0	95.4	n.a.
Gas	72.8	n.a.	n.a.	n.a.

Density/kg m^{-3}: n.a.

Thermal conductivity/W m^{-1} K^{-1}: 15 (est.) [300 K]

Electrical resistivity/Ω m: n.a.

Mass magnetic susceptibility/kg^{-1} m^3: n.a.

Molar volume/cm^3: n.a.

Coefficient of linear thermal expansion/K^{-1}: n.a.

Lattice structure (cell dimensions/pm), space group

n.a.

X-ray diffraction: mass absorption coefficients (μ/ρ)/cm^2 g^{-1}:

n.a.

Discovered in 1939 by Marguerite Perey in Paris, France

[Named after France]

Francium

Nuclear properties

Atomic number: 87

Thermal neutron capture cross-section/barns: n.a.

Number of isotopes (including nuclear isomers): 21

Isotope mass range: 204→224

Key isotopes

Nuclide	Atomic mass	Natural abundance (%)	Half life $T_{1/2}$	Decay mode and energy (MeV)	Nuclear spin I	Nuclear magnetic moment μ	Uses
^{212}Fr		0	19m				
^{223}Fr	223.0198	some	22m*	β^- (1.15); γ			

* longest lived isotope

Electron shell properties

Ground state electron configuration: $[Rn]7s^1$

Term symbol: $^2S_{1/2}$

Electron affinity $(M \rightarrow M^-)$/kJ mol^{-1}: 44 (calc.)

Ionization energies/kJ mol^{-1}

1. $M \rightarrow M^+$	400	6. $M^{5+} \rightarrow M^{6+}$	(6 900)
2. $M^+ \rightarrow M^{2+}$	(2100)	7. $M^{6+} \rightarrow M^{7+}$	(8 100)
3. $M^{2+} \rightarrow M^{3+}$	(3100)	8. $M^{7+} \rightarrow M^{8+}$	(12 300)
4. $M^{3+} \rightarrow M^{4+}$	(4100)	9. $M^{8+} \rightarrow M^{9+}$	(12 800)
5. $M^{4+} \rightarrow M^{5+}$	(5700)	10. $M^{9+} \rightarrow M^{10+}$	(29 300)

Principal lines in atomic spectrum

Wavelength/nm	Species	Sensitivity	Application
n.a.			

Abundance: Minute traces in uranium ores

Biological role: None; toxic due to radioactivity

Gd

Atomic number: 64
Relative atomic mass ($^{12}C = 12.0000$): 157.25

Chemical properties

Silvery-white metal of rare earth group. Obtained from monazite and bastnaesite. Reacts slowly with oxygen and water, dissolves in acids. Limited use in electronics, refractories, and alloys.

Radii/pm: Gd^{3+} 97; atomic 180.2; covalent 161
Electronegativity: 1.20 (Pauling); 1.11 (Allred-Rochow)
Effective nuclear charge: 2.85 (Slater); 8.22 (Clementi); 11.28 (Froese–Fischer)

Standard reduction potentials $E^{\ominus}$/V

	III		0
Acid solution	Gd^{3+}	$\xrightarrow{-2.28}$	Gd
Alkaline solution	$Gd(OH)_3$	$\xrightarrow{-2.82}$	Gd

Oxidation states

Gd^{II} (f^8)	GdI_2
Gd^{III} (f^7)	Gd_2O_3, $Gd(OH)_3$, GdF_3, $GdCl_3$ etc., $[Gd(H_2O)_x]^{3+}$ (aq), Gd^{3+} salts and complexes

Physical properties

Melting point/K: 1586
Boiling point/K: 3539
ΔH_{fusion}/kJ mol^{-1}: 15.5
ΔH_{vap}/kJ mol^{-1}: 301

Thermodynamic properties (298.15 K, 0.1 MPa)

State	$\Delta_f H^{\ominus}$/kJ mol^{-1}	$\Delta_f G^{\ominus}$/kJ mol^{-1}	$S^{\ominus}$/J K^{-1} mol^{-1}	C_p/J K^{-1} mol^{-1}
Solid	0	0	68.07	37.03
Gas	397.5	359.8	193.314	27.547

Density/kg m^{-3}: 7900.4 [298 K]
Thermal conductivity/W m^{-1} K^{-1}: 10.6 [300 K]
Electrical resistivity/Ω m: 134.0×10^{-8} [298 K]
Mass magnetic susceptibility/kg^{-1} m^3: $+6.030 \times 10^{-5}$ (s)
Molar volume/cm^3: 19.90
Coefficient of linear thermal expansion/K^{-1}: 8.6×10^{-6}

Lattice structure (cell dimensions/pm), space group

α-Gd h.c.p. ($a = 363.60$, $c = 578.26$), $P6_3/mmc$
β-Gd b.c.c. ($a = 405$), Im3m

$T(\alpha \rightarrow \beta) = 1535$ K

high pressure form: ($a = 361$, $c = 2603$), $R\bar{3}m$

X-ray diffraction: mass absorption coefficients ((μ/ρ)/cm^2 g^{-1}):
CuK_α 439 MoK_α 64.4

Discovered in 1880 by J.-C. Galissard de Marignac at Geneva, Switzerland. Isolated in 1886 by P.-E. Lecoq de Boisbaudran at Paris, France
[Named after J. Gadolin, a Finnish chemist]

Gadolinium

Nuclear properties

Atomic number: 64
Thermal neutron capture cross-section/barns: $49\,000 \pm 2000$
Number of isotopes (including nuclear isomers): 17
Isotope mass range: $145 \rightarrow 161$

Key isotopes

Nuclide	Atomic mass	Natural abundance (%)	Half life $T_{1/2}$	Decay mode and energy (MeV)	Nuclear spin I	Nuclear magnetic moment μ	Uses
^{152}Gd	151.9195	0.20	1.1×10^{14}y	α (2.24)			
^{153}Gd		0	242d	EC (0.243); γ	3/2		tracer
^{154}Gd	153.9207	2.1	stable				
^{155}Gd	154.9226	14.8	stable		3/2	−0.2584	NMR
^{156}Gd	155.9221	20.6	stable				
^{157}Gd	156.9339	15.7	stable		3/2	−0.3388	NMR
^{158}Gd	157.9241	24.8	stable				
^{160}Gd	159.9071	21.8	stable				

NMR

	^{155}Gd	^{157}Gd
Relative sensitivity ($^1H = 1.00$)	2.79×10^{-4}	5.44×10^{-4}
Absolute sensitivity ($^1H = 1.00$)	4.11×10^{-5}	8.53×10^{-5}
Receptivity ($^{13}C = 1.00$)	0.124	0.292
Magnetogyric ratio/rad T^{-1} s^{-1}	-0.8273×10^7	-1.0792×10^7
Quadrupole moment/m^2	1.6×10^{-28}	2×10^{-28}
Frequency ($^1H = 100$ MHz; 2.3488 T)/MHz	3.819	4.774

Electron shell properties

Ground state electron configuration: $[Xe]4f^7 5d^1 6s^2$
Term symbol: $^9D_2$
Electron affinity ($M \rightarrow M^-$)/kJ mol^{-1}: ≤ 50

Ionization energies/kJ mol^{-1}

1. $M \rightarrow M^+$	592.5	6. $M^{5+} \rightarrow M^{6+}$	
2. $M^+ \rightarrow M^{2+}$	1167	7. $M^{6+} \rightarrow M^{7+}$	
3. $M^{2+} \rightarrow M^{3+}$	1990	8. $M^{7+} \rightarrow M^{8+}$	
4. $M^{3+} \rightarrow M^{4+}$	4250	9. $M^{8+} \rightarrow M^{9+}$	
5. $M^{4+} \rightarrow M^{5+}$		10. $M^{9+} \rightarrow M^{10+}$	

Principal lines in atomic spectrum

Wavelength/nm	Species	Sensitivity	Application
364.620	II		AE
368.413	I		AA
376.841	II		AE
378.305	I		AA
404.870	I		AA

Abundance: Earth's crust 6.1 p.p.m.; seawater 2×10^{-7} p.p.m.
Biological role: None; toxicity unknown, probably low

Ga

Atomic number: 31

Relative atomic mass (^{12}C=12.0000): 69.723

Chemical properties

Soft, silvery-white metal, stable in air and with water. Soluble in acids and alkalis. Longest liquid range of all elements. Semiconductor properties with phosphorus, arsenic, and antimony.

Radii/pm: Ga^{3+} 62; Ga^{+} 113; atomic 122.1; covalent 125

Electronegativity: 1.81 (Pauling); 1.82 (Allred-Rochow)

Effective nuclear charge: 5.00 (Slater); 6.22 (Clementi); 6.72 (Froese–Fischer)

Standard reduction potentials $E^\ominus$/V

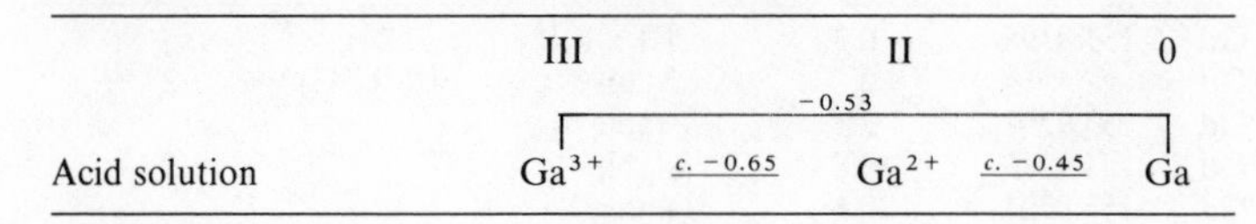

Oxidation states

Ga^{I} (s^2)	Ga_2O, $GaCl_2$ etc., $Ga_2Cl_4 = Ga^{I}[Ga^{III}Cl_4]$
Ga^{II} (s^1)	$[Ga_2Cl_6]^{2-}$
Ga^{III} ($[d^{10}]$)	Ga_2O_3, $Ga(OH)_3$, $[Ga(H_2O)_6]^{3+}$ (aq) GaF_3, Ga_2Cl_6, $GaCl_6^{3-}$

Physical properties

Melting point/K: 302.93

Boiling point/K: 2676

ΔH_{fusion}/kJ mol^{-1}: 5.59

ΔH_{vap}/kJ mol^{-1}: 270.3

Thermodynamic properties (298.15 K, 0.1 MPa)

State	$\Delta_f H^\ominus$/kJ mol^{-1}	$\Delta_f G^\ominus$/kJ mol^{-1}	$S^\ominus$/J K^{-1} mol^{-1}	C_p/J K^{-1} mol^{-1}
Solid	0	0	40.88	25.86
Gas	277.0	238.9	169.06	25.36

Density/kg m^{-3}: 5907 [293 K]; 6113.6 [liquid at m.p.]

Thermal conductivity/W m^{-1} K^{-1}: 40.6 [300 K]

Electrical resistivity/Ω m: 27×10^{-8} [273 K] varies with axis

Mass magnetic susceptibility/kg^{-1} m^3: -3.9×10^{-9} (s)

Molar volume/cm^3: 11.81

Coefficient of linear thermal expansion/K^{-1}:
11.5×10^{-6} (*a* axis); 31.5×10^{-6} (*b* axis); 16.5×10^{-6} (*c* axis)

Lattice structure (cell dimensions/pm), space group

α-Ga orthorhombic (a=451.86, b=765.70, c=452.58), Cmca
β-Ga rhombic (a=290, b=813, c=317), Cmcm (metastable form)
γ-Ga rhombic (a=1060, b=1356, c=519), $Cmc2_1$

$T(\gamma \rightarrow \alpha)$=238 K
High pressure form: (a=279, c=438), I4/mmm

X-ray diffraction: mass absorption coefficients (μ/ρ)/cm^2 g^{-1}:
CuK_α 67.9 MoK_α 60.1

Discovered in 1875 by Paul-Émile Lecoq de Boisbaudran at Paris, France

[Latin, *Gallia* = France]

Gallium

Nuclear properties

Atomic number: 31

Thermal neutron capture cross-section/barns: 3.1 ± 0.3

Number of isotopes (including nuclear isomers): 14

Isotope mass range: 63→76

Key isotopes

Nuclide	Atomic mass	Natural abundance (%)	Half life $T_{1/2}$	Decay mode and energy (MeV)	Nuclear spin I	Nuclear magnetic moment μ	Uses
^{67}Ga		0	78.1h	EC(1.0003); γ	3/2	+1.849	medical tracer
^{69}Ga	68.9257	60.1	stable		3/2	+2.0145	NMR
^{71}Ga	70.9249	39.9	stable		3/2	+2.5597	NMR
^{72}Ga		0	14.1h	β^-(4.00); γ	3	−0.1321	tracer

NMR

	[^{69}Ga]	^{71}Ga
Relative sensitivity (^1H = 1.00)	6.91×10^{-2}	0.14
Absolute sensitivity (^1H = 1.00)	4.17×10^{-2}	5.62×10^{-2}
Receptivity (^{13}C = 1.00)	237	319
Magnetogyric ratio/rad T^{-1} s^{-1}	6.420×10^7	8.158×10^7
Quadrupole moment/m^2	0.178×10^{-28}	0.112×10^{-28}
Frequency (^1H = 100 MHz; 2.3488 T)/MHz	24.003	30.495

Reference: $Ga(H_2O)_6^{3+}$

Electron shell properties

Ground state electron configuration: $[Ar]3d^{10}4s^24p^1$

Term symbol: $^2P_{1/2}$

Electron affinity ($M \rightarrow M^-$)/kJ mol^{-1}: 36 (calc.)

Ionization energies/kJ mol^{-1}

1.	$M \rightarrow M^+$	578.8	6.	$M^{5+} \rightarrow M^{6+}$	(11 400)
2.	$M^+ \rightarrow M^{2+}$	1979	7.	$M^{6+} \rightarrow M^{7+}$	(14 400)
3.	$M^{2+} \rightarrow M^{3+}$	2963	8.	$M^{7+} \rightarrow M^{8+}$	(17 700)
4.	$M^{3+} \rightarrow M^{4+}$	6200	9.	$M^{8+} \rightarrow M^{9+}$	(22 300)
5.	$M^{4+} \rightarrow M^{5+}$	(8700)	10.	$M^{9+} \rightarrow M^{10+}$	(26 100)

Principal lines in atomic spectrum

Wavelength/nm	Species	Sensitivity	Application
287.424	I	U4	AA, AE
294.364	I	U3	AA, AE
403.298*	I	U2	AA, AE
417.206	I	U1	AA, AE
639.656	I		

* Lowest energy transition from ground state to nearest empty orbital

Abundance: Earth's crust 19 p.p.m.; seawater 0.0005 p.p.m.

Biological role: None; non-toxic; stimulatory

Ge

Atomic number: 32
Relative atomic mass (^{12}C=12.0000): 72.61

Chemical properties

Ultrapure it is a silvery-white brittle metalloid element. Minerals are rare; obtained from zinc smelter flue dust. Stable in air and water, unaffected by acids, except nitric, and alkalis. Used in semiconductors, alloys, and special glasses.

Radii/pm: Ge^{2+} 90; atomic 122.5; covalent 122; Ge^{4-} 272

Electronegativity: 2.01 (Pauling); 2.02 (Allred-Rochow)

Effective nuclear charge: 5.65 (Slater); 6.78 (Clementi); 7.92 (Froese–Fischer)

Standard reduction potentials $E^{\ominus}$/V

	IV		II		0		−IV
Acid solution	GeO_2	−0.370	GeO	0.255	Ge	−0.29	GeH_4
	Ge^{4+}	0.00	Ge^{2+}	−0.247 (to Ge)			

[Alkaline solutions contain many different forms]

Covalent bonds

	r/pm	E/kJ mol^{-1}
Ge–H	152.9	288
Ge–C	194	237
Ge–O	165	363
Ge–F	170	464
Ge–Cl	210	340
Ge–Ge	241	163

Oxidation states

Ge^{II} (s^2)	GeO, GeS, GeF_2, $GeCl_2$ etc.
Ge^{IV} (d^{10})	GeO_2, GeH_4 etc., GeF_4, $GeCl_4$ etc., GeF_6^{2-}, $GeCl_6^{2-}$, GeS_2, $[GeO(OH)_3]^-$ (aq)

Physical properties

Melting point/K: 1210.6

Boiling point/K: 3103

ΔH_{fusion}/kJ mol^{-1}: 34.7

ΔH_{vap}/kJ mol^{-1}: 327.6

Thermodynamic properties (298.15 K, 0.1 MPa)

State	$\Delta_f H^{\ominus}$/kJ mol^{-1}	$\Delta_f G^{\ominus}$/kJ mol^{-1}	$S^{\ominus}$/J K^{-1} mol^{-1}	C_p/J K^{-1} mol^{-1}
Solid	0	0	31.09	23.347
Gas	376.6	335.9	167.900	30.731

Density/kg m^{-3}: 5323 [293 K]; 5490 [liquid at m.p.]

Thermal conductivity/W m^{-1} K^{-1}: 59.9 [300 K]

Electrical resistivity/Ω m: 0.46 [295 K]

Mass magnetic susceptibility/kg^{-1} m^3: -1.328×10^{-9} (s)

Molar volume/cm^3: 13.64

Coefficient of linear thermal expansion/K^{-1}: 5.57×10^{-6}

Lattice structure (cell dimensions/pm), space group

Cubic (a=565.754), Fd3m, diamond structure

High pressure forms: (a=488.4; c=269.2), $I4_1$/amd
(a=593; c=698), $P4_32_12$
(a=692), b.c.c.

X-ray diffraction: mass absorption coefficients (μ/ρ)/cm^2 g^{-1}:
CuK_α 75.6 MoK_α 64.8

Discovered in 1886 by C. A. Winkler at Freiberg, Germany

[Latin, *Germania* = Germany]

Germanium

Nuclear properties

Atomic number: 32

Thermal neutron capture cross-section/barns: 2.3 ± 0.3

Number of isotopes (including nuclear isomers): 17

Isotope mass range: 65→78

Key isotopes

Nuclide	Atomic mass	Natural abundance (%)	Half life $T_{1/2}$	Decay mode and energy (MeV)	Nuclear spin I	Nuclear magnetic moment μ	Uses
^{68}Ge		0	287d	EC(*c.* 0.7); γ			tracer
^{70}Ge	69.9243	20.5	stable		0		
^{71}Ge	70.925	0	11.4d	EC(0.235); γ	1/2	+0.546	tracer
^{72}Ge	71.9217	27.4	stable		0		
^{73}Ge	72.9234	7.8	stable		9/2	−0.8792	NMR
^{74}Ge	73.9219	36.5	stable		0		
^{76}Ge	75.9214	7.8	stable		0		
^{77}Ge		0	11.3h	β^-(2.75);γ			tracer

NMR

	^{73}Ge
Relative sensitivity (^{1}H = 1.00)	1.4×10^{-3}
Absolute sensitivity (^{1}H = 1.00)	1.08×10^{-4}
Receptivity (^{13}C = 1.00)	0.617
Magnetogyric ratio/rad T^{-1} s^{-1}	-0.9331×10^7
Quadrupole moment/m^2	-0.2×10^{-28}
Frequency (^{1}H = 100 MHz; 2.3488 T)/MHz	3.488

Reference: $Ge(CH_3)_4$

Electron shell properties

Ground state electron configuration: [Ar] $3d^{10}4s^24p^2$

Term symbol: $^3P_0$

Electron affinity ($M \rightarrow M^-$)/kJ mol^{-1}: 116

Ionization energies/kJ mol^{-1}

1. $M \rightarrow M^+$	762.1	6. $M^{5+} \rightarrow M^{6+}$	(11 900)
2. $M^+ \rightarrow M^{2+}$	1537	7. $M^{6+} \rightarrow M^{7+}$	(15 000)
3. $M^{2+} \rightarrow M^{3+}$	3302	8. $M^{7+} \rightarrow M^{8+}$	(18 200)
4. $M^{3+} \rightarrow M^{4+}$	4410	9. $M^{8+} \rightarrow M^{9+}$	(21 800)
5. $M^{4+} \rightarrow M^{5+}$	9020	10. $M^{9+} \rightarrow M^{10+}$	(27 000)

Principal lines in atomic spectrum

Wavelength/nm	Species	Sensitivity	Application
206.866	I		
259.253	I		AA
265.158*	I		AA, AE
303.906	I	U2	AA, AE
326.949	I	U3	AE

* Lowest energy transition from ground state to nearest empty orbital

Abundance: Earth's crust 1.5 p.p.m.; seawater 1×10^{-4} p.p.m.

Biological role: None; non-toxic; stimulatory

Au

Atomic number: 79
Relative atomic mass (^{12}C=12.0000): 196.96654

Chemical properties

Soft metal with characteristic yellow colour. Highest malleability and ductility of any element. Found in free state. Unaffected by air, water, acids (except HNO_3–HCl), and alkalis. Used as bullion, in jewellery, electronics, and glass (colouring and heat reflecting).

Radii/pm: Au^+ 137; Au^{3+} 91; atomic 144.2; covalent 134
Electronegativity: 2.54 (Pauling); 1.42 (Allred-Rochow)
Effective nuclear charge: 4.20 (Slater); 10.94 (Clementi); 15.94 (Froese–Fischer)

Standard reduction potentials $E^{\ominus}$/V

	III		I		0
Acid solution	Au^{3+}	1.36	Au^+	1.83	Au
		1.52 (Au^{3+} → Au)			
	$AuCl_4^-$	0.926	$AuCl_2^-$	1.154	Au
		1.002 ($AuCl_4^-$ → Au)			
	$Au(SCN)_4^-$	0.623	$Au(SCN)_2^-$	0.662	Au
		0.636 ($Au(SCN)_4^-$ → Au)			

Oxidation states

Au^{-I} (d^{10} s^2)	$[Au(NH_3)_n]^-$ in liquid NH_3
Au^0 (d^{10} s^1)	gold clusters eg. $[Au_8(PPh_3)_8]^{2+}$
Au^I (d^{10})	Au_2S, $[Au(CN)_2]^-$ and other complexes
Au^{II} (d^9)	rare but some complexes known
Au^{III} (d^8)	Au_2O_3, $Au(OH)_4^-$ (aq), $AuCl_4^-$ (aq), $AuCl_3(OH)^-$ (aq), Au_2S_3, AuF_3, Au_2Cl_6, $AuBr_3$, complexes
Au^V (d^6)	AuF_5
Au^{VII} (d^4)	AuF_7

Physical properties

Melting point/K: 1337.58
Boiling point/K: 3080
ΔH_{fusion}/kJ mol^{-1}: 12.7
ΔH_{vap}/kJ mol^{-1}: 343.1

Thermodynamic properties (298.15 K, 0.1 MPa)

State	$\Delta_f H^{\ominus}$/kJ mol^{-1}	$\Delta_f G^{\ominus}$/kJ mol^{-1}	$S^{\ominus}$/J K^{-1} mol^{-1}	C_p/J K^{-1} mol^{-1}
Solid	0	0	47.40	25.418
Gas	336.1	326.3	180.503	20.786

Density/kg m^{-3}: 19 320 [293 K]; 17 280 [liquid at m.p.]
Thermal conductivity/W m^{-1} K^{-1}: 317 [300 K]
Electrical resistivity/Ω m: 2.35×10^{-8} [293 K]
Mass magnetic susceptibility/kg^{-1} m^3: -1.78×10^{-9} (s)
Molar volume/cm^3: 10.19
Coefficient of linear thermal expansion/K^{-1}: 14.16×10^{-6}

Lattice structure (cell dimensions/pm), space group

f.c.c. (a=407.833), Fm3m

X-ray diffraction: mass absorption coefficients (μ/ρ)/cm^2 g^{-1}:
CuK_{α} 208 MoK_{α} 115

Known to ancient civilizations

[Anglo-Saxon, *gold*]

Gold

Nuclear properties

Atomic number: 79
Thermal neutron capture cross-section/barns: 98.8 ± 0.3
Number of isotopes (including nuclear isomers): 21
Isotope mass range: 185→203

Key isotopes

Nuclide	Atomic mass	Natural abundance (%)	Half life $T_{1/2}$	Decay mode and energy (MeV)	Nuclear spin I	Nuclear magnetic moment μ	Uses
^{195}Au		0	183d	EC(0.227); γ	3/2	± 0.147	tracer
^{197}Au	196.9666	100	stable		3/2	+0.144 86	NMR
^{198}Au		0	2.693d	$\beta^-(1.374)$; γ	2	+0.590	tracer, medical
^{199}Au		0	3.15d	$\beta^-(0.46)$; γ	3/2	+0.270	tracer

NMR (difficult to detect)	^{197}Au
Relative sensitivity ($^1H = 1.00$)	2.51×10^{-5}
Absolute sensitivity ($^1H = 1.00$)	2.51×10^{-5}
Receptivity ($^{13}C = 1.00$)	0.06
Magnetogyric ratio/rad T^{-1} s^{-1}	0.357×10^7
Quadrupole moment/m^2	0.58×10^{-28}
Frequency ($^1H = 100$ MHz; 2.3488 T)/MHz	1.712

Electron shell properties

Ground state electron configuration: $[Xe]4f^{14}5d^{10}6s^1$
Term symbol: $^2S_{1/2}$
Electron affinity ($M \rightarrow M^-$)/kJ mol^{-1}: 223

Ionization energies/kJ mol^{-1}

1. $M \rightarrow M^+$	890.1	6. $M^{5+} \rightarrow M^{6+}$	(7000)
2. $M^+ \rightarrow M^{2+}$	1980	7. $M^{6+} \rightarrow M^{7+}$	(9300)
3. $M^{2+} \rightarrow M^{3+}$	(2900)	8. $M^{7+} \rightarrow M^{8+}$	(11000)
4. $M^{3+} \rightarrow M^{4+}$	(4200)	9. $M^{8+} \rightarrow M^{9+}$	(12800)
5. $M^{4+} \rightarrow M^{5+}$	(5600)	10. $M^{9+} \rightarrow M^{10+}$	(14800)

Principal lines in atomic spectrum

Wavelength/nm	Species	Sensitivity	Application
201.200	I		
242.795	I	U1	AA, AE
267.595*	I	U2	AA, AE
274.826	I		AA
312.278	I		AA

* Lowest energy transition from ground state to nearest empty orbital

Abundance: Earth's crust 0.004 p.p.m.; seawater 4×10^{-6} p.p.m.
Biological role: None; non-toxic; stimulatory

Hf	**Atomic number: 72** **Relative atomic mass** ($^{12}C=12.0000$): **178.49**

Chemical properties

Lustrous, silvery, ductile metal. Obtained from zirconium ores. Resists corrosion due to oxide film, but powdered Hf will burn in air. Unaffected by acids (except HF) and alkalis. Used in control rods for nuclear reactors.

Radii/pm: Hf^{4+} 84; atomic 156.4; covalent 144

Electronegativity: 1.3 (Pauling); 1.23 (Allred-Rochow)

Effective nuclear charge: 3.15 (Slater); 9.16 (Clementi); 13.27 (Froese–Fischer)

Standard reduction potentials $E^{\ominus}$/V

IV		0
Hf^{4+}	$\xrightarrow{-1.70}$	Hf
HfO_2	$\xrightarrow{-1.57}$	Hf

Oxidation states

Hf^{I} (d^3)	HfCl?
Hf^{II} (d^2)	$HfCl_2$?
Hf^{III} (d^1)	$HfCl_3$, $HfBr_3$, HfI_3, Hf^{3+} reduces water
Hf^{IV} (d^0, f^{14})	HfO_2, $Hf(OH)^{3+}$ (aq), HfF_4, $HfCl_4$ etc., HfF_6^{2-}, HfF_7^{3-}, HfF_8^{4-}

Physical properties

Melting point/K: 2503

Boiling point/K: 5470

ΔH_{fusion}/kJ mol^{-1}: 25.5

ΔH_{vap}/kJ mol^{-1}: 570.7

Thermodynamic properties (298.15 K, 0.1 MPa)

State	$\Delta_f H^{\ominus}$/kJ mol^{-1}	$\Delta_f G^{\ominus}$/kJ mol^{-1}	$S^{\ominus}$/J K^{-1} mol^{-1}	C_p/J K^{-1} mol^{-1}
Solid	0	0	43.56	25.73
Gas	619.2	576.5	186.892	20.803

Density/kg m^{-3}: 13 310 [293 K]; 12 000 [liquid at m.p.]

Thermal conductivity/W m^{-1} K^{-1}: 23.0 [300 K]

Electrical resistivity/Ω m: 35.1×10^{-8} [293 K]

Mass magnetic susceptibility/kg^{-1} m^3: $+5.3 \times 10^{-9}$ (s)

Molar volume/cm^3: 13.41

Coefficient of linear thermal expansion/K^{-1}: 5.9×10^{-6}

Lattice structure (cell dimensions/pm), space group

α-Hf h.c.p. ($a=319.46$; $c=505.10$), $P6_3/mmc$
β-Hf cubic ($a=362$)
$T(\alpha \rightarrow \beta)=2033$ K

X-ray diffraction: mass absorption coefficients (μ/ρ)/cm^2 g^{-1}:
CuK_α 159 MoK_α 91.7

Discovered in 1923 by D. Coster and G. C. von Hevesey at Copenhagen, Denmark

[Latin, *Hafnia* = Copenhagen]

Hafnium

Nuclear properties

Atomic number: 72

Thermal neutron capture cross-section/barns: 103 ± 3

Number of isotopes (including nuclear isomers): 19

Isotope mass range: $169 \rightarrow 183$

Key isotopes

Nuclide	Atomic mass	Natural abundance (%)	Half life $T_{1/2}$	Decay mode and energy (MeV)	Nuclear spin I	Nuclear magnetic moment μ	Uses
^{172}Hf		0	*c.* 5y	EC(*c.* 1); γ			tracer
^{174}Hf	173.9403	0.2	2×10^{15}y	α(2.55)			
^{175}Hf		0	70d	EC(0.59); γ			tracer
^{176}Hf	175.941 65	5.2	stable				
^{177}Hf	176.9435	18.6	stable		7/2	+0.7902	NMR
^{178}Hf	177.9439	27.1	stable		0		
^{179}Hf	178.9460	13.7	stable		9/2	−0.638	NMR
^{180}Hf	179.9468	35.2	stable		0		
^{181}Hf		0	42.5d	β^-(1.023); γ			tracer
^{182}Hf		0	9×10^6y	β^-(*c.* 0.5)			tracer

NMR (difficult to observe)	^{177}Hf	^{179}Hf
Relative sensitivity (^{1}H = 1.00)	6.38×10^{-4}	2.16×10^{-4}
Absolute sensitivity (^{1}H = 1.00)	1.18×10^{-4}	2.97×10^{-5}
Receptivity (^{13}C = 1.00)	0.88	0.27
Magnetogyric ratio/rad T^{-1} s^{-1}	$+0.945 \times 10^7$	-0.609×10^7
Quadrupole moment/m^2	4.5×10^{-28}	5.1×10^{-28}
Frequency (^{1}H = 100 MHz; 2.3488 T)/MHz	3.120	1.869

Electron shell properties

Ground state electron configuration: [Xe]$4f^{14}5d^26s^2$

Term symbol: $^3F_2$

Electron affinity ($M \rightarrow M^-$)/kJ mol^{-1}: −61

Ionization energies/kJ mol^{-1}

1. $M \rightarrow M^+$	642	6. $M^{5+} \rightarrow M^{6+}$	
2. $M^+ \rightarrow M^{2+}$	1440	7. $M^{6+} \rightarrow M^{7+}$	
3. $M^{2+} \rightarrow M^{3+}$	2250	8. $M^{7+} \rightarrow M^{8+}$	
4. $M^{3+} \rightarrow M^{4+}$	3216	9. $M^{8+} \rightarrow M^{9+}$	
5. $M^{4+} \rightarrow M^{5+}$		10. $M^{9+} \rightarrow M^{10+}$	

Principal lines in atomic spectrum

Wavelength/nm	Species	Sensitivity	Application
202.818	II		
289.826	I		AA, AE
307.288	I		AA, AE
368.224	I		AA, AE
713.182*	I		AE

* Lowest energy transition from ground state to nearest empty orbital

Abundance: Earth's crust 2.8 p.p.m.; seawater <0.008 p.p.m.

Biological role: None; non-toxic

He

Atomic number: 2
Relative atomic mass (^{12}C=12.0000): **4.002602**

Chemical properties

Colourless, odourless gas obtained from natural gas wells. Unreactive chemically. Used in deep-sea diving, weather balloons, and low temperature research instruments.

Radii/pm: atomic 128; van der Waals 122
Electronegativity: n.a. (Pauling); n.a. (Allred-Rochow)
Effective nuclear charge: 1.70 (Slater); 1.69 (Clementi); 1.62 (Froese–Fischer)

Standard reduction potentials $E^{\ominus}$/V

n.a.

Covalent bonds r/pm E/kJ mol^{-1}	**Oxidation states**
n.a.	n.a.

Physical properties

Melting point/K: 0.95 (under pressure)
Boiling point/K: 4.216
ΔH_{fusion}/kJ mol^{-1}: 0.021
ΔH_{vap}/kJ mol^{-1}: 0.082

Thermodynamic properties (298.15 K, 0.1 MPa)

State	$\Delta_f H^{\ominus}$/kJ mol^{-1}	$\Delta_f G^{\ominus}$/kJ mol^{-1}	$S^{\ominus}$/J K^{-1} mol^{-1}	C_p/J K^{-1} mol^{-1}
Gas	0	0	126.150	20.786

Density/kg m^{-3}: n.a. [solid]; 124.8 [liquid at b.p.]; 0.1785 [gas, 273 K]
Thermal conductivity/W m^{-1} K^{-1}: 0.152 [300 K]$_g$
Mass magnetic susceptibility/kg^{-1} m^3: -5.9×10^{-9} (g)
Molar volume/cm^3: 32.07 [4 K]

Lattice structure (cell dimensions/pm), space group [T, p]

α-He h.c.p. (a=353.1; c=569.3), P6$_3$/mmc. [T=1.15 K, p=6.69 MPa]
β-He f.c.c. (a=424.0), Fm3m. [T=16 K, p=127 MPa]
γ-He b.c.c. (a=411), Im3m. [T=1.73 K, p=2.94 MPa]

X-ray diffraction: mass absorption coefficients (μ/ρ)/cm^2 g^{-1}:
CuK$_\alpha$ 0.383 MoK$_\alpha$ 0.207

Isolated in 1895 by Sir William Ramsay at London, UK and independently by P. T. Cleve and N. A. Langlet at Uppsala, Sweden
[Greek, *helios* = sun]

Helium

Nuclear properties

Atomic number: 2
Thermal neutron capture cross-section/barns: *c.* 0.007
Number of isotopes (including nuclear isomers): 5
Isotope mass range: 3→8 except 7

Key isotopes

Nuclide	Atomic mass	Natural abundance (%)	Half life $T_{1/2}$	Decay mode and energy (MeV)	Nuclear spin I	Nuclear magnetic moment μ	Uses
^{3}He	3.016 03	1.38×10^{-4}	stable		1/2	−2.1276	NMR, tracer
^{4}He	4.002 60	99.999 862	stable	0			

NMR (no compounds known)	^{3}He
Relative sensitivity ($^{1}H = 1.00$)	0.44
Absolute sensitivity ($^{1}H = 1.00$)	5.75×10^{-7}
Receptivity ($^{13}C = 1.00$)	0.003 26
Magnetogyric ratio/rad T^{-1} s^{-1}	-20.378×10^{7}
Frequency ($^{1}H = 100$ MHz; 2.3488 T)/MHz	76.178

Electron shell properties

Ground state electron configuration: $1s^2$
Term symbol: $^{1}S_{0}$
Electron affinity ($M \rightarrow M^{-}$)/kJ mol^{-1}: −21 (calc.)

Ionization energies/kJ mol^{-1}

1.	$M \rightarrow M^{+}$	2372.3
2.	$M^{+} \rightarrow M^{2+}$	5250.4

Principal lines in atomic spectrum

Wavelength/nm	Species	Sensitivity	Application
59.1412*(vac)	I		
388.864	I	U2	AE
468.575	II		AE
587.562	I	U3	AE
1083.034	I		

* Lowest energy transition from ground state to nearest empty orbital

Abundance: Atmosphere 5.2 p.p.m. (volume); Earth's crust 0.003 p.p.m.; seawater 7.2×10^{-6} p.p.m.

Biological role: None; non-toxic

Ho

Atomic number: 67

Relative atomic mass ($^{12}C=12.0000$): **164.93032**

Chemical properties

Silvery metal of the rare earth group, obtained from monazite and bastnaesite ores. Slowly attacked by oxygen and water, dissolves in acid. Little used.

Radii/pm: Ho^{3+} 89; atomic 176.6; covalent 158

Electronegativity: 1.23 (Pauling); 1.10 (Allred-Rochow)

Effective nuclear charge: 2.85 (Slater); 8.44 (Clementi); 11.60 (Froese–Fischer)

Standard reduction potentials E°/V

	III		0
Acid solution	Ho^{3+}	$\xrightarrow{-2.33}$	Ho
Alkaline solution	$Ho(OH)_3$	$\xrightarrow{-2.85}$	Ho

Oxidation states

Ho^{III} (f^{10})	Ho_2O_3, $Ho(OH)_3$, $[Ho(H_2O)_x]^{3+}$ (aq), Ho^{3+} salts, HoF_3, $HoCl_3$ etc., $HoCl_6^{3-}$, complexes

Physical properties

Melting point/K: 1747

Boiling point/K: 2968

ΔH_{fusion}/kJ mol^{-1}: 17.2

ΔH_{vap}/kJ mol^{-1}: 303

Thermodynamic properties (298.15 K, 0.1 MPa)

State	$\Delta_f H^{\circ}$/kJ mol^{-1}	$\Delta_f G^{\circ}$/kJ mol^{-1}	S°/J K^{-1} mol^{-1}	C_p/J K^{-1} mol^{-1}
Solid	0	0	75.3	27.15
Gas	300.8	264.8	195.59	20.79

Density/kg m^{-3}: 8795 [298 K]

Thermal conductivity/W m^{-1} K^{-1}: 16.2 [300 K]

Electrical resistivity/Ω m: 87.0×10^{-8} [298 K]

Mass magnetic susceptibility/kg^{-1} m^3: $+5.49 \times 10^{-6}$ (s)

Molar volume/cm^3: 18.75

Coefficient of linear thermal expansion/K^{-1}: 9.5×10^{-6}

Lattice structure (cell dimensions/pm), space group

α-Ho h.c.p. ($a=357.73$; $c=561.58$), $P6_3/mmc$
β-Ho b.c.c. ($a=396$), Im3m
$T(\alpha \to \beta)$ just below melting point
High pressure form: ($a=334$; $c=2410$), $R\bar{3}m$

X-ray diffraction: mass absorption coefficients (μ/ρ)/cm^2 g^{-1}:
CuK_α 128 MoK_α 73.9

Discovered in 1878 by P. T. Cleve at Uppsala, Sweden and independently by M. Delafontaine and J.-L. Soret at Geneva, Switzerland
[Latin, *Holmia* = Stockholm]

Holmium

Nuclear properties

Atomic number: 67
Thermal neutron capture cross-section/barns: 65±2
Number of isotopes (including nuclear isomers): 29
Isotope mass range: 151→170

Key isotopes

Nuclide	Atomic mass	Natural abundance (%)	Half life $T_{1/2}$	Decay mode and energy (MeV)	Nuclear spin I	Nuclear magnetic moment μ	Uses
^{165}Ho	164.9303	100	stable		7/2	+4.12	NMR
^{166}Ho		0	26.9h	β^-(1.847)	0		tracer

NMR

	^{165}Ho
Relative sensitivity (^{1}H = 1.00)	0.18
Absolute sensitivity (^{1}H = 1.00)	0.18
Receptivity (^{13}C = 1.00)	1.16×10^3
Magnetogyric ratio/rad T^{-1} s^{-1}	5.710×10^7
Quadrupole moment/m^2	2.82×10^{-28}
Frequency (^{1}H = 100 MHz; 2.3488 T)/MHz	20.513

Electron shell properties

Ground state electron configuration: [Xe]$4f^{11}6s^2$
Term symbol: $^4I_{15/2}$
Electron affinity ($M \rightarrow M^-$)/kJ mol^{-1}: ≤50

Ionization energies/kJ mol^{-1}

1. $M \rightarrow M^+$	580.7	6. $M^{5+} \rightarrow M^{6+}$	
2. $M^+ \rightarrow M^{2+}$	1139	7. $M^{6+} \rightarrow M^{7+}$	
3. $M^{2+} \rightarrow M^{3+}$	2204	8. $M^{7+} \rightarrow M^{8+}$	
4. $M^{3+} \rightarrow M^{4+}$	4100	9. $M^{8+} \rightarrow M^{9+}$	
5. $M^{4+} \rightarrow M^{5+}$		10. $M^{9+} \rightarrow M^{10+}$	

Principal lines in atomic spectrum

Wavelength/nm	Species	Sensitivity	Application
345.600	II		
374.817	II		AE
405.393	I		AA
410.384	I		AA, AE
416.303	I		

Abundance: Earth's crust 1.3 p.p.m.; seawater 8×10^{-8} p.p.m.
Biological role: None; non-toxic; stimulatory

H

Atomic number: 1
Relative atomic mass (^{12}C=12.0000): 1.00794

Chemical properties

Colourless, odourless gas, insoluble in water. Burns in air, forms explosive mixtures with air. Used in industry for making ammonia, cyclohexane, methanol etc. Obtained from methane CH_4.

Radii/pm: H^- 154; atomic 78; covalent 30; van der Waals 120; H^+ 10^{-5}

Electronegativity: 2.20 (Pauling); n.a. (Allred-Rochow)

Effective nuclear charge: 1.00 (Slater); 1.00 (Clementi); 1.00 (Froese–Fischer)

Standard reduction potentials $E^\ominus$/V

	I		0		−I
Acid solution	H_3O^+	0.00	H_2	−2.25	H^-
Alkaline solution	H_2O	0.828	H_2	−2.25	H^-

Covalent bonds

	r/pm	E/kJ mol^{-1}
H—H	74.14	453.6
H—F	91.7	566
H—Cl	127.4	431
H—Br	140.8	366
H—I	160.9	299
For covalent bonds to hydrogen see other elements		
D—D	74.14	447.3

Oxidation states

H^{-I}	NaH, CaH_2, etc.
H^0	H_2
H^I	H_2O, H_3O^+, etc.. OH^-, HF, HCl, etc., other acids, NH_3 etc., CH_4 etc.

Hydrogen is the most versatile of elements in its range of chemical bonds

Physical properties

Melting point/K: 14.01

Boiling point/K: 20.28

ΔH_{fusion}/kJ mol^{-1}: 0.12

ΔH_{vap}/kJ mol^{-1}: 0.46

Thermodynamic properties (298.15 K, 0.1 MPa)

State	$\Delta_f H^\ominus$/kJ mol^{-1}	$\Delta_f G^\ominus$/kJ mol^{-1}	$S^\ominus$/J K^{-1} mol^{-1}	C_p/J K^{-1} mol^{-1}
Gas (H_2)	0	0	130.684	28.824
Gas (atoms)	217.965	203.247	114.713	20.784

Density/kg m^{-3}: 76.0 [solid, 11 K]; 70.8 [liquid, b.p.]; 0.089 88 [gas, 273 K]

Thermal conductivity/W m^{-1} K^{-1}: 0.1815 [300 K] g

Mass magnetic susceptibility/kg^{-1} m^3: -2.50×10^{-8} (g)

Molar volume/cm^3: 13.26 [11 K]

Lattice structure (cell dimensions/pm), space group

H_2 h.c.p. (a=377.6, c=616.2), $P6_3/mmc$
D_2 h.c.p. (a=360.0, c=585.8), $P6_3/mmc$
H_2 cubic (a=533.8), Fm3m
D_2 cubic (a=509.2), Fm3m
H_2 tetragonal (a=450, c=368), I4
D_2 tetragonal (a=338, c=560)
T (h.c.p.→cubic)=4.5 K

X-ray diffraction: mass absorption coefficients (μ/ρ)/cm^2 g^{-1}:
CuK_α 0.435 MoK_α 0.380

Recognized as an element in 1766 by H. Cavendish at London UK

[Greek, *hydro genes* = water forming]

Hydrogen

Nuclear properties

Atomic number: 1

Thermal neutron capture cross-section/barns: 0.332 ± 0.002

Number of isotopes (including nuclear isomers): 3

Isotope mass range: $1 \rightarrow 3$

Key isotopes

Nuclide	Atomic mass	Natural abundance (%)	Half life $T_{1/2}$	Decay mode and energy (MeV)	Nuclear spin I	Nuclear magnetic moment μ	Uses
^{1}H	1.007 825	99.985	stable		1/2	+2.792 78	NMR
^{2}H	2.0140	0.015	stable		1	+0.857 42	NMR
^{3}H	3.016 05	0	12.262y	β^-(0.01861); no γ	1/2	+2.9789	tracer, NMR

NMR

	^{1}H	^{2}H	^{3}H
Relative sensitivity ($^{1}H = 1.00$)	1.00 (by defn)	9.65×10^{-3}	1.21
Absolute sensitivity ($^{1}H = 1.00$)	1.00 (by defn)	1.45×10^{-6}	0
Receptivity ($^{13}C = 1.00$)	5680	8.2×10^{-6}	—
Magnetogyric ratio/rad T^{-1} s^{-1}	26.7510×10^{7}	4.1064×10^{7}	28.5335×10^{7}
Quadrupole moment/m^2	—	2.73×10^{-31}	—
Frequency ($^{1}H = 100$ MHz; 2.3488 T)/MHz	100.000	15.351	106.663

Reference: $Si(CH_3)_4$

Electron shell properties

Ground state electron configuration: $1s^1$

Term symbol: $^{2}S_{1/2}$

Electron affinity ($M \rightarrow M^-$)/kJ mol^{-1}: 72.8

Ionization energies/kJ mol^{-1}

1. M	$\rightarrow M^+$	1312.0

Principal lines in atomic spectrum

Wavelength/nm	Species	Sensitivity	Application
121.568* (vac)	I		
486.133	I	U3	AE
656.272	I	U2	AE
656.285	I		
1875.10	I		

* Lowest energy transition from ground state to nearest empty orbital

Abundance: Atmosphere 0.53 p.p.m. (volume); Earth's crust 1520 p.p.m.; seawater, constituent

Biological role: Basis of all life (part of DNA molecule)

In

Atomic number: 49

Relative atomic mass (^{12}C=12.0000): 114.82

Chemical properties

Soft, silvery-white metal. Stable in air and with water, dissolves in acids. Used in low melting alloys in safety devices. Semiconductor uses as InAs and InSb in transistors, thermistors etc.

Radii/pm: In^{3+} 92; In^{+} 132; atomic 162.6; covalent 150

Electronegativity: 1.78 (Pauling); 1.49 (Allred-Rochow)

Effective nuclear charge: 5.00 (Slater); 8.47 (Clementi); 9.66 (Froese–Fischer)

Standard reduction potentials $E^{\ominus}$/V

	III		I		0
Acid solution	In^{3+}	−0.444	In^{+}	−0.126	In
	In^{3+} → In: −0.3382				

Covalent bonds

	r/pm	E/kJ mol^{-1}
In–H	185	243
In–C	216	165
In–O	213	109
IN–F	199	*c.*525
In–Cl	240	439
In–In	325.1	*c.* 85

Oxidation states

In^{I} (s^2)	InCl, InBr, InI
In^{II} (s^1)	$[In_2Cl_6]^{2-}$
	$[In_2Br_2]^{2-}$, $[In_2I_6]^{2-}$
In^{III} (d^{10})	In_2O_3, $In(OH)_3$,
	$[In(H_2O)_6]^{3+}$ (aq)
	InF_3, $InCl_3$ etc.,
	$InCl_5^{2-}$, $InCl_6^{3-}$,
	complexes

Physical properties

Melting point/K: 429.32

Boiling point/K: 2353

ΔH_{fusion}/kJ mol^{-1}: 3.27

ΔH_{vap}/kJ mol^{-1}: 231.8

Thermodynamic properties (298.15 K, 0.1 MPa)

State	$\Delta_f H^{\ominus}$/kJ mol^{-1}	$\Delta_f G^{\ominus}$/kJ mol^{-1}	$S^{\ominus}$/J K^{-1} mol^{-1}	C_p/J K^{-1} mol^{-1}
Solid	0	0	57.82	26.74
Gas	243.30	208.71	173.79	20.84

Density/kg m^{-3}: 7310 [298 K]; 7032 [liquid at m.p.]

Thermal conductivity/W m^{-1} K^{-1}: 81.6 [300 K]

Electrical resistivity/Ω m: 8.37×10^{-8} [293 K]

Mass magnetic susceptibility/kg^{-1} m^3: -7.0×10^{-9} (s)

Molar volume/cm^3: 15.71

Coefficient of linear thermal expansion/K^{-1}: 33×10^{-6}

Lattice structure (cell dimensions/pm), space group

Face centred tetragonal (a=325.30, c=494.55), I4/mmm

X-ray diffraction: mass absorption coefficients (μ/ρ)/cm^2 g^{-1}:
CuK_α 243 MoK_α 29.3

Index

The elements are listed alphabetically in the main part of the book. The following properties are given for each element where known. Some properties are listed in separate tables, pp. 217–49, for all the elements.

Discovered in 1863 by Ferdinand Reich and Hieronymous Richter at Freiberg, Germany

[Named after the indigo line in its spectrum]

Indium

Nuclear properties

Atomic number: 49

Thermal neutron capture cross-section/barns: 194±2

Number of isotopes (including nuclear isomers): 34

Isotope mass range: 106→124

Key isotopes

Nuclide	Atomic mass	Natural abundance (%)	Half life $T_{1/2}$	Decay mode and energy (MeV)	Nuclear spin I	Nuclear magnetic moment μ	Uses
^{111}In			2.81d	EC; γ	9/2	+5.53	medical tracer
^{113}In	112.9043	4.3	stable		9/2	+5.5229	NMR
^{113m}In			100m	IT(0.3916)	1/2	−0.210	medical tracer
$^{114m_1}In$			50d	IT, EC(0.191); γ	5	+4.7	tracer
^{115}In	114.9041	95.7	6×10^{14}y	β^-; no γ	9/2	+5.5348	NMR

NMR

	[^{113}In]	^{115}In
Relative sensitivity ($^1H=1.00$)	0.34	0.34
Absolute sensitivity ($^1H=1.00$)	1.47×10^{-2}	0.33
Receptivity ($^{13}C=1.00$)	83.8	1890
Magnetogyric ratio/rad T^{-1} s^{-1}	5.8493×10^7	5.8618×10^7
Quadrupole moment/m^2	1.14×10^{-28}	1.16×10^{-28}
Frequency ($^1H=100$ MHz; 2.3488 T)/MHz	21.866	21.914

Reference: $In(H_2O)_6^{3+}$

Electron shell properties

Ground state electron configuration: [Kr]$4d^{10}5s^25p^1$

Term symbol: $^2P_{1/2}$

Electron affinity ($M\rightarrow M^-$)/kJ mol^{-1}: 34

Ionization energies/kJ mol^{-1}

1. $M \rightarrow M^+$	558.3	6. $M^{5+}\rightarrow M^{6+}$	(9 500)
2. $M^+\rightarrow M^{2+}$	1820.6	7. $M^{6+}\rightarrow M^{7+}$	(11 700)
3. $M^{2+}\rightarrow M^{3+}$	2704	8. $M^{7+}\rightarrow M^{8+}$	(13 900)
4. $M^{3+}\rightarrow M^{4+}$	5200	9. $M^{8+}\rightarrow M^{9+}$	(17 200)
5. $M^{4+}\rightarrow M^{5+}$	(7400)	10. $M^{9+}\rightarrow M^{10+}$	(19 700)

Principal lines in atomic spectrum

Wavelength/nm	Species	Sensitivity	Application
303.936	I	U4	AA, AE
325.609	I	U3	AA, AE
325.856	I	U5	AE
410.177*	I	U2	AA, AE
451.132	I	U1	AA, AE

* Lowest energy transition from ground state to nearest empty orbital

Abundance: Earth's crust 0.24 p.p.m.; seawater 0.02 p.p.m.

Biological role: None; teratogenic; stimulatory

I

Atomic number: 53
Relative atomic mass ($^{12}C=12.0000$): 126.90447

Chemical properties

Black, shiny solid, I_2, obtained from certain brines rich in iodide or from iodates. Sublimes easily. Compounds used in food supplements, dyes, catalysts and photography.

Radii/pm: I^- 220; Covalent 133.3; van der Waals 215

Electronegativity: 2.66 (Pauling); 2.21 (Allred-Rochow)

Effective nuclear charge: 7.60 (Slater); 11.61 (Clementi); 14.59 (Froese–Fischer)

Standard reduction potentials $E^{\ominus}$/V

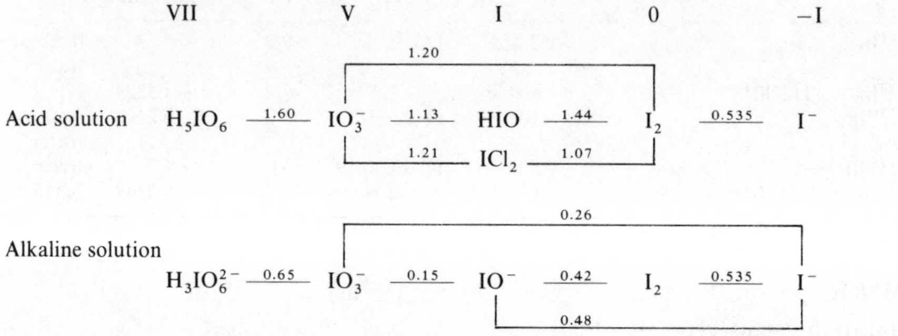

	VII		V		I		0		−I
Acid solution	H_5IO_6	1.60	IO_3^-	1.13	HIO	1.44	I_2	0.535	I^-
			IO_3^-	1.21	ICl_2^-	1.07	I_2		
			IO_3^-	1.20			I_2		
Alkaline solution	$H_3IO_6^{2-}$	0.65	IO_3^-	0.15	IO^-	0.42	I_2	0.535	I^-
			IO_3^-	0.26					I^-
					IO^-	0.48			I^-

Covalent bonds

	r/pm	E/kJ mol^{-1}
I–H	160.9	299
I–C	213	218
I–O	195	234
I–F	191	280
I–Cl	232	208
I–I	266.6	151
I–Si	243	234
I–P	252	184

Oxidation states

I^{-I}	I^-(aq), HI, KI, etc
I^0	I_2, I_3^-, I_5^-, etc.
I^I	I_n^+, ICl_2^-, etc.
I^{III}	I_4O_9 $(=I^{3+}(IO_3^-)_3)$ ICl_3
I^V	I_2O_5, HIO_3, IO_3^-(aq) IF_5, IF_6^-
I^{VII}	H_5IO_6, $H_4IO_6^-$(aq) etc., HIO_4, IO_4^-(aq), IF_7

Physical properties

Melting point/K: 386.7

Boiling point/K: 457.50

ΔH_{fusion}/kJ mol^{-1}: 15.27 ΔH_{vap}/kJ mol^{-1}: 41.67

Thermodynamic properties (298.15 K, 0.1 MPa)

State	$\Delta_f H^{\ominus}$/kJ mol^{-1}	$\Delta_f G^{\ominus}$/kJ mol^{-1}	$S^{\ominus}$/J K^{-1} mol^{-1}	C_p/J K^{-1} mol^{-1}
Solid	0	0	116.135	54.438
Gas (I_2)	62.438	19.327	260.69	36.90
Gas (atoms)	106.838	70.250	180.791	20.786

Density/kg m^{-3}: 4930 [293 K]

Thermal conductivity/W m^{-1} K^{-1}: 0.449 [300 K]

Electrical resistivity/Ω m: 1.37×10^7 [293 K]

Mass magnetic susceptibility/kg^{-1} m^3: -4.40×10^{-9} (s)

Molar volume/cm^3: 25.74

Coefficient of linear thermal expansion/K^{-1}: n.a.

Lattice structure (cell dimensions/pm), space group

Orthorhombic ($a=7.26.47$, $b=478.57$, $c=979.08$), Cmca

X-ray diffraction: mass absorption coefficients (μ/ρ)/cm^2 g^{-1}:
CuK_α 294 MoK_α 37.1

Discovered in 1811 by B. Courtois at Paris, France

[Greek, *iodes* = violet]

Iodine

Nuclear properties

Atomic number: 53
Thermal neutron capture cross-section/barns: 6.2 ± 0.2
Number of isotopes (including nuclear isomers): 24
Isotope mass range: $117 \rightarrow 139$

Key isotopes

Nuclide	Atomic mass	Natural abundance (%)	Half life $T_{1/2}$	Decay mode and energy (MeV)	Nuclear spin I	Nuclear magnetic moment μ	Uses
^{123}I		0	13.3h	EC(*c.* 1.4); γ	5/2		tracer
^{125}I		0	60.2d	EC(0.149); γ	5/2	+3.0	tracer medical
^{127}I	126.9004	100	stable		5/2	+2.8091	NMR
^{129}I		0	1.7×10^7y	β^-(0.189); γ	7/2	+2.6174	
^{131}I		0	8.070d	β^-(0.970); γ	7/2	+2.738	tracer, medical

NMR

	^{127}I
Relative sensitivity ($^1H = 1.00$)	9.34×10^{-2}
Absolute sensitivity ($^1H = 1.00$)	9.34×10^{-2}
Receptivity ($^{13}C = 1.00$)	530
Magnetogyric ratio/rad T^{-1} s^{-1}	5.3525×10^7
Quadrupole moment/m^2	-0.79×10^{-28}
Frequency ($^1H = 100$ MHz; 2.3488 T)/MHz	20.007

Reference: NaI (aq)

Electron shell properties

Ground state electron configuration: $[Kr]4d^{10}5s^25p^5$
Term symbol: $^2P_{3/2}$
Electron affinity ($M \rightarrow M^-$)/kJ mol^{-1}: 295.3

Ionization energies/kJ mol^{-1}

1.	$M \rightarrow M^+$	1008.4	6. $M^{5+} \rightarrow M^{6+}$	(7400)
2.	$M^+ \rightarrow M^{2+}$	1845.9	7. $M^{6+} \rightarrow M^{7+}$	(8700)
3.	$M^{2+} \rightarrow M^{3+}$	3200	8. $M^{7+} \rightarrow M^{8+}$	(16400)
4.	$M^{3+} \rightarrow M^{4+}$	(4100)	9. $M^{8+} \rightarrow M^{9+}$	(19300)
5.	$M^{4+} \rightarrow M^{5+}$	(5000)	10. $M^{9+} \rightarrow M^{10+}$	(22100)

Principal lines in atomic spectrum

Wavelength/nm	Species	Sensitivity	Application
183.036* (vac)	I		
516.119	II		AE
546.461	II		AE
804.374	I		

* Lowest energy transition from ground state to nearest empty orbital

Abundance: Earth's crust 0.46 p.p.m.; seawater 0.05 p.p.m.
Biological role: Essential; I_2 vapour harmful

Ir

Atomic number: 77
Relative atomic mass (^{12}C=12.0000): **192.22**

Chemical properties

Hard, lustrous, silvery metal of the platinum group. Stable to air and water, inert to all acids but fused NaOH will attack it. Used in special alloys and spark plugs.

Radii/pm: Ir^{2+} 89; Ir^{3+} 75; Ir^{4+} 66; atomic 135.7; covalent 126

Electronegativity: 2.20 (Pauling); 1.55 (Allred-Rochow)

Effective nuclear charge: 3.90 (Slater); 10.57 (Clementi); 15.33 (Froese–Fischer)

Standard reduction potentials $E^\ominus$/V

IV		III		0
IrO_2	0.223	Ir^{3+}	1.156	Ir
IrO_2 → Ir: 0.926				
$IrCl_6^{2-}$	0.867	$IrCl_6^{3-}$	0.86	Ir

Oxidation states

Ir^{-I}	(d^{10})	rare $[Ir(CO)_3(PPh_3)]^-$
Ir^0	(d^9)	rare $[Ir_4(CO)_{12}]$
Ir^I	(d^8)	rare $[Ir(CO)Cl(PPh_3)_2]$
Ir^{II}	(d^7)	$IrCl_2$
Ir^{III}	(d^6)	IrF_3, $IrCl_3$ etc., $[IrCl_6]^{3-}$(aq)
Ir^{IV}	(d^5)	IrO_2, IrF_4, IrS_2, $[IrCl_6]^2$(aq)
Ir^V	(d^4)	IrF_5, $[IrF_6]^-$
Ir^{VI}	(d^3)	IrF_6

Physical properties

Melting point/K: 2683
Boiling point/K: 4403
ΔH_{fusion}/kJ mol^{-1}: 26.4
ΔH_{vap}/kJ mol^{-1}: 612.1

Thermodynamic properties (298.15 K, 0.1 MPa)

State	$\Delta_f H^\ominus$/kJ mol^{-1}	$\Delta_f G^\ominus$/kJ mol^{-1}	$S^\ominus$/J K^{-1} mol^{-1}	C_p/J K^{-1} mol^{-1}
Solid	0	0	35.48	25.10
Gas	665.3	617.9	193.578	20.786

Density/kg m^{-3}: 22 420 [290 K]; 20 000 [liquid at m.p.]
Thermal conductivity/W m^{-1} K^{-1}: 147 [300 K]
Electrical resistivity/Ω m: 5.3×10^{-8} [293 K]
Mass magnetic susceptibility/kg^{-1} m^3: $+1.67 \times 10^{-9}$ (s)
Molar volume/cm^3: 8.57
Coefficient of linear thermal expansion/K^{-1}: 6.8×10^{-6}

Lattice structure (cell dimensions/pm), space group

f.c.c. (a=383.89), Fm3m

X-ray diffraction: mass absorption coefficients (μ/ρ)/cm^2 g^{-1}:
CuK$_\alpha$ 193 MoK$_\alpha$ 110

Discovered in 1803 by S. Tennant at London, UK

[Latin, *iris* = rainbow]

Iridium

Nuclear properties

Atomic number: 77
Thermal neutron capture cross-section/barns: 425±15
Number of isotopes (including nuclear isomers): 25
Isotope mass range: 182→198

Key isotopes

Nuclide	Atomic mass	Natural abundance (%)	Half life $T_{1/2}$	Decay mode and energy (MeV)	Nuclear spin I	Nuclear magnetic moment μ	Uses
^{191}Ir	190.9609	37.3	stable		3/2	+0.1454	NMR
^{192}Ir		0	74.2d	β^-(1.453); EC; β^+; γ	4	+1.90	tracer medical
^{193}Ir	192.9633	62.7	stable		3/2	+0.1583	NMR

NMR (never used)	^{191}Ir	^{193}Ir
Relative sensitivity (^{1}H = 1.00)	2.53×10^{-5}	3.27×10^{-5}
Absolute sensitivity (^{1}H = 1.00)	9.43×10^{-6}	2.05×10^{-5}
Receptivity (^{13}C = 1.00)	0.023	0.050
Magnetogyric ratio/rad T^{-1} s^{-1}	0.539×10^7	0.391×10^7
Quadrupole moment/m^2	1.5×10^{-28}	1.4×10^{-28}
Frequency (^{1}H = 100 MHz; 2.3488 T)/MHz	1.718	1.871

Electron shell properties

Ground state electron configuration: [Xe]$4f^{14}5d^76s^2$
Term symbol: $^4F_{9/2}$
Electron affinity ($M \rightarrow M^-$)/kJ mol^{-1}: 190

Ionization energies/kJ mol^{-1}

1.	$M \rightarrow M^+$	880	6.	$M^{5+} \rightarrow M^{6+}$	(6900)
2.	$M^+ \rightarrow M^{2+}$	(1680)	7.	$M^{6+} \rightarrow M^{7+}$	(8500)
3.	$M^{2+} \rightarrow M^{3+}$	(2600)	8.	$M^{7+} \rightarrow M^{8+}$	(10000)
4.	$M^{3+} \rightarrow M^{4+}$	(3800)	9.	$M^{8+} \rightarrow M^{9+}$	(11700)
5.	$M^{4+} \rightarrow M^{5+}$	(5500)	10.	$M^{9+} \rightarrow M^{10+}$	

Principal lines in atomic spectrum

Wavelength/nm	Species	Sensitivity	Application
208.882	I		AA
263.971	I		AA
322.078	I	U1	AE
351.365	I	U2	AE
380).012*			

* Lowest energy transition from ground state to nearest empty orbital

Abundance: Earth's crust 0.001 p.p.m.; seawater n.a. (minute)
Biological role: None; low toxicity

Fe

Atomic number: 26
Relative atomic mass (^{12}C=12.0000): 55.847

Chemical properties

When pure, iron is lustrous, silvery, and soft (workable). Chief ores are haematite (Fe_2O_3), magnesite (Fe_3O_4), and siderite ($FeCO_3$). Most important of all metals, used principally as steels. Rusts in damp air, dissolves in dilute acids.

Radii/pm: Fe^{2+} 82; Fe^{3+} 67: atomic (α form) 124.1; covalent 116.5

Electronegativity: 1.83 (Pauling); 1.64 (Allred-Rochow)

Effective nuclear charge: 3.75 (Slater); 5.43 (Clementi); 7.40 (Froese–Fischer)

Standard reduction potentials $E^{\ominus}$/V

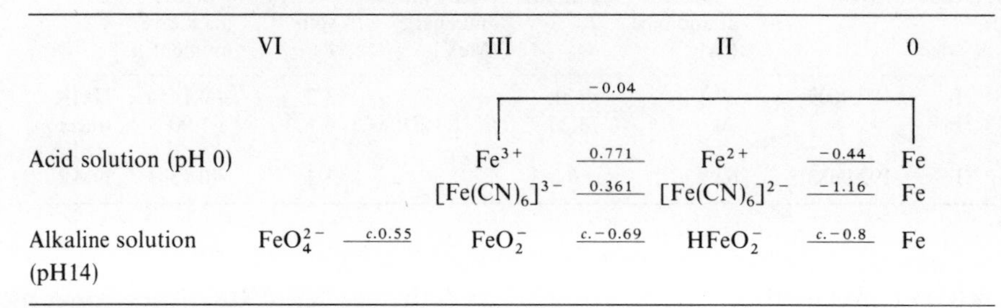

VI
III
II
0
−0.04
Acid solution (pH 0)
Fe3+
0.771
Fe2+
−0.44
Fe
[Fe(CN)6]3−
0.361
[Fe(CN)6]2−
−1.16
Fe
Alkaline solution (pH14)
FeO4 2−
c.0.55
FeO2 −
c.−0.69
HFeO2 −
c.−0.8
Fe

Oxidation states

Fe^{-II}	(d^{10})	rare $Fe(CO)_4^{2-}$
Fe^{-I}	(d^9)	rare $Fe_2(CO)_8^{2-}$
Fe^0	(d^8)	rare $Fe(CO)_5$
Fe^I	(d^7)	rare $[Fe(NO)(H_2O)_5]^{2+}$
Fe^{II}	(d^6)	FeO, FeS_2 (=$Fe^{II}S_2^{2-}$) $Fe(OH)_2$, $[Fe(H_2O)_6]^{2+}$(aq), FeF_2, $Fe(C_5H_5)_2$, etc.
Fe^{III}	(d^5)	Fe_2O_3, Fe_3O_4 (=$Fe^{II}O.Fe_2^{III}O_3$), FeF_3, $FeCl_3$, FeO(OH), $[Fe(H_2O)_6]^{3+}$(aq), etc.
Fe^{IV}	(d^4)	rare, some complexes
Fe^V	(d^3)	FeO_4^{3-} ?
Fe^{IV}	(d^2)	FeO_4^{2-}

Physical properties*

*Very much affected by impurities such as carbon

Melting point/K: 1808

Boiling point/K: 3023

ΔH_{fusion}/kJ mol^{-1}: 14.9 ΔH_{vap}/kJ mol^{-1}: 340.2

Thermodynamic properties (298.15 K, 0.1 MPa)

State	$\Delta_f H^{\ominus}$/kJ mol^{-1}	$\Delta_f G^{\ominus}$/kJ mol^{-1}	$S^{\ominus}$/J K^{-1} mol^{-1}	C_p/J K^{-1} mol^{-1}
Solid	0	0	27.28	25.10
Gas	416.3	370.7	180.490	25.677

Density/kg m^{-3}: 7874 [293 K]; 7035 [liquid at m.p.]

Thermal conductivity/W m^{-1} K^{-1}: 80.2 [300 K]

Electrical resistivity/Ω m: 9.71×10^{-8} [293 K]

Mass magnetic susceptibility/kg^{-1} m^3: ferromagnetic

Molar volume/cm^3: 7.09

Coefficient of linear thermal expansion/K^{-1}: 12.3×10^{-6}

Lattice structure (cell dimensions/pm), space group

α-Fe b.c.c. (a=286.645), Im3m
β-Fe not true allotrope
γ-Fe c.c.p. (a=364.68), Fm3m
δ-Fe b.c.c. (a=293.22), Im3m

$T(\alpha \rightarrow \gamma)$=1183 K
$T(\gamma \rightarrow \delta)$=1663 K

X-ray diffraction: mass absorption coefficients (μ/ρ)/cm^2 g^{-1}:
CuK_α 308 MoK_α 38.5

Known to ancient civilizations

[Anglo-Saxon, *iron*; Latin, *ferrum*]

Iron

Nuclear properties

Atomic number: 26

Thermal neutron capture cross-section/barns: 2.56±0.05

Number of isotopes (including nuclear isomers): 10

Isotope mass range: 52→61

Key isotopes

Nuclide	Atomic mass	Natural abundance (%)	Half life $T_{1/2}$	Decay mode and energy (MeV)	Nuclear spin I	Nuclear magnetic moment μ	Uses
^{52}Fe		0	8.2h	β^+, EC(2.37); γ			tracer
^{54}Fe	53.9396	5.8	stable				
^{55}Fe	54.938	0	2.6y	EC(0.232); no γ			tracer medical
^{56}Fe	55.9349	91.7	stable				
^{57}Fe	56.9354	2.2	stable		1/2	+0.090 42	NMR
^{58}Fe	57.9333	0.3	stable				
^{59}Fe	58.935	0	45.1d	β^-(1.57 max); γ	3/2	±1.1	tracer, medical
^{60}Fe		0	3×10^5y	β^-(0.14)			

NMR

	^{57}Fe
Relative sensitivity (^1H=1.00)	3.37×10^{-5}
Absolute sensitivity (^1H=1.00)	7.38×10^{-7}
Receptivity (^{13}C=1.00)	4.2×10^{-3}
Magnetogyric ratio/rad T^{-1} s^{-1}	0.8661×10^7
Frequency (^1H=100 MHz; 2.3488 T)/MHz	3.231

Reference: $Fe(CO)_5$

Electron shell properties

Ground state electron configuration: [Ar]$3d^64s^2$

Term symbol: $^5D_4$

Electron affinity ($M\rightarrow M^-$)/kJ mol^{-1}: 44

Ionization energies/kJ mol^{-1}

1. $M\rightarrow M^+$	759.3	6. $M^{5+}\rightarrow M^{6+}$	9 600
2. $M^+\rightarrow M^{2+}$	1561	7. $M^{6+}\rightarrow M^{7+}$	12 100
3. $M^{2+}\rightarrow M^{3+}$	2957	8. $M^{7+}\rightarrow M^{8+}$	14 575
4. $M^{3+}\rightarrow M^{4+}$	5290	9. $M^{8+}\rightarrow M^{9+}$	22 678
5. $M^{4+}\rightarrow M^{5+}$	7240	10. $M^{9+}\rightarrow M^{10+}$	25 290

Principal lines in atomic spectrum

Wavelength/nm	Species	Sensitivity	Application
248.327	I		AA
248.814	I		AA
371.994	I	U1	AE
373.713	I	U2	AAAE
374.556	I	U3	AE
835.991	I		AA

* Lowest energy transition from ground state to nearest empty orbital

Abundance: Earth's crust 62 000 p.p.m.; seawater 0.01 p.p.m.

Biological role: Essential; non-toxic

Atomic number: 36
Relative atomic mass ($^{12}C=12.0000$): 83.80

Chemical properties

Colourless, odourless gas obtained from liquid air. Chemically inert to everything but fluorine. ^{86}Kr has orange-red line in atomic spectrum which is fundamental standard of length: 1 metre = 1 650 763.73 wavelengths.

Radii/pm: covalent 189; van der Waals 198; Kr^+ 169

Electronegativity: n.a. (Pauling); n.a. (Allred-Rochow)

Effective nuclear charge: 8.25 (Slater); 9.77 (Clementi); 11.79 (Froese–Fischer)

Standard reduction potentials $E^{\ominus}$/V

n.a. [KrF_2 decomposes in water]

Covalent bonds	r/pm	E/kJ mol^{-1}
Kr–F	188.9	50

Oxidation states	
Kr^0	clathrates $Kr_8(H_2O)_{46}$, Kr (quinol)$_3$
Kr^{II}	KrF_2, $[KrF]^+[AsF_6]^-$

Physical properties

Melting point/K: 116.6

Boiling point/K: 120.85

ΔH_{fusion}/kJ mol^{-1}: 1.64

ΔH_{vap}/kJ mol^{-1}: 9.05

Thermodynamic properties (298.15 K, 0.1 MPa)

State	$\Delta_f H^{\ominus}$/kJ mol^{-1}	$\Delta_f G^{\ominus}$/kJ mol^{-1}	$S^{\ominus}$/J K^{-1} mol^{-1}	C_p/J K^{-1} mol^{-1}
Gas	0	0	164.082	20.786

Density/kg m^{-3}: 2823 [solid, m.p.]; 2413 [liquid b.p.]; 3.7493 [gas, 273 K]

Thermal conductivity/W m^{-1} K^{-1}: 0.009 49 [300 K]$_g$

Mass magnetic susceptibility/kg^{-1} m^3: -4.32×10^{-9} (g)

Molar volume/cm^3: 29.68 [116 K]

Lattice structure (cell dimensions/pm), space group
f.c.c. (80 K) ($a=572.1$), Fm3m

X-ray diffraction: mass absorption coefficients (μ/ρ)/cm^2 g^{-1}:
CuK_α 108 MoK_α 84.9

Discovered in 1898 by Sir William Ramsay and M. W. Travers at London, UK

[Greek, *kryptos* = hidden]

Krypton

Nuclear properties

Atomic number: 36

Thermal neutron capture cross-section/barns: 24.1 ± 1.0

Number of isotopes (including nuclear isomers): 23

Isotope mass range: 74→94

Key isotopes

Nuclide	Atomic mass	Natural abundance (%)	Half life $T_{1/2}$	Decay mode and energy (MeV)	Nuclear spin I	Nuclear magnetic moment μ	Uses
^{78}Kr	77.9204	0.35	stable				
^{80}Kr	79.9164	2.25	stable				
^{82}Kr	81.9135	11.6	stable		0		
^{83}Kr	82.914	11.5	stable		9/2	−0.970	NMR
^{84}Kr	83.912	57.0	stable		0		
^{85}Kr	84.913	0	10.76y	β^-(0.67); γ	9/2	± 1.005	tracer
^{86}Kr	85.911	17.3	stable		0		

NMR (few compounds)	^{83}Kr
Relative sensitivity ($^1H = 1.00$)	1.88×10^{-3}
Absolute sensitivity ($^1H = 1.00$)	2.17×10^{-4}
Receptivity ($^{13}C = 1.00$)	1.23
Magnetogyric ratio/rad T^{-1} s^{-1}	-1.029×10^7
Quadrupole moment/m^2	0.15×10^{-28}
Frequency ($^1H = 100$ MHz; 2.3488 T)/MHz	3.847

Electron shell properties

Ground state electron configuration: $[Ar]3d^{10}4s^24p^6$

Term symbol: $^1S_0$

Electron affinity ($M \rightarrow M^-$)/kJ mol^{-1}: −39 (calc.)

Ionization energies/kJ mol^{-1}

1. $M \rightarrow M^+$	1350.7	6. $M^{5+} \rightarrow M^{6+}$	7570
2. $M^+ \rightarrow M^{2+}$	2350	7. $M^{6+} \rightarrow M^{7+}$	10710
3. $M^{2+} \rightarrow M^{3+}$	3565	8. $M^{7+} \rightarrow M^{8+}$	12200
4. $M^{3+} \rightarrow M^{4+}$	5070	9. $M^{8+} \rightarrow M^{9+}$	22229
5. $M^{4+} \rightarrow M^{5+}$	6240	10. $M^{9+} \rightarrow M^{10+}$	(28900)

Principal lines in atomic spectrum

Wavelength/nm	Species	Sensitivity	Application
123.584* (vac)	I		
557.029	I	U3	AE
587.092	I	U2	AE
877.675	I		

* Lowest energy transition from ground state to nearest empty orbital

Abundance: Atmosphere 1.14 p.p.m. (volume); Earth's crust *c.* 0; seawater 0.0003 p.p.m.

Biological role: None; non-toxic

La

Atomic number: 57

Relative atomic mass (^{12}C=12.0000): 138.9055

Chemical properties

Soft, silvery-white metal; rapidly tarnishes in air and burns easily. Reacts with water to give hydrogen gas. Used in optical glass and for flints. La^{3+} is used as a biological tracer for calcium Ca^{2+}.

Radii/pm: La^{3+} 122; atomic 187.7; covalent 169

Electronegativity: 1.10 (Pauling); 1.08 (Allred-Rochow)

Effective nuclear charge: 2.85 (Slater); 9.31 (Clementi); 10.43 (Froese–Fischer)

Standard reduction potentials $E^{\ominus}$/V

	III		0
Acid solution	La^{3+}	$\xrightarrow{-2.38}$	La
Alkaline solution	$La(OH)_3$	$\xrightarrow{-2.80}$	La

Oxidation states

La^{III} ([Xe])	La_2O_3, $La(OH)_3$, $[La(H_2O)_x]^{3+}$(aq) LaF_3, $LaCl_3$ etc., La^{3+} salts, LaOCl, $[La(NCS)_6]^{3-}$, complexes LaH_2–LaH_3 is probably $La^{3+}H^-$

Physical properties

Melting point/K: 1194

Boiling point/K: 3730

ΔH_{fusion}/kJ mol^{-1}: 10.04

ΔH_{vap}/kJ mol^{-1}: 402.1

Thermodynamic properties (298.15 K, 0.1 MPa)

State	$\Delta_f H^{\ominus}$/kJ mol^{-1}	$\Delta_f G^{\ominus}$/kJ mol^{-1}	$S^{\ominus}$/J K^{-1} mol^{-1}	C_p/J K^{-1} mol^{-1}
Solid	0	0	56.9	27.11
Gas	431.0	393.56	182.377	22.753

Density/kg m^{-3}: 6145 [298 K]

Thermal conductivity/W m^{-1} K^{-1}: 13.5 [300 K]

Electrical resistivity/Ω m: 57×10^{-8} [298 K]

Mass magnetic susceptibility/kg^{-1} m^3: $+1.1 \times 10^{-8}$ (s)

Molar volume/cm^3: 22.60

Coefficient of linear thermal expansion/K^{-1}: 4.9×10^{-6}

Lattice structure (cell dimensions/pm), space group

α-La hexagonal (a=377.0, c=121.59), $P6_3/mmc$
β-La f.c.c. (a=529.6), Fm3m
γ-La b.c.c. (a=426), Im3m

$T(\alpha\rightarrow\beta)$= 583 K
$T(b\rightarrow\gamma)$=1137 K

X-ray diffraction: mass absorption coefficients (μ/ρ)/cm^2 g^{-1}:
CuK_α 341 MoK_α 45.8

Discovered in 1839 by C. G. Mosander at Stockholm, Sweden

[Greek, *lanthanein* = to lie hidden]

Lanthanum

Nuclear properties

Atomic number: 57

Thermal neutron capture cross-section/barns: 8.9 ± 0.2

Number of isotopes (including nuclear isomers): 19

Isotope mass range: 126→144

Key isotopes

Nuclide	Atomic mass	Natural abundance (%)	Half life $T_{1/2}$	Decay mode and energy (MeV)	Nuclear spin I	Nuclear magnetic moment μ	Uses
^{138}La	137.9068	0.09	stable		5	+3.707	NMR
^{139}La	138.9061	99.91			7/2	+2.778	NMR
^{140}La		0	40.22h	β^-(3.769); γ	3	+0.73	tracer

NMR

	[^{138}La]	^{139}La
Relative sensitivity ($^1H = 1.00$)	9.19×10^{-2}	5.92×10^{-2}
Absolute sensitivity ($^1H = 1.00$)	8.18×10^{-5}	5.91×10^{-2}
Receptivity ($^{13}C = 1.00$)	0.43	336
Magnetogyric ratio/rad T^{-1} s^{-1}	3.5295×10^{7}	3.7787×10^{7}
Quadrupole moment/m^2	-0.47×10^{-28}	0.21×10^{-28}
Frequency ($^1H = 100$ MHz; 2.3488 T)/MHz	13.193	14.126

References: 0.01M $LaCl_3$

Electron shell properties

Ground state electron configuration: [Xe]$5d^16s^2$

Term symbol: $^2D_{3/2}$

Electron affinity ($M \rightarrow M^-$)/kJ mol^{-1}: 53

Ionization energies/kJ mol^{-1}

1. $M \rightarrow M^+$	538.1	6. $M^{5+} \rightarrow M^{6+}$	(7600)
2. $M^+ \rightarrow M^{2+}$	1067	7. $M^{6+} \rightarrow M^{7+}$	(9600)
3. $M^{2+} \rightarrow M^{3+}$	1850	8. $M^{7+} \rightarrow M^{8+}$	(11000)
4. $M^{3+} \rightarrow M^{4+}$	4819	9. $M^{8+} \rightarrow M^{9+}$	(12400)
5. $M^{4+} \rightarrow M^{5+}$	(6400)	10. $M^{9+} \rightarrow M^{10+}$	(15900)

Principal lines in atomic spectrum

Wavelength/nm	Species	Sensitivity	Application
394.910	II	V2	
418.732	I		AA
550.134	I		AA
593.065	I	U2	AE
624.993	I	U1	AE
645.599*	I		

* Lowest energy transition from ground state to nearest empty orbital

Abundance: Earth's crust 35 p.p.m.; seawater 0.0003 p.p.m.

Biological role: None; moderately toxic

Lr

Atomic number: **103**
Relative atomic mass (^{12}C=12.0000): **(260)**

Chemical properties

Radioactive metal, of which only a few atoms have ever been produced by bombarding ^{252}Cf with boron nuclei.

Radii/pm: Lr^{2+} 112; Lr^{3+} 94; Lr^{4+} 83

Electronegativity: 1.3 (Pauling)

Effective nuclear charge: 1.80 (Slater)

Standard reduction potentials $E^{\ominus}$/V

III		0
Lr^{3+}	$\xrightarrow{-2.06}$	Lr

Oxidation states

Lr^{III} (f^{14})	Lr^{3+} (aq)

Physical properties

Melting point/K: n.a.
Boiling point/K: n.a.
ΔH_{fusion}/kJ mol^{-1}: n.a.
ΔH_{vap}/kJ mol^{-1}: n.a.

Thermodynamic properties (298.15 K, 0.1 MPa)

State	$\Delta_f H^{\ominus}$/kJ mol^{-1}	$\Delta_f G^{\ominus}$/kJ mol^{-1}	$S^{\ominus}$/J K^{-1} mol^{-1}	C_p/J K^{-1} mol^{-1}
Solid	0	0	n.a.	n.a.
Gas	n.a.	n.a.	n.a.	n.a.

Density/kg m^{-3}: n.a.
Thermal conductivity/W m^{-1} K^{-1}: 10 (est.) [300 K]
Electrical resistivity/Ω m: n.a.
Mass magnetic susceptibility/kg^{-1} m^{3}: n.a.
Molar volume/cm^{3}: n.a.
Coefficient of linear thermal expansion/K^{-1}: n.a.

Lattice structure (cell dimensions/pm), space group

n.a.

X-ray diffraction: mass absorption coefficients (μ/ρ)/cm^{2} g^{-1}:
n.a.

Prepared in 1961 by A. Ghiorso, T. Sikkeland, A. E. Larsh, and R. M. Latimer at Berkeley, California, USA

[Named after Ernest O. Lawrence]

Lawrencium

Nuclear properties

Atomic number: 103

Thermal neutron capture cross-section/barns: n.a.

Number of isotopes (including nuclear isomers): 6

Isotope mass range: 255→260

Key isotopes

Nuclide	Atomic mass	Natural abundance (%)	Half life $T_{1/2}$	Decay mode and energy (MeV)	Nuclear spin I	Nuclear magnetic moment μ	Uses
^{260}Lr		0	*c.* 3m	α(8.03)			

Electron shell properties

Ground state electron configuration: $[Rn]5f^{14}6d^{1}7s^{2}$

Term symbol: $^{2}D_{5/2}$

Electron affinity ($M \rightarrow M^{-}$)/kJ mol^{-1}: n.a.

Ionization energies/kJ mol^{-1}

1. $M \rightarrow M^{+}$	n.a.	6. $M^{5+} \rightarrow M^{6+}$	
2. $M^{+} \rightarrow M^{2+}$		7. $M^{6+} \rightarrow M^{7+}$	
3. $M^{2+} \rightarrow M^{3+}$		8. $M^{7+} \rightarrow M^{8+}$	
4. $M^{3+} \rightarrow M^{4+}$		9. $M^{8+} \rightarrow M^{9+}$	
5. $M^{4+} \rightarrow M^{5+}$		10. $M^{9+} \rightarrow M^{10+}$	

Principal lines in atomic spectrum

Wavelength/nm	Species	Sensitivity	Application
n.a.			

Abundance: Earth's crust nil; seawater nil

Biological role: None; never encountered, but toxic due to radioactivity

Pb

Atomic number: 82
Relative atomic mass ($^{12}C=12.0000$): 207.2

Chemical properties

Soft, weak, ductile, dull grey metal. Main ore is galena PbS. Tarnishes in moist air but stable to oxygen and water, dissolves in nitric acid. Used in batteries, cables, paints, glass, solder, petrol, radiation shielding etc.

Radii/pm: Pb^{2+} 132; Pb^{4+} 84; atomic 175.0; covalent 154; Pb^{4-} 215

Electronegativity: 2.33 (Pauling); 1.55 (Allred-Rochow)

Effective nuclear charge: 5.65 (Slater); 12.39 (Clementi); 15.33 (Froese–Fischer)

Standard reduction potentials $E^{\ominus}$/V

	IV		II		0		−II
Acid solution	Pb^{4+}	1.69	Pb^{2+}	−0.1251	Pb	−1.507	PbH_2

[Alkaline solutions contain many different forms]

Covalent bonds	r/pm	E/kJ mol^{-1}
Pb–H	184	180
Pb–C	229	130
Pb–O	192	398
Pb–F	213	314
Pb–Cl	247	244
Pb–Pb	350	100

Oxidation states

Pb^{II}	PbO, PbF_2, $PbCl_2$ etc., $PbOH^+(aq)$, $Pb(H_2O)_n^{2+}(aq)$ salts, complexes.
Pb^{IV}	PbO_2, Pb_3O_4 $(=2PbO.PbO_2)$ PbF_4, $PbCl_4$, $PbBr_4$, $PbCl_6^{2-}$, $Pb(OH)_6^{2-}(aq)$ [Pb^{4+} does not exist in water], organo lead compounds, complexes

Physical properties

Melting point/K: 600.65
Boiling point/K: 2013
ΔH_{fusion}/kJ mol^{-1}: 5.121
ΔH_{vap}/kJ mol^{-1}: 177.8

Thermodynamic properties (298.15 K, 0.1 MPa)

State	$\Delta_f H^{\ominus}$/kJ mol^{-1}	$\Delta_f G^{\ominus}$/kJ mol^{-1}	$S^{\ominus}$/J K^{-1} mol^{-1}	C_p/J K^{-1} mol^{-1}
Solid	0	0	64.81	26.44
Gas	195.0	161.9	175.373	20.786

Density/kg m^{-3}: 11 350 [293 K]; 10 678 [liquid, m.p.]
Thermal conductivity/W m^{-1} K^{-1}: 35.3 [300 K]
Electrical resistivity/Ω m: 20.648×10^{-8} [293 K]
Mass magnetic susceptibility/kg^{-1} m^3: -1.39×10^{-9} (s)
Molar volume/cm^3: 18.26
Coefficient of linear thermal expansion/K^{-1}: 29.1×10^{-9}

Lattice structure (cell dimensions/pm), space group

f.c.c. (a=495.00), Fm3m

X-ray diffraction: mass absorption coefficients (μ/ρ)/cm^2 g^{-1}:
CuK_α 232 MoK_α 120

Known to ancient civilizations

[Anglo-Saxon, *lead* Latin *plumbum*]

Lead

Nuclear properties

Atomic number: 82

Thermal neutron capture cross-section/barns: 0.18±0.01

Number of isotopes (including nuclear isomers): 29

Isotope mass range: 194→214

Key isotopes

Nuclide	Atomic mass	Natural abundance (%)	Half life $T_{1/2}$	Decay mode and energy (MeV)	Nuclear spin I	Nuclear magnetic moment μ	Uses
^{204}Pb	203.9731	1.4	stable				
^{205}Pb		0	3.0×10^{7}y	EC(*c*.0.035)			
^{206}Pb	205.9745	24.1	stable		0		
^{207}Pb	206.9759	22.1	stable		1/2	+0.5783	NMR
^{208}Pb	207.9766	52.3	stable		0		
^{210}Pb		trace	20.4y	β^-(0.061) 81%; α; γ			tracer
^{214}Pb		trace	10.64h	β^-(0.58); γ			

NMR

	^{207}Pb
Relative sensitivity (^{1}H=1.00)	9.16×10^{-3}
Absolute sensitivity (^{1}H=1.00)	2.07×10^{-3}
Receptivity (^{13}C=1.00)	11.8
Magnetogyric ratio/rad T^{-1} s^{-1}	5.5797×10^{7}
Frequency (^{1}H=100 MHz; 2.3488 T)/MHz	20.921

Reference: $Pb(CH_3)_4$

Electron shell properties

Ground state electron configuration: $[Xe]4f^{14}5d^{10}6s^{2}6p^{2}$

Term symbol: $^3P_0$

Electron affinity ($M\rightarrow M^-$)/kJ mol^{-1}: 35.2

Ionization energies/kJ mol^{-1}

1. $M \rightarrow M^+$	715.5	6. $M^{5+}\rightarrow M^{6+}$	(8 100)
2. $M^+\rightarrow M^{2+}$	1450.4	7. $M^{6+}\rightarrow M^{7+}$	(9 900)
3. $M^{2+}\rightarrow M^{3+}$	3081.5	8. $M^{7+}\rightarrow M^{8+}$	(11 800)
4. $M^{3+}\rightarrow M^{4+}$	4083	9. $M^{8+}\rightarrow M^{9+}$	(13 700)
5. $M^{4+}\rightarrow M^{5+}$	6640	10. $M^{9+}\rightarrow M^{10+}$	(16 700)

Principal lines in atomic spectrum

Wavelength/nm	Species	Sensitivity	Application
216.999	I		AA, AE
261.418	I		AA, AE
283.305*	I		AA
368.346	I	U2	AA, AE
405.781	I	U2	AE

* Lowest energy transition from ground state to nearest empty orbital

Abundance: Earth's crust 13 p.p.m.; seawater 0.003 p.p.m.

Biological role: None; toxic, teratogenic, carcinogenic

Li	Atomic number: 3 Relative atomic mass ($^{12}C=12.0000$): **6.941**

Chemical properties

Soft, white, silvery metal. Reacts slowly with oxygen and water. Large deposits known of spodumene, $LiAlSi_2O_6$. Used in alloys (with Al and Mg), greases, glass, medicine, and nuclear bombs.

Radii/pm: Li^+ 78; atomic 152; covalent 123

Electronegativity: 0.98 (Pauling); 0.97 (Allred-Rochow)

Effective nuclear charge: 1.30 (Slater); 1.28 (Clementi); 1.55 (Froese–Fischer)

Standard reduction potentials $E^{\ominus}$/V

I		0
Li^+	−3.040	Li

Oxidation states

Li^{-1} (s^2)	Li solutions in liquid ammonia
Li^{I} ([He])	Li_2O, LiOH, LiH, $LiAlH_4$, LiF, LiCl etc., $Li(H_2O)_4^+$(aq), Li_2CO_3, salts of Li^+, some complexes, $(LiCH_3)_4$

Physical properties

Melting point/K: 453.69

Boiling point/K: 1620

ΔH_{fusion}/kJ mol^{-1}: 4.60

ΔH_{vap}/kJ mol^{-1}: 147.7

Thermodynamic properties (298.15 K, 0.1 MPa)

State	$\Delta_f H^{\ominus}$/kJ mol^{-1}	$\Delta_f G^{\ominus}$/kJ mol^{-1}	$S^{\ominus}$/J K^{-1} mol^{-1}	C_p/J K^{-1} mol^{-1}
Solid	0	0	29.12	24.77
Gas	159.37	126.66	138.77	20.786

Density/kg m^{-3}: 534 [293 K]; 515 [liquid m.p.]

Thermal conductivity/W m^{-1} K^{-1}: 84.7 [300 K]

Electrical resistivity/Ω m: 8.55×10^{-8} [273 K]

Mass magnetic susceptibility/kg^{-1} m^3: $+2.56 \times 10^{-8}$ (s)

Molar volume/cm^3: 13.00

Coefficient of linear thermal expansion/K^{-1}: 56×10^{-6}

Lattice structure (cell dimensions/pm), space group

α-Li b.c.c. ($a=351.00$), Im3m
β-Li f.c.c. ($a=437.9$), Fm3m

$T(\alpha \rightarrow \beta)=77$ K

X-ray diffraction: mass absorption coefficients (μ/ρ)/cm^2 g^{-1}:
CuK_α 0.716 MoK_α 0.217

Discovered in 1817 by J. A. Arfvedson at Stockholm, Sweden.
Isolated by W. T. Brande 1821
[Greek, *lithos* = stone]

Lithium

Nuclear properties

Atomic number: 3
Thermal neutron capture cross-section/barns: 71 ± 1
Number of isotopes (including nuclear isomers): 5
Isotope mass range: 5→9

Key isotopes

Nuclide	Atomic mass	Natural abundance (%)	Half life $T_{1/2}$	Decay mode and energy (MeV)	Nuclear spin I	Nuclear magnetic moment μ	Uses
^{6}Li	6.015 12	7.5	stable		1	+0.822 03	NMR
^{7}Li	7.016 00	92.5	stable		3/2	+3.256 36	NMR

NMR

	[^{6}Li]	^{7}Li
Relative sensitivity (^{1}H = 1.00)	8.50×10^{3}	0.29
Absolute sensitivity (^{1}H = 1.00)	6.31×10^{-4}	0.27
Receptivity (^{13}C = 1.00)	3.58	1540
Magnetogyric ratio/rad T^{-1} s^{-1}	3.9366×10^{7}	10.3964×10^{7}
Quadrupole moment/m_2	-8×10^{-32}	-4.5×10^{-30}
Frequency (^{1}H = 100 MHz; 2.3488 T)/MHz	14.716	38.863

Reference: LiCl (aq)

Electron shell properties

Ground state electron configuration: [He]$2s^1$
Term symbol: $^{2}S_{1/2}$
Electron affinity ($M \rightarrow M^-$)/kJ mol^{-1}: 59.8

Ionization energies/kJ mol^{-1}

1. $M \rightarrow M^+$	513.3
2. $M^+ \rightarrow M^{2+}$	7 298.0
3. $M^{2+} \rightarrow M^{3+}$	11 814.8

Principal lines in atomic spectrum

Wavelength/nm	Species	Sensitivity	Application
323.261	I	U2	AA, AE
460.300	I	U4	AE
610.362	I	U3	AA, AE
670.784*	I	U1	AA, AE

* Lowest energy transition from ground state to nearest empty orbital

Abundance: Earth's crust 18 p.p.m.; seawater 0.2 p.p.m.
Biological role: None; non-toxic; teratogenic; stimulatory; anti-depressant

Lu

Atomic number: 71
Relative atomic mass (^{12}C=12.0000): 174.967

Chemical properties

Hardest, densest, and one of the rarest of the lanthanide (rare earth) metals. Obtained from euxenite, monazite, and bastnaesite. Little used except in research.

Radii/pm: Lu^{3+} 85; atomic 173.4; covalent 156
Electronegativity: 1.27 (Pauling); 1.14 (Allred-Rochow)
Effective nuclear charge: 3.00 (Slater); 8.80 (Clementi); 12.68 (Froese–Fischer)

Standard reduction potentials $E^{\ominus}$/V

	III		0
Acid solution	Lu^{3+}	$\xrightarrow{-2.30}$	Lu
Alkaline solution	$Lu(OH)_3$	$\xrightarrow{-2.83}$	Lu

Oxidation states

Lu^{III} (f^{14})	Lu_2O_3, $Lu(OH)_3$, $[Lu(H_2O)_x]^{3+}$ (aq), Lu^{3+} salts, LuF_3, $LuCl_3$ etc., $LuCl_6^{3-}$, complexes

Physical properties

Melting point/K: 1936
Boiling point/K: 3668
ΔH_{fusion}/kJ mol^{-1}: 19.2
ΔH_{vap}/kJ mol^{-1}: 428

Thermodynamic properties (298.15 K, 0.1 MPa)

State	$\Delta_f H^{\ominus}$/kJ mol^{-1}	$\Delta_f G^{\ominus}$/kJ mol^{-1}	$S^{\ominus}$/J K^{-1} mol^{-1}	C_p/J K^{-1} mol^{-1}
Solid	0	0	50.96	26.86
Gas	427.6	387.8	184.800	20.861

Density/kg m^{-3}: 9840 [298 K]
Thermal conductivity/W m^{-1} K^{-1}: 16.4 [300 K]
Electrical resistivity/Ω m: 79.0×10^{-8} [298 K]
Mass magnetic susceptibility/kg^{-1} m^3: $+1.3 \times 10^{-9}$ (s)
Molar volume/cm^3: 17.78
Coefficient of linear thermal expansion/K^{-1}: 8.12×10^{-6}

Lattice structure (cell dimensions/pm), space group

α-Lu h.c.p. (a=350.31, c=555.09), $P6_3/mmc$
β-Lu b.c.c. (a=390), Im3m

X-ray diffraction: mass absorption coefficients ((μ/ρ)/cm^2 g^{-1}):
CuK$_\alpha$ 153 MoK$_\alpha$ 88.2

Discovered in 1907 by G. Urbain at Paris, France and independently by C. James at the University of New Hampshire, USA

[Latin, *Lutetia* = Paris]

Lutetium

Nuclear properties

Atomic number: 71

Thermal neutron capture cross-section/barns: 75 ± 2

Number of isotopes (including nuclear isomers): 22

Isotope mass range: 167→180

Key isotopes

Nuclide	Atomic mass	Natural abundance (%)	Half life $T_{1/2}$	Decay mode and energy (MeV)	Nuclear spin I	Nuclear magnetic moment μ	Uses
^{175}Lu	174.9409	97.39	stable		7/2	+2.203	NMR
^{176}Lu		2.61	2.2×10^{10}y	$\beta^-(1.02)$; γ	7	+3.18	
^{177}Lu		0	6.74d	$\beta^-(0.497)$; γ	7/2	+2.24	tracer

NMR (not used)	^{175}Lu
Relative sensitivity ($^1H = 1.00$)	3.12×10^{-2}
Absolute sensitivity ($^1H = 1.00$)	3.03×10^{-2}
Receptivity ($^{13}C = 1.00$)	156
Magnetogyric ratio/rad T^{-1} s^{-1}	3.05×10^7
Quadrupole moment/m^2	5.68×10^{-28}
Frequency (^1H = 100 MHz; 2.3488 T)/MHz	11.407

Electron shell properties

Ground state electron configuration: $[Xe]4f^{14}5d^16s^2$

Term symbol: $^2D_{3/2}$

Electron affinity ($M \rightarrow M^-$)/kJ mol^{-1}: $\leqslant 50$

Ionization energies/kJ mol^{-1}

1. $M \rightarrow M^+$	523.5	6. $M^{5+} \rightarrow M^{6+}$	
2. $M^+ \rightarrow M^{2+}$	1340	7. $M^{6+} \rightarrow M^{7+}$	
3. $M^{2+} \rightarrow M^{3+}$	2022	8. $M^{7+} \rightarrow M^{8+}$	
4. $M^{3+} \rightarrow M^{4+}$	4360	9. $M^{8+} \rightarrow M^{9+}$	
5. $M^{4+} \rightarrow M^{5+}$		10. $M^{9+} \rightarrow M^{10+}$	

Principal lines in atomic spectrum

Wavelength/nm	Species	Sensitivity	Application
261.542	II		
291.139	II		AE
331.211	I		AA, AE
335.956	I		AE
347.248	II		AE
451.857	I		AA, AE

Abundance: Earth's crust 0.8 p.p.m.; seawater 4×10^{-8} p.p.m.

Biological role: None; low toxicity

Mg

Atomic number: 12

Relative atomic mass (^{12}C=12.0000): 24.3050

Chemical properties

Silvery white, lustrous, relatively soft metal. Obtained by electrolysis of fused $MgCl_2$. Burns in air and reacts with hot water. Used as bulk metal and in lightweight alloys for engines, also as a sacrificial electrode to protect other metals.

Radii/pm: Mg^{2+} 78; atomic 160; covalent 136

Electronegativity: 1.31 (Pauling); 1.23 (Allred-Rochow)

Effective nuclear charge: 2.85 (Slater); 3.31 (Clementi); 4.15 (Froese–Fischer)

Standard reduction potentials E°/V

	II		I		0
Acid solution	Mg^{2+}	−2.054	Mg^+	−2.657	Mg
	Mg^{2+} → Mg: −2.356				
Alkaline solution	$Mg(OH)_2$	−2.687			Mg

Oxidation states

Mg^{II}	MgO, MgO_2, $Mg(OH)_2$, $[Mg(H_2O)_6]^{2+}$(aq), MgH_2, $MgCO_3$, Mg^{2+} salts, MgF_2, $MgCl_2$, etc., CH_3MgI, complexes

Physical properties

Melting point/K: 922.0

Boiling point/K: 1363

ΔH_{fusion}/kJ mol^{-1}: 9.04

ΔH_{vap}/kJ mol^{-1}: 127.6

Thermodynamic properties (298.15 K, 0.1 MPa)

State	$\Delta_f H^\circ$/kJ mol^{-1}	$\Delta_f G^\circ$/kJ mol^{-1}	S°/J K^{-1} mol^{-1}	C_p/J K^{-1} mol^{-1}
Solid	0	0	32.68	24.89
Gas	147.70	113.10	148.650	20.786

Density/kg m^{-3}: 1738 [293 K]; 1585 [liquid at m.p.]

Thermal conductivity/W m^{-1} K^{-1}: 156 [300 K]

Electrical resistivity/Ω m: 4.45×10^{-8} [293 K]

Mass magnetic susceptibility/kg^{-1} m^3: $+6.8 \times 10^{-9}$ (s)

Molar volume/cm^3: 13.98

Coefficient of linear thermal expansion/K^{-1}: 26.1×10^{-6}

Lattice structure (cell dimensions/pm), space group

h.c.p. (a=320.94; c=521.03), $P6_3/mmc$

X-ray diffraction: mass absorption coefficients (μ/ρ)/cm^2 g^{-1}:
CuK_α 38.6 MoK_α 4.11

Recognized as an element in 1755 by Joseph Black at Edinburgh, Scotland; isolated by Sir Humphry Davy in 1808

[Greek, *Magnesia*, district of Thessaly]

Magnesium

Nuclear properties

Atomic number: 12

Thermal neutron capture cross-section/barns: 0.064 ± 0.002

Number of isotopes (including nuclear isomers): 8

Isotope mass range: $20 \rightarrow 28$

Key isotopes

Nuclide	Atomic mass	Natural abundance (%)	Half life $T_{1/2}$	Decay mode and energy (MeV)	Nuclear spin I	Nuclear magnetic moment μ	Uses
^{24}Mg	23.98504	78.99	stable		0		
^{25}Mg	24.98584	10.00	stable		5/2	−0.8564	NMR
^{26}Mg	25.98259	11.01	stable		0		
^{28}Mg*		0	21.2h	β^-(1.84); γ			

* longest lived radioactive isotope

NMR

	^{25}Mg
Relative sensitivity ($^1H = 1.00$)	2.67×10^{-3}
Absolute sensitivity ($^1H = 1.00$)	2.71×10^{-4}
Receptivity ($^{13}C = 1.00$)	1.54
Magnetogyric ratio/rad T^{-1} s^{-1}	1.6375×10^7
Quadrupole moment/m_2	0.22×10^{-28}
Frequency ($^1H = 100$ MHz; 2.3488 T)/MHz	6.1195
Reference: $MgCl_2$(aq)	

Electron shell properties

Ground state electron configuration: $[Ne]3s^2$

Term symbol: $^1S_0$

Electron affinity ($M \rightarrow M^-$)/kJ mol^{-1}: −21

Ionization energies/kJ mol^{-1}

1. $M \rightarrow M^+$	737.7	6. $M^{5+} \rightarrow M^{6+}$	17995
2. $M^+ \rightarrow M^{2+}$	1450.7	7. $M^{6+} \rightarrow M^{7+}$	21703
3. $M^{2+} \rightarrow M^{3+}$	7732.6	8. $M^{7+} \rightarrow M^{8+}$	25656
4. $M^{3+} \rightarrow M^{4+}$	10540	9. $M^{8+} \rightarrow M^{9+}$	31642
5. $M^{4+} \rightarrow M^{5+}$	13630	10. $M^{9+} \rightarrow M^{10+}$	35461

Principal lines in atomic spectrum

Wavelength/nm	Species	Sensitivity	Application
202.582	I		AA
279.553	II	V1	AA, AE
285.213	I	U1	AA, AE
383.230	I	U3	AA
383.826	I	U2	AE
457.110*			

* Lowest energy transition from ground state to nearest empty orbital

Abundance: Earth's crust 27640 p.p.m.; seawater 1300 p.p.m.

Biological role: Essential; non-toxic

Mn

Atomic number: 25

Relative atomic mass ($^{12}C = 12.0000$): **54.93805**

Chemical properties

Hard, brittle, silvery metal. Occurs as pyrolusite MnO_2. Reactive when impure and will burn in oxygen. Surface oxidation occurs in air; will react with water, dissolves in dilute acids. Most Mn is used in steel production.

Radii/pm: Mn^{2+} 91; Mn^{3+} 70; Mn^{4+} 52; atomic 124; covalent 117

Electronegativity: 1.55 (Pauling); 1.60 (Allred-Rochow)

Effective nuclear charge: 3.60 (Slater); 5.23 (Clementi); 7.17 (Froese–Fischer)

Standard reduction potentials $E^\ominus$/V

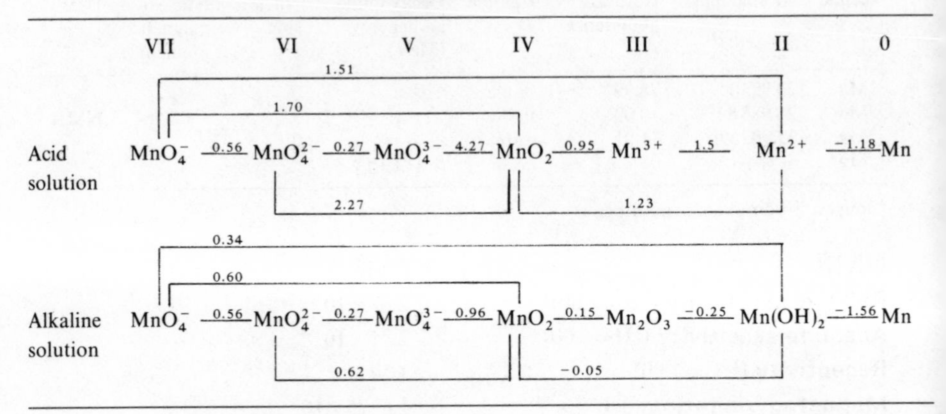

Oxidation states

Mn^{-III}	(d^{10})	$[Mn(NO)_3(CO)]$
Mn^{-II}	(d^9)	some complexes known
Mn^{-I}	(d^8)	$[Mn(CO)_5]^-$
Mn^0	(d^7)	$Mn_2(CO)_{10}$
Mn^I	(d^6)	$[Mn(CN)_6]^-$
Mn^{II}	(d^5)	MnO, Mn_3O_4 ($=Mn^{II}Mn_2^{III}O_4$), $[Mn(H_2O)_6]^{2+}$(aq), MnF_2, $MnCl_2$ etc., salts, complexes
Mn^{III}	(d^4)	Mn_2O_3 $[Mn(H_2O)_6]^{3+}$(aq) unstable; MnF_3, $MnCl_5^{2-}$
Mn^{IV}	(d^3)	MnO_2, MnF_4, MnF_6^{2-}
Mn^V	(d^2)	MnO_4^{3-}
Mn^{VI}	(d^1)	MnO_4^{2-}
Mn^{VII}	(d^0, [Ar])	Mn_2O_7, MnO_4^-

Physical properties

Melting point/K: 1517

Boiling point/K: 2235

ΔH_{fusion}/kJ mol^{-1}: 14.4 ΔH_{vap}/kJ mol^{-1}: 220.5

Thermodynamic properties (298.15 K, 0.1 MPa)

State	$\Delta_f H^\ominus$/kJ mol^{-1}	$\Delta_f G^\ominus$/kJ mol^{-1}	$S^\ominus$/J K^{-1} mol^{-1}	C_p/J K^{-1} mol^{-1}
Solid	0	0	32.01	26.32
Gas	280.7	238.5	173.70	20.79

Density/kg m^{-3}: 7440 (α) [293 K]; 6430 [liquid at m.p.]

Thermal conductivity/W m^{-1} K^{-1}: 7.82 [300 K]

Electrical resistivity/Ω m: 185.0×10^{-8} [298 K]

Mass magnetic susceptibility/kg^{-1} m^3: $+1.21 \times 10^{-7}$ (s)

Molar volume/cm^3: 7.38

Coefficient of linear thermal expansion/K^{-1}: 22×10^{-6}

Lattice structure (cell dimensions/pm), space group

α-Mn b.c.c. ($a = 891.39$), I$\bar{4}$3m
γ-Mn f.c.c. ($a = 386.3$), Fm3m
β-Mn b.c.c. ($a = 631.45$), P$4_1$32
δ-Mn b.c.c. ($a = 308.1$), Im3m

$T(\alpha \to \beta) = 973$ K; $T(\beta \to \gamma) = 1352$ K; $T(\gamma \to \delta) = 1413$ K

X-ray diffraction: mass absorption coefficients (μ/ρ)/cm^2 g^{-1}:
CuK_α 285 MoK_α 34.7

Isolated in 1774 by J. G. Grahn at Stockholm, Sweden

[Latin, *magnes* = magnet]

Manganese

Nuclear properties

Atomic number: 25

Thermal neutron capture cross-section/barns: 13.3 ± 0.1

Number of isotopes (including nuclear isomers): 11

Isotope mass range: $50 \rightarrow 58$

Key isotopes

Nuclide	Atomic mass	Natural abundance (%)	Half life $T_{1/2}$	Decay mode and energy (MeV)	Nuclear spin I	Nuclear magnetic moment μ	Uses
^{53}Mn	52.9413	0	2×10^6y	EC(0.598); no γ	7/2	± 5.02	
^{54}Mn	53.9402	0	303d	EC(1.379); γ	3	$+3.278$	tracer
^{55}Mn	54.9381	100	stable		5/2	$+3.449$	NMR
^{56}Mn		0	2.576h	β^-(3.702); γ	3	$+3.223$	tracer

NMR

	^{55}Mn
Relative sensitivity (^{1}H = 1.00)	0.18
Absolute sensitivity (^{1}H = 1.00)	0.18
Receptivity (^{13}C = 1.00)	994
Magnetogyric ratio/rad T^{-1} s^{-1}	6.6195×10^7
Quadrupole moment/m^2	0.55×10^{-28}
Frequency (^{1}H = 100 MHz; 2.3488 T)/MHz	24.664

Reference: $KMnO_4$(aq)

Electron shell properties

Ground state electron configuration: [Ar]$3d^5 4s^2$

Term symbol: $^6S_{5/2}$

Electron affinity ($M \rightarrow M^-$)/kJ mol^{-1}: -94

Ionization energies/kJ mol^{-1}

1.	$M \rightarrow M^+$	717.4
2.	$M^+ \rightarrow M^{2+}$	1509.0
3.	$M^{2+} \rightarrow M^{3+}$	3248.4
4.	$M^{3+} \rightarrow M^{4+}$	4940
5.	$M^{4+} \rightarrow M^{5+}$	6990
6.	$M^{5+} \rightarrow M^{6+}$	9 200
7.	$M^{6+} \rightarrow M^{7+}$	11 508
8.	$M^{7+} \rightarrow M^{8+}$	18 956
9.	$M^{8+} \rightarrow M^{9+}$	21 400
10.	$M^{9+} \rightarrow M^{10+}$	23 960

Principal lines in atomic spectrum

Wavelength/nm	Species	Sensitivity	Application
279.482	I		AA
279.827	I		AA, AE
403.076	I	U2	AA, AE
403.307	I	U2	AE
539.467*	I		

* Lowest energy transition from ground state to nearest empty orbital

Abundance: Earth's crust 1060 p.p.m.; seawater 0.002 p.p.m.

Biological role: Essential; non-toxic, suspected carcinogenic

Md

Atomic number: 101
Relative atomic mass ($^{12}C=12.0000$): **(258)**

Chemical properties

Radioactive metal of which only a few atoms have been made by bombarding ^{253}Es with α particles (^4He)

Radii/pm: Md^{2+} 114; Md^{3+} 96; Md^{4+} 84
Electronegativity: 1.3 (Pauling); n.a. (Allred-Rochow)
Effective nuclear charge: 1.65 (Slater)

Standard reduction potentials $E^\ominus$/V

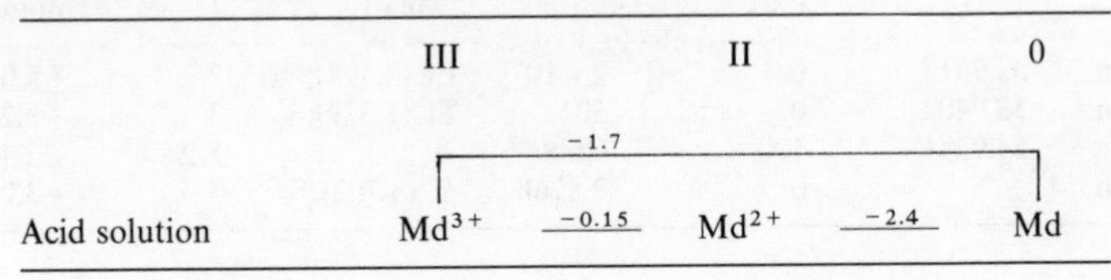

Oxidation states

Md^{II} (f^{13})	stable ?
Md^{III} (f^{12})	$[Md(H_2O)_x]^{3+}$(aq)

Physical properties

Melting point/K: n.a.
Boiling point/K: n.a.
ΔH_{fusion}/kJ mol^{-1}: n.a.
ΔH_{vap}/kJ mol^{-1}: n.a.

Thermodynamic properties (298.15 K, 0.1 MPa)

State	$\Delta_f H^\ominus$/kJ mol^{-1}	$\Delta_f G^\ominus$/kJ mol^{-1}	$S^\ominus$/J K^{-1} mol^{-1}	C_p/J K^{-1} mol^{-1}
Solid	0	0	n.a.	n.a.
Gas	n.a.	n.a.	n.a.	n.a.

Density/kg m^{-3}: n.a.
Thermal conductivity/W m^{-1} K^{-1}: 10 (est.) [300 K]
Electrical resistivity/Ω m: n.a.
Mass magnetic susceptibility/kg^{-1} m^3: n.a.
Molar volume/cm^3: n.a.
Coefficient of linear thermal expansion/K^{-1}: n.a.

Lattice structure (cell dimensions/pm), space group

n.a.

X-ray diffraction: mass absorption coefficients (μ/ρ)/cm^2 g^{-1}:
n.a.

Prepared in 1955 by A. Ghiorso, B. G. Harvey, G. R. Choppin, S. G. Thompson, and G. T. Seaborg at Berkeley, California, USA
[Named after Dmitri Mendeleyev]

Mendelevium

Nuclear properties

Atomic number: 101
Thermal neutron capture cross-section/barns: n.a.
Number of isotopes (including nuclear isomers): 10
Isotope mass range: 248→258

Key isotopes

Nuclide	Atomic mass	Natural abundance (%)	Half life $T_{1/2}$	Decay mode and energy (MeV)	Nuclear spin I	Nuclear magnetic moment μ	Uses
^{258}Md		0	54d	EC; α(6.79); SF			

Electron shell properties

Ground state electron configuration: $[Rn]5f^{13}7s^2$
Term symbol: $^2F_{7/2}$
Electron affinity ($M \rightarrow M^-$)/kJ mol^{-1}: n.a.

Ionization energies/kJ mol^{-1}

1. $M \rightarrow M^+$	635	6. $M^{5+} \rightarrow M^{6+}$	
2. $M^+ \rightarrow M^{2+}$		7. $M^{6+} \rightarrow M^{7+}$	
3. $M^{2+} \rightarrow M^{3+}$		8. $M^{7+} \rightarrow M^{8+}$	
4. $M^{3+} \rightarrow M^{4+}$		9. $M^{8+} \rightarrow M^{9+}$	
5. $M^{4+} \rightarrow M^{5+}$		10. $M^{9+} \rightarrow M^{10+}$	

Principal lines in atomic spectrum

Wavelength/nm	Species	Sensitivity	Application
n.a.			

Abundance: Earth's crust nil; seawater nil
Biological role: None; toxic due to radiation

Hg

Atomic number: 80
Relative atomic mass ($^{12}C=12.0000$): 200.59

Chemical properties

Liquid silvery metal obtained by roasting cinnabar, HgS. Stable with air and water, unreactive to acids (except conc HNO_3) and alkalis. Used in chlorine and NaOH manufacture, street lights, fungicides, electrical apparatus, etc.

Radii/pm: Hg^+ 127; Hg^{2+} 112; atomic 160; covalent 144

Electronegativity: 2.00 (Pauling); 1.44 (Allred-Rochow)

Effective nuclear charge: 4.35 (Slater); 11.15 (Clementi); 16.22 (Froese–Fischer)

Standard reduction potentials $E^\ominus$/V

	II		I		0
		(0.8535, II → 0)			
Acid solution	Hg^{2+}	−0.9110	Hg_2^{2+}	0.7960	Hg
Alkaline solution	HgO		0.0977		Hg

Oxidation states

Hg^I ($d^{10}s^1$)	Hg_2F_2, Hg_2Cl_2 etc., Hg_2^{2+} salts, most insoluble, $Hg_2(NO_3)_2 \cdot 2H_2O$ soluble.
Hg^{II} (d^{10})	HgO, HgS, HgF_2, $HgCl_2$, etc., Hg^{2+} salts, $[Hg(H_2O)_6]^{2+}$ (aq) HgN(OH), complexes eg $[Hg(SCN)_4]^{2-}$, $Hg(CH_3)_2$, etc.

Physical properties

Melting point/K: 234.28
Boiling point/K: 629.73
ΔH_{fusion}/kJ mol^{-1}: 2.331
ΔH_{vap}/kJ mol^{-1}: 59.11

Thermodynamic properties (298.15 K, 0.1 MPa)

State	$\Delta_f H^\ominus$/kJ mol^{-1}	$\Delta_f G^\ominus$/kJ mol^{-1}	$S^\ominus$/J K^{-1} mol^{-1}	C_p/J K^{-1} mol^{-1}
Solid	0	0	76.02	27.983
Gas	61.317	31.820	174.96	20.786

Density/kg m^{-3}: 13 546 [293 K]
Thermal conductivity/W m^{-1} K^{-1}: 8.34 [300 K]
Electrical resistivity/Ω m: 94.1×10^{-8} [273 K]
Mass magnetic susceptibility/kg^{-1} m^3: -2.095×10^{-9} (l)
Molar volume/cm^3: 14.81
Coefficient of cubical thermal expansion/K^{-1}: 18.1×10^{-5}

Lattice structure (cell dimensions/pm), space group

α-Hg rhombohedral ($a=299.25$, $\alpha=70°\ 44.6'$), $R\bar{3}m$
β-Hg tetragonal ($a=399.5$, $c=282.5$), I4/mmm
α→β high pressure

X-ray diffraction: mass absorption coefficients (μ/ρ)/cm^2 g^{-1}:
CuK_α 216 MoK_α 117

Known to ancient civilizations

[Named after planet Mercury. Latin, *hydragyrum*=liquid silver]

Mercury

Nuclear properties

Atomic number: 80

Thermal neutron capture cross-section/barns: 375 ± 5

Number of isotopes (including nuclear isomers): 26

Isotope mass range: 185→206

Key isotopes

Nuclide	Atomic mass	Natural abundance (%)	Half life $T_{1/2}$	Decay mode and energy (MeV)	Nuclear spin I	Nuclear magnetic moment μ	Uses
^{196}Hg	195.9658	0.2	stable				
^{197}Hg		0	65h	EC(0.42); γ	1/2	+0.5241	tracer, medical
^{198}Hg	197.9668	10.1	stable		0		
^{199}Hg	198.9683	17.0	stable		1/2	+0.502 71	NMR
^{200}Hg	199.9682	23.1	stable		0		
^{201}Hg	200.9703	13.2	stable		3/2	−0.556 71	NMR
^{202}Hg	201.9706	29.6	stable		0		
^{203}Hg		0	46.9	β^-(0.492); γ	5/2	+0.86	tracer, medical
^{204}Hg	203.9735	6.8	stable	0			

NMR

	^{199}Hg	[^{201}Hg]
Relative sensitivity (^1H=1.00)	5.67×10^{-3}	1.44×10^{-3}
Absolute sensitivity (^1H=1.00)	9.54×10^{-4}	1.90×10^{-4}
Receptivity (^{13}C=1.00)	5.42	1.08
Magnetogyric ratio/rad T^{-1} s^{-1}	4.7912×10^7	-1.7686×10^7
Quadrupole moment/m^2	—	0.5×10^{-28}
Frequency (^1H=100 MHz; 2.3488 T)/MHz	17.827	6.599

Reference: $Hg(CH_3)_2$

Electron shell properties

Ground state electron configuration: [Xe]$4f^{14}5d^{10}6s^2$

Term symbol: $^1S_0$

Electron affinity ($M \rightarrow M^-$)/kJ mol^{-1}: −18

Ionization energies/kJ mol^{-1}

1. $M \rightarrow M^+$	1007.0	6. $M^{5+} \rightarrow M^{6+}$ (7400)
2. $M^+ \rightarrow M^{2+}$	1809.7	7. $M^{6+} \rightarrow M^{7+}$ (9100)
3. $M^{2+} \rightarrow M^{3+}$	3300	8. $M^{7+} \rightarrow M^{8+}$ (11600)
4. $M^{3+} \rightarrow M^{4+}$	(4400)	9. $M^{8+} \rightarrow M^{9+}$ (13400)
5. $M^{4+} \rightarrow M^{5+}$	(5900)	10. $M^{9+} \rightarrow M^{10+}$ (15300)

Principal lines in atomic spectrum

Wavelength/nm	Species	Sensitivity	Application
253.652*	I	U2	AA, AE
365.015	I	U3	AE
365.483	I	U4	AE
366.328	I	U5	AE
435.833	I		AE

* Lowest energy transition from ground state to nearest empty orbital

Abundance: Earth's crust 0.08 p.p.m.; seawater 3×10^{-5} p.p.m.

Biological role: None; toxic, methyl mercury very toxic; teratogenic

Mo

Atomic number: 42
Relative atomic mass ($^{12}C = 12.0000$): 95.94

Chemical properties

Metal is lustrous, silvery, and fairly soft when pure. Usually obtained as grey powder. Key ore is molybdenite, MoS_2, but is also obtained as by-product of copper production. Used in alloys, electrodes, and catalysts.

Radii/pm: Mo^{2+} 92; Mo^{6+} 6.2; atomic 136.2; covalent 129

Electronegativity: 2.16 (Pauling); 1.30 (Allred-Rochow)

Effective nuclear charge: 3.45 (Slater); 6.98 (Clementi); 9.95 (Froese–Fischer)

Standard reduction potentials $E^{\ominus}$/V

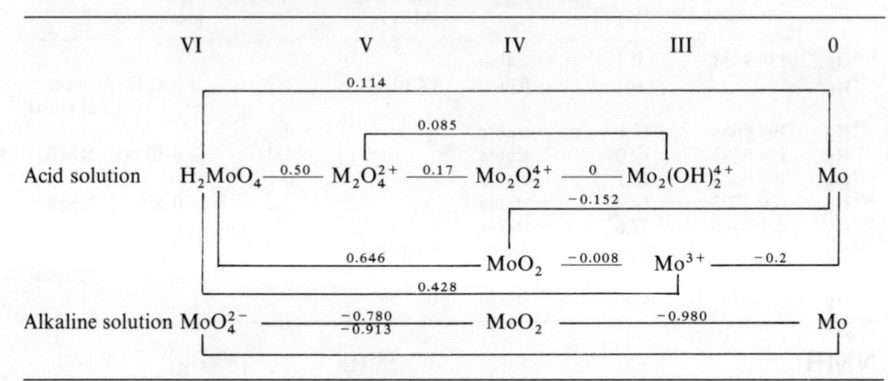

Oxidation states

Mo^{-II}	(d^8)	rare $[Mo(CO)_5]^{2-}$
Mo^0	(d^6)	rare $Mo(CO)_6$
Mo^I	(d^5)	rare $[Mo(C_6H_6)_2]^+$
Mo^{II}	(d^4)	Mo_6Cl_{12}, $[Mo_2Cl_8]^{4-}$, Mo_2^{4+}(aq)
Mo^{III}	(d^3)	MoF_3, $MoCl_3$ etc., $[Mo(H_2O)_6]^{3+}$ (aq)
Mo^{IV}	(d^2)	MoO_2, MoS_2, MoF_4, $MoCl_4$, etc.
Mo^V	(d^1)	Mo_2O_5, MoF_5, $MoCl_5$
Mo^{VI}	(d^0, [Kr])	MoO_3, MoO_4^{2-}(aq), MoF_6, MoF_8^{2-}, $MoOF_4$

Physical properties

Melting point/K: 2890

Boiling point/K: 4885

ΔH_{fusion}/kJ mol^{-1}: 27.6 $\qquad$ ΔH_{vap}/kJ mol^{-1}: 589.9

Thermodynamic properties (298.15 K, 0.1 MPa)

State	$\Delta_f H^{\ominus}$/kJ mol^{-1}	$\Delta_f G^{\ominus}$/kJ mol^{-1}	$S^{\ominus}$/J K^{-1} mol^{-1}	C_p/J K^{-1} mol^{-1}
Solid	0	0	28.66	24.06
Gas	658.1	612.5	181.950	20.786

Density/kg m^{-3}: 10 220 [293 K]; 9330 [liquid at m.p.]

Thermal conductivity/W m^{-1} K^{-1}: 138 [300 K]

Electrical resistivity/Ω m: 5.2×10^{-8} [273 K]

Mass magnetic susceptibility/kg^{-1} m^3: $+1.2 \times 10^{-8}$ (s)

Molar volume/cm^3: 9.39

Coefficient of linear thermal expansion/K^{-1}: 5.43×10^{-6}

Lattice structure (cell dimensions/pm), space group

b.c.c. ($a = 314.700$), Im3m

X-ray diffraction: mass absorption coefficients (μ/ρ)/cm^2 g^{-1}:
CuK_α 162 $\quad$ MoK_α 18.4

Isolated in 1781 by P. J. Hjelm at Uppsala, Sweden

[Greek, *molybdos* = lead]

Molybdenum

Nuclear properties

Atomic number: 42

Thermal neutron capture cross-section/barns: 2.65 ± 0.05

Number of isotopes (including nuclear isomers): 20

Isotope mass range: 88→105

Key isotopes

Nuclide	Atomic mass	Natural abundance (%)	Half life $T_{1/2}$	Decay mode and energy (MeV)	Nuclear spin I	Nuclear magnetic moment μ	Uses
^{92}Mo	91.9063	14.84	stable		0		
^{94}Mo	93.9047	9.25	stable		0		
^{95}Mo	94.905 84	15.92	stable		5/2	−0.9135	NMR
^{96}Mo	95.9046	16.68	stable		0		
^{97}Mo	96.9058	9.55	stable		5/2	−0.9327	NMR
^{98}Mo	97.9055	24.13	stable		0		
^{99}Mo		0	66.69h	β^-(1.37); γ			tracer
^{100}Mo	99.9076	9.63	stable		0		

NMR

	^{95}Mo	[^{97}Mo]
Relative sensitivity ($^1H = 1.00$)	3.23×10^{-3}	3.43×10^{-3}
Absolute sensitivity ($^1H = 1.00$)	5.07×10^{-4}	3.24×10^{-4}
Receptivity ($^{13}C = 1.00$)	2.88	1.84
Magnetogyric ratio/rad T^{-1} s^{-1}	1.7433×10^7	-1.7799×10^7
Quadrupole moment/m^2	0.12×10^{-28}	1.1×10^{-28}
Frequency ($^1H = 100$ MHz; 2.3488 T)/MHz	6.514	6.652

Reference: MoO_4^{2-} (aq)

Electron shell properties

Ground state electron configuration: [Kr]$4d^5 5s^1$

Term symbol: $^7S_3$

Electron affinity ($M \rightarrow M^-$)/kJ mol^{-1}: 114

Ionization energies/kJ mol^{-1}

1. $M \rightarrow M^+$	685.0	6. $M^{5+} \rightarrow M^{6+}$	6 560
2. $M^+ \rightarrow M^{2+}$	1558	7. $M^{6+} \rightarrow M^{7+}$	12 230
3. $M^{2+} \rightarrow M^{3+}$	2621	8. $M^{7+} \rightarrow M^{8+}$	14 800
4. $M^{3+} \rightarrow M^{4+}$	4480	9. $M^{8+} \rightarrow M^{9+}$	(16 800)
5. $M^{4+} \rightarrow M^{5+}$	5900	10. $M^{9+} \rightarrow M^{10+}$	(19 700)

Principal lines in atomic spectrum

Wavelength/nm	Species	Sensitivity	Application
202.030	II		
313.259	I		AA
317.035	I		AA
379.825	I	U1	AA, AE
386.411	I	U2	AE
390.296	I	U3	AE

Abundance: Earth's crust 1.2 p.p.m.; seawater 0.01 p.p.m.

Biological role: Essential; moderately toxic; teratogenic

Nd

Atomic number: 60

Relative atomic mass (^{12}C=12.0000): 144.24

Chemical properties

Silvery-white metal of the lanthanide (rare earth) group. Obtained from monazite and bastnaesite. Tarnishes in air, reacts slowly with cold water, rapidly with hot. Used in alloys, flints, glazes, and glass.

Radii/pm: Nd^{3+} 104; atomic 182.1; covalent 164

Electronegativity: 1.14 (Pauling); 1.07 (Allred-Rochow)

Effective nuclear charge: 2.85 (Slater); 9.31 (Clementi); 10.83 (Froese–Fischer)

Standard reduction potentials $E^{\ominus}$/V

	IV		III		II		0
Acid solution	Nd^{4+}	4.9	Nd^{3+}	−2.6	Nd^{2+}	−2.2	Nd
			Nd^{3+}	−2.32			Nd
Alkaline solution	$[NdO_2]$	2.5	$Nd(OH)_3$	−2.78			Nd

Oxidation states

Nd^{II} (f^4)	NdO, $NdCl_2$, NdI_2
Nd^{III} (f^3)	Nd_2O_3, $Nd(OH)_3$, $[Nd(H_2O)_x]^{3+}$ (aq), NdF_3, $NdCl_3$, etc., Nd^{3+} salts, complexes
Nd^{IV} (f^2)	Cs_3NdF_7

Physical properties

Melting point/K: 1294

Boiling point/K: 3341

ΔH_{fusion}/kJ mol^{-1}: 7.113

ΔH_{vap}/kJ mol^{-1}: 328

Thermodynamic properties (298.15 K, 0.1 MPa)

State	$\Delta_f H^{\ominus}$/kJ mol^{-1}	$\Delta_f G^{\ominus}$/kJ mol^{-1}	$S^{\ominus}$/J K^{-1} mol^{-1}	C_p/J K^{-1} mol^{-1}
Solid	0	0	71.5	27.45
Gas	327.6	292.4	189.406	22.092

Density/kg m^{-3}: 7007 [293 K]

Thermal conductivity/W m^{-1} K^{-1}: 16.5 [300 K]

Electrical resistivity/Ω m: 64.0×10^{-8} [293 K]

Mass magnetic susceptibility/kg^{-1} m^3: $+4.902 \times 10^{-7}$ (s)

Molar volume/cm^3: 20.59

Coefficient of linear thermal expansion/K^{-1}: 6.7×10^{-6}

Lattice structure (cell dimensions/pm), space group

α-Nd hexagonal (a=365.79, c=1179.92), $P6_3/mmc$
β-Nd b.c.c. (a=413), Im3m

$T(\alpha \rightarrow \beta)$=1135 K

High pressure form: f.c.c. (a=480), Fm3m

X-ray diffraction: mass absorption coefficients (μ/ρ)/cm^2 g^{-1}: CuK_α 374 MoK_α 53.2

Separated in 1885 by Baron Auer von Welsbach at Vienna, Austria
[Greek, *neos didymos* = new twin]

Neodymium

Nuclear properties

Atomic number: 60
Thermal neutron capture cross-section/barns: 49 ± 20
Number of isotopes (including nuclear isomers): 16
Isotope mass range: 138 → 151

Key isotopes

Nuclide	Atomic mass	Natural abundance (%)	Half life $T_{1/2}$	Decay mode and energy (MeV)	Nuclear spin I	Nuclear magnetic moment μ	Uses
^{142}Nd	141.9075	27.16	stable				
^{143}Nd	142.9096	12.18	stable		7/2	−1.063	NMR
^{144}Nd	143.9099	23.80	2.4×10^{15}y	α(1.83)			
^{145}Nd	144.9122	8.29	stable		7/2	−0.654	NMR
^{146}Nd	145.9172	17.19	stable				
^{147}Nd		0	11.06d	β^-(0.91); γ	5/2	±0.55	tracer
^{148}Nd	147.9165	5.75	stable				
^{150}Nd	149.9207	5.63	stable				

NMR

	^{143}Nd	^{145}Nd
Relative sensitivity (^{1}H = 1.00)	3.38×10^{-3}	7.86×10^{-4}
Absolute sensitivity (^{1}H = 1.00)	4.11×10^{-4}	6.52×10^{-5}
Receptivity (^{13}C = 1.00)	2.43	0.393
Magnetogyric ratio/rad T^{-1} s^{-1}	-1.474×10^7	-0.913×10^7
Quadrupole moment/m^2	-0.48×10^{-28}	-0.25×10^{-28}
Frequency (^{1}H = 100 MHz; 2.3488 T)/MHz	5.437	3.346

Electron shell properties

Ground state electron configuration: [Xe]$4f^4 6s^2$
Term symbol: $^5I_4$
Electron affinity ($M \rightarrow M^-$)/kJ mol^{-1}: ⩽50

Ionization energies/kJ mol^{-1}

1. $M \rightarrow M^+$	529.6	6. $M^{5+} \rightarrow M^{6+}$	
2. $M^+ \rightarrow M^{2+}$	1035	7. $M^{6+} \rightarrow M^{7+}$	
3. $M^{2+} \rightarrow M^{3+}$	2130	8. $M^{7+} \rightarrow M^{8+}$	
4. $M^{3+} \rightarrow M^{4+}$	3899	9. $M^{8+} \rightarrow M^{9+}$	
5. $M^{4+} \rightarrow M^{5+}$		10. $M^{9+} \rightarrow M^{10+}$	

Principal lines in atomic spectrum

Wavelength/nm	Species	Sensitivity	Application
395.115	II		AE
417.732	II		AE
430.357	II		AE
463.424	I		AA
492.453	I		AE

Abundance: Earth's crust 40 p.p.m.; seawater 3×10^{-6} p.p.m.
Biological role: None; moderately toxic, eye irritant

Ne

Atomic number: 10
Relative atomic mass ($^{12}C = 12.0000$): 20.1797

Chemical properties

Colourless, odourless gas, obtained from liquid air. Chemically inert towards everything including fluorine gas. Used in ornamental lighting ('neon' signs)

Radii/pm: van der Waals 160
Electronegativity: n.a. (Pauling); n.a. (Allred-Rochow)
Effective nuclear charge: 5.85 (Slater); 5.76 (Clementi); 5.18 (Froese–Fischer)

Standard reduction potentials $E^{\ominus}$/V

n.a.

Covalent bonds	r/pm	E/kJ mol^{-1}
n.a.		

Oxidation states
n.a.

Physical properties

Melting point/K: 24.48
Boiling point/K: 27.10
ΔH_{fusion}/kJ mol^{-1}: 0.324
ΔH_{vap}/kJ mol^{-1}: 1.736

Thermodynamic properties (298.15 K, 0.1 MPa)

State	$\Delta_f H^{\ominus}$/kJ mol^{-1}	$\Delta_f G^{\ominus}$/kJ mol^{-1}	$S^{\ominus}$/J K^{-1} mol^{-1}	C_p/J K^{-1} mol^{-1}
Gas	0	0	146.328	20.786

Density/kg m^{-3}: 1444 [solid, m.p.]; 1207.3 [liquid, b.p.]; 0.899 94 [gas, 273 K]
Thermal conductivity/W m^{-1} K^{-1}: 0.0493 [300 K]$_g$
Mass magnetic susceptibility/kg^{-1} m^3: -4.2×10^{-9} (g);
Molar volume/cm^3: 13.97 [24 K]

Lattice structure (cell dimensions/pm), space group

f.c.c. ($a = 445.462$), Fm3m
h.c.p. ($a = 314.5$, $c = 514$), $P6_3/mmc$ [3 K]

X-ray diffraction: mass absorption coefficients (μ/ρ)/cm^2 g^{-1}:
CuK$_\alpha$ 22.9 MoK$_\alpha$ 2.47

Discovered in 1898 by Sir William Ramsay and M. W. Travers at London, UK

[Greek, *neos* = new]

Neon

Nuclear properties

Atomic number: 10

Thermal neutron capture cross-section/barns: 0.038

Number of isotopes (including nuclear isomers): 8

Isotope mass range: 17→24

Key isotopes

Nuclide	Atomic mass	Natural abundance (%)	Half life $T_{1/2}$	Decay mode and energy (MeV)	Nuclear spin I	Nuclear magnetic moment μ	Uses
^{20}Ne	19.992 44	90.51	stable		0		
^{21}Ne	20.993 95	0.27	stable		3/2	−0.661 76	NMR
^{22}Ne	21.991 38	9.22	stable		0		

NMR (no known compounds)	^{21}Ne
Relative sensitivity ($^1H = 1.00$)	2.50×10^{-3}
Absolute sensitivity ($^1H = 1.00$)	6.43×10^{-6}
Receptivity ($^{13}C = 1.00$)	0.0359
Magnetogyric ratio/rad T^{-1} s^{-1}	-2.1118×10^7
Quadrupole moment/m^2	9×10^{-30}
Frequency ($^1H = 100$ MHz; 2.3488 T)/MHz	7.894

Electron shell properties

Ground state electron configuration: $[He]2s^22p^6$

Term symbol: $^1S_0$

Electron affinity ($M \rightarrow M^-$)/kJ mol^{-1}: −29 (calc.)

Ionization energies/kJ mol^{-1}

1.	$M \rightarrow M^+$	2 080.6	6.	$M^{5+} \rightarrow M^{6+}$	15 238
2.	$M^+ \rightarrow M^{2+}$	3 952.2	7.	$M^{6+} \rightarrow M^{7+}$	19 998
3.	$M^{2+} \rightarrow M^{3+}$	6 122	8.	$M^{7+} \rightarrow M^{8+}$	23 069
4.	$M^{3+} \rightarrow M^{4+}$	9 370	9.	$M^{8+} \rightarrow M^{9+}$	115 377
5.	$M^{4+} \rightarrow M^{5+}$	12 177	10.	$M^{9+} \rightarrow M^{10+}$	131 429

Principal lines in atomic spectrum

Wavelength/nm	Species	Sensitivity	Application
74.370*	I		
540.056	I		AE
585.249	I		AE
640.225	I		AE
865.438	I		

* Lowest energy transition from ground state to nearest empty orbital

Abundance: Atmosphere 18 p.p.m. (volume); Earth's crust 0.04 p.p.m.; seawater 0.45 p.p.m.

Biological role: None; non-toxic

Np

Atomic number: 93

Relative atomic mass ($^{12}C = 12.0000$): **237.0482**

Chemical properties

Radioactive silvery metal obtained in kg quantities as ^{237}Np from uranium fuel elements. Attacked by oxygen, steam, and acids, but not alkalis.

Radii/pm: Np^{3+} 110; Np^{4+} 95; Np^{5+} 88; Np^{6+} 82; atomic 131

Electronegativity: 1.36 (Pauling); 1.22 (Allred-Rochow)

Effective nuclear charge: 1.80 (Slater)

Standard reduction potentials $E^{\ominus}$/V

	VII		VI		V		IV		III		0
Acid solution	NpO_3^+	2.04	NpO_2^{2+}	1.24	NpO_2^+	0.66	Np^{4+}	0.18	Np^{3+}	−1.79	Np
			NpO_2^{2+} → Np^{4+}: 0.95				Np^{4+} → Np: −1.30				
Alkaline solution	NpO_5^{3-}	0.58	$NpO_2(OH)_2$	0.6	NpO_2OH	0.3	NpO_2	−2.1	$Np(OH)_3$	−2.2	Np

Oxidation states

Np^{II}	(f^4d^1)	NpO
Np^{III}	(f^4)	NpF_3, $NpCl_3$, etc., $[NpCl_6]^{3-}$, $(Np(H_2O)_x]^{3+}$ (aq),
Np^{IV}	(f^3)	NpO_2, $[Np(H_2O)_x]^{4+}$ (aq), NpF_4, $NpCl_4$, $NpBr_4$, $[NpCl_6]^{2-}$, complexes
Np^{V}	(f^2)	Np_2O_5, NpF_5, $CsNpF_6$, Na_3NpF_8, NpO_2^+ (aq)
Np^{VI}	(f^1)	$NpO_3 \cdot H_2O$, NpO_2^{2+} (aq), NpF_6
Np^{VII}	([Rn])	Li_5NpO_6

Physical properties

Melting point/K: 913

Boiling point/K: 4175

ΔH_{fusion}/kJ mol^{-1}: 9.46

ΔH_{vap}/kJ mol^{-1}: 336.6

Thermodynamic properties (298.15 K, 0.1 MPa)

State	$\Delta_f H^{\ominus}$/kJ mol^{-1}	$\Delta_f G^{\ominus}$/kJ mol^{-1}	$S^{\ominus}$/J K^{-1} mol^{-1}	C_p/J K^{-1} mol^{-1}
Solid	0	0	n.a.	n.a.
Gas	n.a.	n.a.	n.a.	n.a.

Density/kg m^{-3}: 20 250 [293 K]

Thermal conductivity/W m^{-1} K^{-1}: 6.3 [300 K]

Electrical resistivity/Ω m: 122×10^{-8} [293 K]

Mass magnetic susceptibility/kg^{-1} m^3: n.a.

Molar volume/cm^3: 11.71

Coefficient of linear thermal expansion/K^{-1}: 27.5×10^{-6}

Lattice structure (cell dimensions/pm), space group

α-Np orthorhombic ($a = 472.3$, $b = 488.7$, $c = 666.3$), Pmcn
β-Np tetragonal ($a = 489.7$, $c = 338.8$), $P42_12$
γ-Np cubic ($a = 352$), Im3m

$T(\alpha \rightarrow \beta) = 551$ K
$T(\beta \rightarrow \gamma) = 850$ K

X-ray diffraction: mass absorption coefficients (μ/ρ)/cm^2 g^{-1}: n.a.

Prepared in 1940 by E. M. McMillan and P. Abelson at Berkeley, California, USA

[Named after planet Neptune]

Neptunium

Nuclear properties

Atomic number: 93

Thermal neutron capture cross-section/barns: 170 ± 5 (^{237}Np)

Number of isotopes (including nuclear isomers): 15

Isotope mass range: $229 \rightarrow 241$

Key isotopes

Nuclide	Atomic mass	Natural abundance (%)	Half life $T_{1/2}$	Decay mode and energy (MeV)	Nuclear spin I	Nuclear magnetic moment μ	Uses
^{237}Np	237.0480	0	2.14×10^6y	α(4.956); γ	5/2	+2.4	tracer

NMR

	^{237}Np
Relative sensitivity ($^1H = 1.00$)	—
Absolute sensitivity ($^1H = 1.00$)	—
Receptivity ($^{13}C = 1.00$)	—
Magnetogyric ratio/rad T^{-1} s^{-1}	3.1×10^7
Quadrupole moment/m^2	4.2×10^{-28}
Frequency ($^1H = 100$ MHz; 2.3488 T)/MHz	11.25

Electron shell properties

Ground state electron configuration: $[Rn]5f^4 6d^1 7s^2$

Term symbol: $^6L_{11/2}$

Electron affinity ($M \rightarrow M^-$)/kJ mol^{-1}: n.a.

Ionization energies/kJ mol^{-1}

1. $M \rightarrow M^+$	597	6. $M^{5+} \rightarrow M^{6+}$	
2. $M^+ \rightarrow M^{2+}$		7. $M^{6+} \rightarrow M^{7+}$	
3. $M^{2+} \rightarrow M^{3+}$		8. $M^{7+} \rightarrow M^{8+}$	
4. $M^{3+} \rightarrow M^{4+}$		9. $M^{8+} \rightarrow M^{9+}$	
5. $M^{4+} \rightarrow M^{5+}$		10. $M^{9+} \rightarrow M^{10+}$	

Principal lines in atomic spectrum

Wavelength/nm	Species	Wavelength	Species
901.618	I	1214.818	I
1009.199	I	1237.742	I
1081.745	I	1240.799	I
1169.515	I	1383.433	I
1177.664	I		

Abundance: Earth's crust: trace amounts in uranium minerals; seawater nil

Biological role: None; toxic due to radioactivity

Ni

Atomic number: 28
Relative atomic mass (^{12}C=12.0000): 58.69

Chemical properties

Silvery-white metal, lustrous, malleable, and ductile. Obtained from garnierite $(Ni,Mg)_6Si_4O_{10}(OH)_2$ and pentlandite $(Ni,Fe)_9S_8$ ores. Resists corrosion, soluble in acids, except concentrated HNO_3, unaffected by alkalis. Used in alloys, coins, metal plating, and catalysts.

Radii/pm: Ni^{2+} 78; Ni^{3+} 62; atomic 124.6; covalent 115
Electronegativity: 1.91 (Pauling); 1.75 (Allred-Rochow)
Effective nuclear charge: 4.05 (Slater); 5.71 (Clementi); 7.86 (Froese–Fischer)

Standard reduction potentials $E^{\ominus}$/V

	VI		IV		II		0
		(VI → II: >1.6)					
Acid solution	NiO_4^{2-}	>1.8	NiO_2	1.593	Ni^{2+}	−0.257	Ni
Alkaline solution	NiO_4^{2-}	>0.4	NiO_2	0.490	$Ni(OH)_2$	−0.72	Ni

Oxidation states

Ni^{-I}	($d^{10}s^1$)	$[Ni_2(CO)_6]^{2-}$
Ni^0	(d^{10})	$Ni(CO)_4$, $K_4[Ni(CN)_4]$
Ni^I	(d^9)	$K_2[Ni(CN)_4]$
Ni^{II}	(d^8)	NiO, $Ni(OH)_2$, $[Ni(H_2O)_6]^{2+}$ (aq), NiF_2, $NiCl_2$, etc., salts, NiAs, $[NiCl_4]^{2-}$, complexes, $[Ni(C_5H_5)_2]$
Ni^{III}	(d^7)	NiO(OH), NiF_3 ?, NiF_6^{3-}
Ni^{IV}	(d^6)	NiO_2 ?, NiF_6^{2-}
Ni^{VI}	(d^4)	K_2NiO_4 ?

Physical properties

Melting point/K: 1726
Boiling point/K: 3005
ΔH_{fusion}/kJ mol^{-1}: 17.6
ΔH_{vap}/kJ mol^{-1}: 374.8

Thermodynamic properties (298.15 K, 0.1 MPa)

State	$\Delta_f H^{\ominus}$/kJ mol^{-1}	$\Delta_f G^{\ominus}$/kJ mol^{-1}	$S^{\ominus}$/J K^{-1} mol^{-1}	C_p/J K^{-1} mol^{-1}
Solid	0	0	29.87	26.07
Gas	429.7	384.5	182.193	23.359

Density/kg m^{-3}: 8902 [298 K]; 7780 [liquid at m.p.]
Thermal conductivity/W m^{-1} K^{-1}: 90.7 [300 K]
Electrical resistivity/Ω m: 6.84×10^{-8} [293 K]
Mass magnetic susceptibility/kg^{-1} m^3: ferromagnetic
Molar volume/cm^3: 6.59
Coefficient of linear thermal expansion/K^{-1}: 13.3×10^{-6}

Lattice structure (cell dimensions/pm), space group

f.c.c. (a=352.38), Fm3m
'Hexagonal' nickel* (a=266, c=432), $P6_3/mmc$

*impure form of nickel

X-ray diffraction: mass absorption coefficients (μ/ρ)/cm^2 g^{-1}:
CuK$_\alpha$ 45.7 MoK$_\alpha$ 46.6

Discovered in 1751 by A. F. Cronstedt at Stockholm, Sweden

[German, *Nickel* = Satan]

Nickel

Nuclear properties

Atomic number: 28

Thermal neutron capture cross-section/barns: 4.51 ± 0.1

Number of isotopes (including nuclear isomers): 9

Isotope mass range: 56→64

Key isotopes

Nuclide	Atomic mass	Natural abundance (%)	Half life $T_{1/2}$	Decay mode and energy (MeV)	Nuclear spin I	Nuclear magnetic moment μ	Uses
^{58}Ni	57.9353	68.27	stable				
^{59}Ni	58.934	0	8×10^4y	EC(1.072); no γ			
^{60}Ni	59.9332	26.10	stable				
^{61}Ni	60.9310	1.13	stable		3/2	−0.7498	NMR
^{62}Ni	61.9283	3.59	stable				
^{63}Ni	62.930	0	92y	β^-(0.067); no γ			tracer
^{64}Ni	63.9280	0.91	stable				

NMR

	^{61}Ni
Relative sensitivity ($^1H = 1.00$)	3.57×10^{-3}
Absolute sensitivity ($^1H = 1.00$)	4.25×10^{-5}
Receptivity ($^{13}C = 1.00$)	0.242
Magnetogyric ratio/rad T^{-1} s^{-1}	-2.3948×10^7
Quadrupole moment/m^2	0.16×10^{-28}
Frequency ($^1H = 100$ MHz; 2.3488 T)/MHz	8.936

Electron shell properties

Ground state electron configuration: [Ar]$3d^8 4s^2$

Term symbol: $^3F_4$

Electron affinity ($M \rightarrow M^-$)/kJ mol^{-1}: 156

Ionization energies/kJ mol^{-1}

1. $M \rightarrow M^+$	736.7	6. $M^{5+} \rightarrow M^{6+}$	10400
2. $M^+ \rightarrow M^{2+}$	1753.0	7. $M^{6+} \rightarrow M^{7+}$	12800
3. $M^{2+} \rightarrow M^{3+}$	3393	8. $M^{7+} \rightarrow M^{8+}$	15600
4. $M^{3+} \rightarrow M^{4+}$	5300	9. $M^{8+} \rightarrow M^{9+}$	18600
5. $M^{4+} \rightarrow M^{5+}$	7280	10. $M^{9+} \rightarrow M^{10+}$	21660

Principal lines in atomic spectrum

Wavelength/nm	Species	Sensitivity	Application
231.096	I		AA
232.003	I		AA
336.957*	I		
341.477	I	U1	AA, AE
349.296	I	U2	AE
352.454	I		AE

* Lowest energy transition from ground state to nearest empty orbital

Abundance: Earth's crust 99 p.p.m.; seawater 0.0005 p.p.m.

Biological role: Uncertain; carcinogenic; stimulatory

Nb

Atomic number: 41

Relative atomic mass (^{12}C=12.0000): 92.90638

Chemical properties

Shiny silvery metal, soft when pure. Found as columbite, $(Fe,Mn)Nb_2O_6$, but obtained as by-product of tin extraction. Resists corrosion due to oxide film, attacked by hot, concentrated acids but resists fused alkalis. Used in stainless steels.

Radii/pm: Nb^{4+} 74; Nb^{5+} 69; atomic 142.9; covalent 134

Electronegativity: 1.6 (Pauling); 1.23 (Allred-Rochow)

Effective nuclear charge: 3.30 (Slater); 6.70 (Clementi); 9.60 (Froese–Fischer)

Standard reduction potentials $E^\ominus$/V

	V		III		0
Acid solution	Nb_2O_5	$\xrightarrow{-0.1}$	Nb^{3+}	$\xrightarrow{-1.1}$	Nb
			−0.65 (Nb_2O_5 → Nb)		

Oxidation states

Nb^{-III}	(d^8)	$[Nb(CO)_5]^{3-}$
Nb^{-I}	(d^6)	$[Nb(CO)_5]^-$
Nb^I	(d^4)	$[(C_5H_5)Nb(CO)_4]$
Nb^{II}	(d^3)	NbO
Nb^{III}	(d^2)	$LiNbO_2$, $NbCl_3$, $NbBr_3$, NbI_3, $[Nb(CN)_8]^{5-}$
Nb^{IV}	(d^1)	NbO_2, NbF_4, $NbCl_4$ etc., $NbOCl_2$
Nb^V	(d^0, f^{14})	Nb_2O_5, $[HNb_6O_{19}]^-$ (aq), NbF_5, $NbCl_5$, etc., NbO_2F, $NbOCl_3$

Physical properties

Melting point/K: 2741

Boiling point/K: 5015

ΔH_{fusion}/kJ mol^{-1}: 27.2

ΔH_{vap}/kJ mol^{-1}: 680.19

Thermodynamic properties (298.15 K, 0.1 MPa)

State	$\Delta_f H^\ominus$/kJ mol^{-1}	$\Delta_f G^\ominus$/kJ mol^{-1}	$S^\ominus$/J K^{-1} mol^{-1}	C_p/J K^{-1} mol^{-1}
Solid	0	0	36.40	24.60
Gas	725.9	681.1	186.256	30.158

Density/kg m^{-3}: 8570 [293 K]; 7830 [liquid at m.p.]

Thermal conductivity/W m^{-1} K^{-1}: 53.7 [300 K]

Electrical resistivity/Ω m: 12.5×10^{-8} [273 K]

Mass magnetic susceptibility/kg^{-1} m^3: $+2.76\times10^{-8}$ (s)

Molar volume/cm^3: 10.84

Coefficient of linear thermal expansion/K^{-1}: 7.07×10^{-6}

Lattice structure (cell dimensions/pm), space group

b.c.c. (a=329.86), Im3m

X-ray diffraction: mass absorption coefficients (μ/ρ)/cm^2 g^{-1}: CuK_α 153 MoK_α 17.1

Discovered in 1801 by C. Hatchett at London, UK

[Greek, *Niobe* = daughter of Tantalus]

Niobium

Nuclear properties

Atomic number: 41
Thermal neutron capture cross-section/barns: 1.15 ± 0.05
Number of isotopes (including nuclear isomers): 24
Isotope mass range: 88 → 101

Key isotopes

Nuclide	Atomic mass	Natural abundance (%)	Half life $T_{1/2}$	Decay mode and energy (MeV)	Nuclear spin I	Nuclear magnetic moment μ	Uses
^{93}Nb	92.9060	100	stable		9/2	+6.167	NMR
^{94}Nb	93.9057	0	2×10^4y	β^-(2.06); γ			

NMR

	^{93}Nb
Relative sensitivity (^{1}H = 1.00)	0.48
Absolute sensitivity (^{1}H = 1.00)	0.48
Receptivity (^{13}C = 1.00)	2740
Magnetogyric ratio/rad T^{-1} s^{-1}	6.5476×10^7
Quadrupole moment/m^2	-0.2×10^{-28}
Frequency (^{1}H = 100 MHz; 2.3488 T)/MHz	24.442

Reference: NbF_6^- (conc. HF)

Electron shell properties

Ground state electron configuration: [Kr]$4d^4 5s^1$
Term symbol: $^6D_{1/2}$
Electron affinity ($M \rightarrow M^-$)/kJ mol^{-1}: 109

Ionization energies/kJ mol^{-1}

1. $M \rightarrow M^+$	664	6. $M^{5+} \rightarrow M^{6+}$	9899
2. $M^+ \rightarrow M^{2+}$	1382	7. $M^{6+} \rightarrow M^{7+}$	12100
3. $M^{2+} \rightarrow M^{3+}$	2416	8. $M^{7+} \rightarrow M^{8+}$	
4. $M^{3+} \rightarrow M^{4+}$	3695	9. $M^{8+} \rightarrow M^{9+}$	
5. $M^{4+} \rightarrow M^{5+}$	4877	10. $M^{9+} \rightarrow M^{10+}$	

Principal lines in atomic spectrum

Wavelength/nm	Species	Sensitivity	Application
334.370	I		AA
358.027	I		AA
405.894	I	U1	AA, AE
407.973	I	U2	AA, AE
410.092	I	U3	AE
525.162*	I		

* Lowest energy transition from ground state to nearest empty orbital

Abundance: Earth's crust 20 p.p.m.; seawater 1×10^{-5} p.p.m.
Biological role: None; moderately toxic

N

Atomic number: 7

Relative atomic mass ($^{12}C = 12.0000$): 14.00674

Chemical properties

Colourless, odourless gas (N_2) obtained from liquid air. Generally unreactive at room temperature. Extensive inorganic and organic chemistry. Used in fertilizers, acids (HNO_3), explosives, plastics, dyes, etc.

Radii/pm: N^{3+} 16; atomic 71; covalent 70 (single bond); van der Waals 154; N^{3-} 171

Electronegativity: 3.04 (Pauling); 3.07 (Allred-Rochow)

Effective nuclear charge: 3.90 (Slater); 3.83 (Clementi); 3.46 (Froese–Fischer)

Standard reduction potentials $E^\ominus$/V

Acid solution

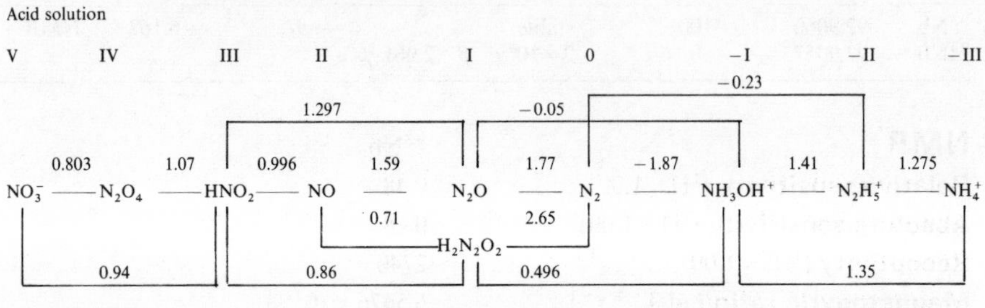

Alkaline solution

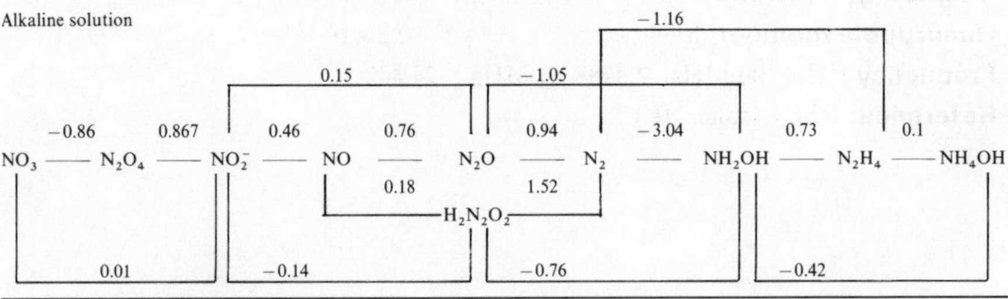

Covalent bonds	r/pm	E/kJ mol^{-1}
N—H	101	390
N—N	147	160
N=N	125	415
N≡N	110	946
N—Cl	195	193

Oxidation states

N^{-III}	NH_3, NH_4^+ (aq)	N^{III}	HNO_2, NO_2^- (aq), NF_3
N^{-II}	N_2H_4, $N_2H_5^+$ (aq)	N^{IV}	$N_2O_4 \rightleftharpoons 2NO_2$
N^{-I}	NH_2OH	N^{V}	HNO_3, NO_3^- (aq)
N^0	N_2		
N^{II}	NO		

Physical properties

Melting point/K: 63.29 — ΔH_{fusion}/kJ mol^{-1}: 0.720

Boiling point/K: 77.4 — ΔH_{vap}/kJ mol^{-1}: 5.577

Thermodynamic properties (298.15 K, 0.1 MPa)

State	$\Delta_f H^\ominus$/kJ mol^{-1}	$\Delta_f G^\ominus$/kJ mol^{-1}	$S^\ominus$/J K^{-1} mol^{-1}	C_p/J K^{-1} mol^{-1}
Gas (N_2)	0	0	191.61	29.125
Gas (atoms)	472.704	455.563	153.298	20.786

Density/kg m^{-3}: 1026 [solid, 21 K]; 880 [liquid, b.p.]; 1.2506 [gas, 273 K]

Thermal conductivity/W m^{-1} K^{-1}: 0.025 98 [300 K]$_g$

Mass magnetic susceptibility/kg^{-1} m^3: -5.4×10^{-9} (g)

Molar volume/cm^3: 13.65 [21 K]

Lattice structure (cell dimensions/pm), space group

α-N_2 cubic ($a = 564.4$), $P2_13$
β-N_2 h.c.p. ($a = 404.2$, $c = 660.1$), $P6_3/mmc$
$T(\alpha \rightarrow \beta) = 35$ K

X-ray diffraction: mass absorption coefficients (μ/ρ)/cm^2 g^{-1}: CuK_α 7.52 MoK_α 0.916

Discovered by D. Rutherford in 1772 at Edinburgh, Scotland

[Greek, *nitron genes* = nitre forming]

Nitrogen

Nuclear properties

Atomic number: 7

Thermal neutron capture cross-section/barns: 0.075 ± 0.008

Number of isotopes (including nuclear isomers): 8

Isotope mass range: $12 \rightarrow 18$

Key isotopes

Nuclide	Atomic mass	Natural abundance (%)	Half life $T_{1/2}$	Decay mode and energy (MeV)	Nuclear spin I	Nuclear magnetic moment μ	Uses
^{14}N	14.003 07	99.63	stable		1	+0.4036	NMR
^{15}N	15.000 11	0.37	stable		1/2	−0.2831	NMR

No long-lived radioactive isotopes: ^{13}N has $T_{1/2} = 9.97$ m

NMR

	[^{14}N]	^{15}N
Relative sensitivity ($^1H = 1.00$)	1.01×10^{-3}	1.04×10^{-3}
Absolute sensitivity ($^1H = 1.00$)	1.01×10^{-3}	3.85×10^{-6}
Receptivity ($^{13}C = 1.00$)	5.69	0.0219
Magnetogyric ratio/rad T^{-1} s^{-1}	1.9331×10^7	-2.7116×10^7
Quadrupole moment/m^2	1.6×10^{-30}	—
Frequency ($^1H = 100$ MHz; 2.3488 T)/MHz	7.224	10.133

Reference: $MeNO_2$ or NO_3^-

Electron shell properties

Ground state electron configuration: $[He]2s^22p^3$

Term symbol: $^4S_{3/2}$

Electron affinity ($M \rightarrow M^-$)/kJ mol^{-1}: −7

Ionization energies/kJ mol^{-1}

1. $M \rightarrow M^+$ 1402.3	6. $M^{5+} \rightarrow M^{6+}$ 53 265.6
2. $M^+ \rightarrow M^{2+}$ 2856.1	7. $M^{6+} \rightarrow M^{7+}$ 64 358.7
3. $M^{2+} \rightarrow M^{3+}$ 4578.0	
4. $M^{3+} \rightarrow M^{4+}$ 7474.9	
5. $M^{4+} \rightarrow M^{5+}$ 9440.0	

Principal lines in atomic spectrum

Wavelength/nm	Species	Sensitivity	Application
120.002* (vac)	I		
399.500	II		
409.995	I	U3	AE
410.998	I	U2	AE
567.956	II	V2	AE

* Lowest energy transition from ground state to nearest empty orbital

Abundance: Atmosphere 780 900 p.p.m. (volume); Earth's crust 19 p.p.m.; seawater 0.8 p.p.m.

Biological role: Basis of life (part of DNA); nitrogen cycle in nature

No

Atomic number: 102

Relative atomic mass ($^{12}C=12.0000$): **(259)**

Chemical properties

Radioactive metal, only a few atoms of which have ever been made. Obtained by bombardment of ^{246}Cm with carbon nuclei.

Radii/pm: No^{2+} 113; No^{3+} 95; No^{4+} 83

Electronegativity: 1.3 (Pauling); n.a. (Allred-Rochow)

Effective nuclear charge: 1.65 (Slater)

Standard reduction potentials $E^{\ominus}$/V

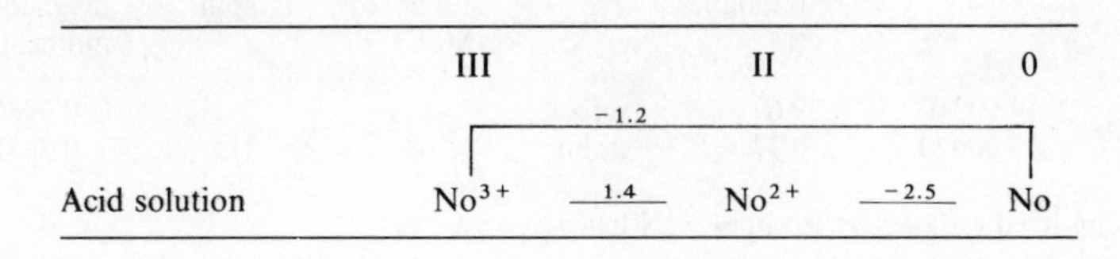

Oxidation states

No^{II} (f^{14})	$[No(H_2O)_x]^{2+}$ (aq)
No^{III} (f^{13})	$[No(H_2O)_x]^{3+}$ (aq)

Physical properties

Melting point/K: n.a.

Boiling point/K: n.a.

ΔH_{fusion}/kJ mol^{-1}: n.a.

ΔH_{vap}/kJ mol^{-1}: n.a.

Thermodynamic properties (298.15 K, 0.1 MPa)

State	$\Delta_f H^{\ominus}$/kJ mol^{-1}	$\Delta_f G^{\ominus}$/kJ mol^{-1}	$S^{\ominus}$/J K^{-1} mol^{-1}	C_p/J K^{-1} mol^{-1}
Solid	0	0	n.a.	n.a.
Gas	n.a.	n.a.	n.a.	n.a.

Density/kg m^{-3}: n.a.

Thermal conductivity/W m^{-1} K^{-1}: 10 (est.) [300 K]

Electrical resistivity/Ω m: n.a.

Mass magnetic susceptibility/kg^{-1} m^3: n.a.

Molar volume/cm^3: n.a.

Coefficient of linear thermal expansion/K^{-1}: n.a.

Lattice structure (cell dimensions/pm), space group

n.a.

X-ray diffraction: mass absorption coefficients (μ/ρ)/cm^2 g^{-1}: n.a.

Conclusively identified in 1958 by A. Ghiorso, T. Sikkeland, J. R. Walton, and G. T. Seaborg at Berkeley, California, USA

[Named after Alfred Nobel]

Nobelium

Nuclear properties

Atomic number: 102

Thermal neutron capture cross-section/barns: n.a.

Number of isotopes (including nuclear isomers): 9

Isotope mass range: 251→259

Key isotopes

Nuclide	Atomic mass	Natural abundance (%)	Half life $T_{1/2}$	Decay mode and energy (MeV)	Nuclear spin I	Nuclear magnetic moment μ	Uses
^{259}No		0	58m	α(7.50); SF			

Electron shell properties

Ground state electron configuration: $[Rn]5f^{14}7s^2$

Term symbol: $^1S_0$

Electron affinity ($M \rightarrow M^-$)/kJ mol^{-1}: n.a.

Ionization energies/kJ mol^{-1}

1. $M \rightarrow M^+$	642	6. $M^{5+} \rightarrow M^{6+}$	
2. $M^+ \rightarrow M^{2+}$		7. $M^{6+} \rightarrow M^{7+}$	
3. $M^{2+} \rightarrow M^{3+}$		8. $M^{7+} \rightarrow M^{8+}$	
4. $M^{3+} \rightarrow M^{4+}$		9. $M^{8+} \rightarrow M^{9+}$	
5. $M^{4+} \rightarrow M^{5+}$		10. $M^{9+} \rightarrow M^{10+}$	

Principal lines in atomic spectrum

Wavelength/nm	Species	Sensitivity	Application
n.a.			

Abundance: Earth's crust nil; seawater nil

Biological role: None; toxic due to radioactivity

Os

Atomic number: 76

Relative atomic mass (^{12}C=12.0000): 190.2

Chemical properties

Lustrous silvery metal of platinum group. Found in free state but obtained from wastes of nickel refining. Unaffected by air, water, and acids, but dissolves in molten alkalis. Smells, due to formation of volatile OsO_4. Used in alloys and catalysts.

Radii/pm: Os^{2+} 89; Os^{3+} 81; Os^{4+} 67; atomic 135; covalent 126

Electronegativity: 2.2 (Pauling); 1.52 (Allred-Rochow)

Effective nuclear charge: 3.75 (Slater); 10.32 (Clementi); 14.90 (Froese–Fischer)

Standard reduction potentials $E^{\ominus}$/V

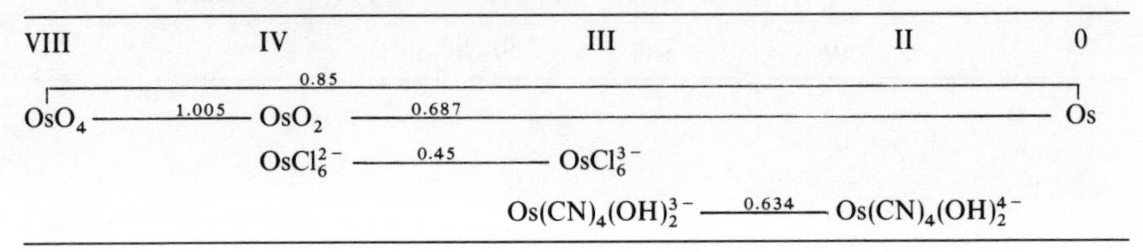

Oxidation states

Os^{-II} (d^{10})	$[Os(CO)_4]^{2-}$
Os^{0} (d^{8})	$Os(CO)_5$, $Os_2(CO)_9$
Os^{I} (d^{7})	OsI
Os^{II} (d^{6})	$OsCl_2$, OsI_2
Os^{III} (d^{5})	$OsCl_3$, $OsBr_3$, OsI_3, complexes
Os^{IV} (d^{4})	OsO_2, OsO_2 (aq), OsF_4, $OsCl_4$, $OsBr_4$, $OsCl_6^{2-}$, complexes
Os^{V} (d^{3})	OsF_5, $OsCl_5$
Os^{VI} (d^{2})	OsO_3?, OsF_6
Os^{VII} (d^{1})	OsF_7
Os^{VIII} (d^{0}, f^{14})	OsO_4, $[OsO_4(OH)_2]^{2-}$ (aq)

Physical properties

Melting point/K: 3327

Boiling point/K: 5300

ΔH_{fusion}/kJ mol^{-1}: 29.3

ΔH_{vap}/kJ mol^{-1}: 738.06

Thermodynamic properties (298.15 K, 0.1 MPa)

State	$\Delta_f H^{\ominus}$/kJ mol^{-1}	$\Delta_f G^{\ominus}$/kJ mol^{-1}	$S^{\ominus}$/J K^{-1} mol^{-1}	C_p/J K^{-1} mol^{-1}
Solid	0	0	32.6	24.7
Gas	791	745	192.573	20.786

Density/kg m^{-3}: 22 590 [293 K]; 20 100 [liquid at m.p.]

Thermal conductivity/W m^{-1} K^{-1}: 87.6 [300 K]

Electrical resistivity/Ω m: 8.12×10^{-8} [273 K]

Mass magnetic susceptibility/kg^{-1} m^{3}: $+6.5 \times 10^{-10}$ (s)

Molar volume/cm^{3}: 8.43

Coefficient of linear thermal expansion/K^{-1}:
4.3×10^{-6} (*a* axis); 6.1×10^{-6} (*b* axis); 6.8×10^{-6} (*c* axis)

Lattice structure (cell dimensions/pm), space group

h.c.p. (a=273.43; c=432.00), $P6_3/mmc$

X-ray diffraction: mass absorption coefficients (μ/ρ)/cm^{2} g^{-1}:
CuK_α 186 MoK_α 106

Discovered in 1803 by S. Tennant at London, UK

[Greek, *osme* = smell]

Osmium

Nuclear properties

Atomic number: 76

Thermal neutron capture cross-section/barns: 15.3 ± 0.7

Number of isotopes (including nuclear isomers): 19

Isotope mass range: 181→195

Key isotopes

Nuclide	Atomic mass	Natural abundance (%)	Half life $T_{1/2}$	Decay mode and energy (MeV)	Nuclear spin I	Nuclear magnetic moment μ	Uses
^{184}Os	183.9526	0.020	stable				
^{185}Os		0	96.6d	EC(0.982); γ			tracer
^{186}Os	185.9539	1.58	stable				
^{187}Os	186.9560	1.6	stable		1/2	+0.0643	NMR
^{188}Os	187.9560	13.3	stable				
^{189}Os	188.9586	16.1	stable		3/2	+0.6565	NMR
^{190}Os	189.9586	26.4	stable				
^{191}Os		0	15.0d	β^-(0.310); γ			tracer
^{192}Os		41.0	stable				

NMR (Only OsO_4 studied)	^{187}Os	^{189}Os
Relative sensitivity ($^1H = 1.00$)	1.22×10^{-5}	2.34×10^{-3}
Absolute sensitivity ($^1H = 1.00$)	2.00×10^{-7}	3.76×10^{-4}
Receptivity ($^{13}C = 1.00$)	1.14×10^{-3}	2.13
Magnetogyric ratio/rad T^{-1} s^{-1}	0.6105×10^7	2.0773×10^7
Quadrupole moment/rad T^{-1} s^{-1}	—	0.8×10^{-28}
Frequency ($^1H = 100$ MHz; 2.3488 T)/MHz	2.282	7.758

Reference: OsO_4

Electron shell properties

Ground state electron configuration: $[Xe]4f^{14}5d^66s^2$

Term symbol: $^5D_4$

Electron affinity ($M \rightarrow M^-$)/kJ mol^{-1}: 139

Ionization energies/kJ mol^{-1}

1. $M \rightarrow M^+$ 840	6. $M^{5+} \rightarrow M^{6+}$ (6600)
2. $M^+ \rightarrow M^{2+}$ (1600)	7. $M^{6+} \rightarrow M^{7+}$ (8100)
3. $M^{2+} \rightarrow M^{3+}$ (2400)	8. $M^{7+} \rightarrow M^{8+}$ (9500)
4. $M^{3+} \rightarrow M^{4+}$ (3900)	9. $M^{8+} \rightarrow M^{9+}$
5. $M^{4+} \rightarrow M^{5+}$ (5200)	10. $M^{9+} \rightarrow M^{10+}$

Principal lines in atomic spectrum

Wavelength/nm	Species	Sensitivity	Application
201.814	I		
202.026	I		
290.906	I	U2	AA, AE
305.866	I		AA, AE
442.047*	I		AE

* Lowest energy transition from ground state to nearest empty orbital

Abundance: Earth's crust 0.005 p.p.m.; seawater n.a. ($<1 \times 10^{-10}$ p.p.m.?)

Biological role: None; very toxic

O

Atomic number: 8
Relative atomic mass (^{12}C=12.0000): **15.9994**

Chemical properties

Colourless, odourless gas. Obtained from liquid air. Very reactive and forms oxides of all elements except He, Ne, Ar, and Kr. Moderately soluble in water (30.8 cm^3 per dm^3) at 293 K. Used in steel making, metal cutting, chemical industry.

Radii/pm: O^+ 22; O^{2-} 132; covalent (single bonds) 66; van der Waals 140

Electronegativity: 3.44 (Pauling); 3.50 (Allred-Rochow)

Effective nuclear charge: 4.55 (Slater); 4.45 (Clementi); 4.04 (Froese–Fischer)

Standard reduction potentials $E^{\ominus}$/V

	0		−I		−II			
Acid solutions	O_2	0.695	H_2O_2	1.763	H_2O	O_3	1.246	O_2
	(O_2 → H_2O: 1.229)							
Alkaline solutions	O_2	−0.0649	HO_2^-	0.867	OH^-	O_3	2.075	O_2
	(O_2 → OH^-: 0.401)							

Covalent bonds

	r/pm	E/kJ mol^{-1}
O—O	148	146
O=O (O_2)	120.8	498
N—O	146	200
N=O	115	678
N≡O (NO)	106	1063

For other covalent bonds to oxygen see other elements

Oxidation states

O^{-II}	H_2O, H_3O^+, OH^-, oxides, etc.
O^{-I}	H_2O_2, peroxides
O^0	O_2, O_3
O^I	O_2F_2
O^{II}	OF_2

Physical properties

Melting point/K: 54.8

Boiling point/K: 90.188

ΔH_{fusion}/kJ mol^{-1}: 0.444

ΔH_{vap}/kJ mol^{-1}: 6.82

Thermodynamic properties (298.15 K, 0.1 MPa)

State	$\Delta_f H^{\ominus}$/kJ mol^{-1}	$\Delta_f G^{\ominus}$/kJ mol^{-1}	$S^{\ominus}$/J K^{-1} mol^{-1}	C_p/J K^{-1} mol^{-1}
Gas (O_2)	0	0	205.138	29.355
Gas (atoms)	249.170	231.731	161.055	21.912

Density/kg m^{-3}: 2000 [solid, m.p.]; 1140 [liquid, b.p.]; 1.429 [gas, 273 K]

Thermal conductivity/W m^{-1} K^{-1}: 0.2674 [300 K]

Mass magnetic susceptibility/kg^{-1} m^3: $+1.355 \times 10^{-6}$ (g)

Molar volume/cm^3: 8.00 [54 K]

Lattice structure (cell dimensions/pm), space group

α-O_2 rhombic (a=540.3; b=342.9; c=508.6; β=132.53°), C2/m
β-O_2 rhombohedral (a=330.7; c=1125.6), R$\bar{3}$m
γ-O_2 cubic (a=683), Pm3n

$T(\alpha \rightarrow \beta)$=23.8 K
$T(\beta \rightarrow \gamma)$=43.8 K

X-ray diffraction: mass absorption coefficients (μ/ρ)/cm^2 g^{-1}:
CuK_α 11.5 MoK_α 1.31

Discovered in 1774 by J. Priestley at Leeds, UK, and independently by C. W. Scheele at Uppsala, Sweden

[Greek, *oxy genes* = acid forming]

Oxygen

Nuclear properties

Atomic number: 8

Thermal neutron capture cross-section/barns: 0.178×10^{-3} (^{16}O)

Number of isotopes (including nuclear isomers): 8

Isotope mass range: 13→20

Key isotopes

Nuclide	Atomic mass	Natural abundance (%)	Half life $T_{1/2}$	Decay mode and energy (MeV)	Nuclear spin I	Nuclear magnetic moment μ	Uses
^{16}O	15.994 91	99.762	stable		0		
^{17}O	16.9991	0.038	stable		5/2	−1.8937	NMR
^{18}O	17.9992	0.200	stable		0		

No long-lived radioactive isotopes: ^{15}O has $T_{1/2} = 124$ s

NMR

	^{17}O
Relative sensitivity ($^1H = 1.00$)	2.91×10^{-2}
Absolute sensitivity ($^1H = 1.00$)	1.08×10^{-5}
Receptivity ($^{13}C = 1.00$)	0.061
Magnetogyric ratio/rad $T^{-1} s^{-1}$	-3.6264×10^7
Quadrupole moment/m^2	-2.6×10^{-30}
Frequency ($^1H = 100$ MHz; 2.3488 T)/MHz	13.557

Reference: H_2O

Electron shell properties

Ground state electron configuration: $[He]2s^22p^4$

Term symbol: $^3P_2$

Electron affinity ($M \rightarrow M^-$)/kJ mol^{-1}: 141

Ionization energies/kJ mol^{-1}

1. $M \rightarrow M^+$	1313.9	6. $M^{5+} \rightarrow M^{6+}$ 13 326.2
2. $M^+ \rightarrow M^{2+}$	3388.2	7. $M^{6+} \rightarrow M^{7+}$ 71 333.3
3. $M^{2+} \rightarrow M^{3+}$	5300.3	8. $M^{7+} \rightarrow M^{8+}$ 84 076.3
4. $M^{3+} \rightarrow M^{4+}$	7469.1	
5. $M^{4+} \rightarrow M^{5+}$	10 989.3	

Principal lines in atomic spectrum

Wavelength/nm	Species	Sensitivity	Application
135.852*			
777.193	I	U2	AE
777.414	I	U1	AE
777.543	I	U3	AE
844.636			

* Lowest energy transition from ground state to nearest empty orbital

Abundance: Atmosphere 209 500 p.p.m. (volume); Earth's crust 455 000 p.p.m.; seawater, constituent

Biological role: Basis of life (part of DNA)

Pd

Atomic number: 46

Relative atomic mass ($^{12}C = 12.0000$): 106.42

Chemical properties

Silvery-white metal, lustrous, malleable, ductile. Extracted from Cu–Zn ores. Resists corrosion, dissolves in oxidising acids and fused alkalis. Readily absorbs hydrogen gas. Main use is as catalyst.

Radii/pm: Pd^{2+} 86; Pd^{4+} 64; atomic 137.6; covalent 128

Electronegativity: 2.20 (Pauling); 1.35 (Allred-Rochow)

Effective nuclear charge: 4.05 (Slater); 7.84 (Clementi); 11.11 (Froese–Fischer)

Standard reduction potentials $E^{\ominus}$/V

	VI		IV		II		0
Acid solution			PdO_2	1.263	Pd^{2+}	0.915	Pd
Alkaline solution	'PdO_3'	2.03	PdO_2	1.283	$Pd(OH)_2$	−0.19	Pd

Oxidation states

Pd^0 (d^{10})	$[Pd(PPh_3)_3]$, $[Pd(PF_3)_4]$
$\mathbf{Pd^{II}}$ (d^8)	PdO, $[Pd(H_2O)_4]^{2+}$ (aq); PdF_3, $PdCl_2$ etc., $PdCl_4^{2-}$, salts, complexes
Pd^{IV} (d^6)	PdO_2, PdF_4, $PdCl_6^{2-}$

Physical properties

Melting point/K: 1825

Boiling point/K: 3413

ΔH_{fusion}/kJ mol^{-1}: 17.2

ΔH_{vap}/kJ mol^{-1}: 361.5

Thermodynamic properties (298.15 K, 0.1 MPa)

State	$\Delta_f H^{\ominus}$/kJ mol^{-1}	$\Delta_f G^{\ominus}$/kJ mol^{-1}	$S^{\ominus}$/J K^{-1} mol^{-1}	C_p/J K^{-1} mol^{-1}
Solid	0	0	37.57	25.98
Gas	378.2	339.7	167.05	20.786

Density/kg m^{-3}: 12 020 [293 K]; 10 379 [liquid at m.p.]

Thermal conductivity/W m^{-1} K^{-1}: 71.8 [300 K]

Electrical resistivity/Ω m: 10.8×10^{-8} [293 K]

Mass magnetic susceptibility/kg^{-1} m^3: $+6.702 \times 10^{-8}$ (s)

Molar volume/cm^3: 8.85

Coefficient of linear thermal expansion/K^{-1}: 11.2×10^{-6}

Lattice structure (cell dimensions/pm), space group

f.c.c. ($a = 389.08$), Fm3m

X-ray diffraction: mass absorption coefficients (μ/ρ)/cm^2 g^{-1}:
CuK_{α} 206 MoK_{α} 24.1

Discovered in 1803 by W. H. Wollaston at London, UK

[Named after the asteroid Pallas]

Palladium

Nuclear properties

Atomic number: 46

Thermal neutron capture cross-section/barns: 6.0±1.0

Number of isotopes (including nuclear isomers): 21

Isotope mass range: 95→115

Key isotopes

Nuclide	Atomic mass	Natural abundance (%)	Half life $T_{1/2}$	Decay mode and energy (MeV)	Nuclear spin I	Nuclear magnetic moment μ	Uses
^{102}Pd	101.9049	1.02	stable				
^{103}Pd		0	17d	EC(0.56); γ			tracer
^{104}Pd	103.9036	11.14	stable				
^{105}Pd	104.9046	22.33	stable		5/2	−0.642	NMR
^{106}Pd	105.9032	27.33	stable				
^{108}Pd	107.903 89	26.46	stable				
^{109}Pd		0	13.47h	β^-(1.115); γ			tracer
^{110}Pd		11.72	stable				

NMR

^{105}Pd (Only K_2PdCl_6 recorded)

Relative sensitivity (^1H=1.00) 1.12×10^{-3}

Absolute sensitivity (^1H=1.00) 2.49×10^{-4}

Receptivity (^{13}C=1.00) 1.41

Magnetogyric ratio/rad T^{-1} s^{-1} -0.756×10^7

Quadrupole moment/m^2 $+0.8 \times 10^{-28}$

Frequency (^1H=100 MHz; 2.3488 T)/MHz 4.576

Electron shell properties

Ground state electron configuration: [Kr]$4d^{10}$

Term symbol: $^1S_0$

Electron affinity ($M \rightarrow M^-$)/kJ mol^{-1}: 98.4

Ionization energies/kJ mol^{-1}

1. $M \rightarrow M^+$	805	6. $M^{5+} \rightarrow M^{6+}$	(8 700)
2. $M^+ \rightarrow M^{2+}$	1875	7. $M^{6+} \rightarrow M^{7+}$	(10 700)
3. $M^{2+} \rightarrow M^{3+}$	3177	8. $M^{7+} \rightarrow M^{8+}$	(12 700)
4. $M^{3+} \rightarrow M^{4+}$	(4700)	9. $M^{8+} \rightarrow M^{9+}$	(15 000)
5. $M^{4+} \rightarrow M^{5+}$	(6300)	10. $M^{9+} \rightarrow M^{10+}$	(17 200)

Principal lines in atomic spectrum

Wavelength/nm	Species	Sensitivity	Application
244.791	I		AA
247.642	I		AA
276.308*	I		AA
340.458	I	U1	AA, AE
342.124	I	U2	AE
363.470	I	U3	AE

* Lowest energy transition from ground state to nearest empty orbital

Abundance: Earth's crust 0.015 p.p.m.; seawater n.a. ($<1 \times 10^{10}$ p.p.m.?)

Biological role: None; non-toxic

P

Atomic number: 15

Relative atomic mass ($^{12}C = 12.0000$): 30.973762

Chemical properties

White phosphorus (P_4) is soft and flammable, red phosphorus is powdery and usually non-flammable. Neither form reacts with water or dilute acid but alkalis react to form phosphine gas. Occurs as vast deposits of apatite, $Ca_5(PO_4)_3F$. Used in fertilizers, insecticides, etc.

Radii/pm: P^{3+} 44; atomic 93 (white) 115 (red); covalent 110; van der Waals 190; P^{3-} 212

Electronegativity: 2.19 (Pauling); 2.06 (Allred-Rochow)

Effective nuclear charge: 4.80 (Slater); 4.89 (Clementi); 5.28 (Froese–Fischer)

Standard reduction potentials $E^{\ominus}$/V

	V	'IV'	III	'I'	0	−II	−III
Acid solution	H_3PO_4 —(−0.933)—	$H_4P_2O_6$ —(0.380)—	H_3PO_3 —(−0.499)—	H_3PO_2 —(−0.365)—	P —(−0.100)—	P_2H_4 —(−0.006)—	PH_3
Alkaline solution	PO_4^{3-} —(−1.12)—		HPO_3^{2-} —(−1.57)—	$H_2PO_2^-$ —(−2.05)—	P —(−0.89)—		PH_3

Acid solution: H_3PO_4 → H_3PO_3 −0.276; H_3PO_3 → P −0.502; P → PH_3 −0.063

Alkaline solution: HPO_3^{2-} → P −1.73; $H_2PO_2^-$ → PH_3 −1.18

Covalent bonds

Bond	r/pm	E/kJ mol^{-1}
P—H	144	328
P—C	185	264
P—O	164	407
P=O	145	560
P—F	157	490
P—Cl	204	319
P—P	222	209

Oxidation states

P^{-III}	PH_3, Ca_3P_2
P^{-II}	P_2H_4
P^{0}	P_4
P^{II}	P_2I_4
P^{III}	P_4O_6, H_3PO_3 (aq) etc., H_3PO_2 (aq), $H_2PO_3^-$ PCl_3, etc.
P^{V}	P_4O_{10}, H_3PO_4 (aq, $H_2PO_4^-$ (aq) etc., PF_5, PCl_5, $POCl_3$, phosphates

Physical properties

Melting point/K: 317.3 (P_4); 683 (red) under pressure

Boiling point/K: 553 (P_4)

ΔH_{fusion}/kJ mol^{-1}: 2.51 (P_4)

ΔH_{vap}/kJ mol^{-1}: 51.9 (P_4)

Thermodynamic properties (298.15 K, 0.1 MPa)

State	$\Delta_f H^{\ominus}$/kJ mol^{-1}	$\Delta_f G^{\ominus}$/kJ mol^{-1}	$S^{\ominus}$/J K^{-1} mol^{-1}	C_p/J K^{-1} mol^{-1}
Solid (P_4)	0	0	41.09	23.840
Solid (red)	−17.6	−12.1	22.80	21.21
Gas	314.64	278.25	163.193	20.786

Density/kg m^{-3}: 1820 (P_4); 2200 (red); 2690 (black) [293 K];

Thermal conductivity/W m^{-1} K^{-1}: 0.235 (P_4); 12.1 (black) [300 K]

Electrical resistivity/Ω m: 1×10^9 (P_4) [293 K]

Mass magnetic susceptibility/kg^{-1} m^3: -1.1×10^{-8} (P_4); -8.4×10^{-9} (red)

Molar volume/cm^3: 17.02 (P_4)

Coefficient of linear thermal expansion/K^{-1}: 124.5×10^{-6} (P_4)

Lattice structure (cell dimensions/pm), space group

α-P_4 white, cubic ($a = 1851$), $I\bar{4}3m$
β-P_4 white, rhombohedral ($a = 337.7$; $c = 880.6$), $R\bar{3}m$ [high pressure form]
γ-P_4 white, cubic ($a = 237.7$), Pm3m [high pressure form]
Red, cubic ($a = 1131$) Pm3m or $P\bar{4}3$
Hittorf's phosphorus (red), monoclinic ($a = 921$; $b = 915$; $c = 2260$; $\beta = 106.1°$), P2/c
Black, orthorhombic ($a = 331.36$; $b = 1047.8$; $c = 437.63$), Cmca

X-ray diffraction: mass absorption coefficients (μ/ρ)/cm^2 g^{-1}:
CuK_α 74.1 MoK_α 7.89

Discovered in 1669 by Hennig Brandt at Hamburg, Germany

[Greek, *phosphoros* = bringer of light]

Phosphorus

Nuclear properties

Atomic number: 15

Thermal neutron capture cross-section/barns: 0.190

Number of isotopes (including nuclear isomers): 7

Isotope mass range: 28→34

Key isotopes

Nuclide	Atomic mass	Natural abundance (%)	Half life $T_{1/2}$	Decay mode and energy (MeV)	Nuclear spin I	Nuclear magnetic moment μ	Uses
^{31}P	30.973 76	100	stable		1/2	+1.1317	NMR
^{32}P	31.9739	0	14.3d	β^-(1.710); no γ	1	−0.2523	tracer, medical
^{33}P	32.9717	0	25d	β^-(0.248); no γ			

NMR

	^{31}P
Relative sensitivity (^1H = 1.00)	6.63×10^{-2}
Absolute sensitivity (^1H = 1.00)	6.63×10^{-2}
Receptivity (^{13}C = 1.00)	377
Magnetogyric ratio/rad T^{-1} s^{-1}	10.8289×10^7
Frequency (^1H = 100 MHz; 2.3488 T)/MHz	40.481

References: 85% H_3PO_4

Electron shell properties

Ground state electron configuration: $[Ne]3s^23p^3$

Term symbol: $^4S_{3/2}$

Electron affinity ($M \rightarrow M^-$)/kJ mol^{-1}: 71.7

Ionization energies/kJ mol^{-1}

1. $M \rightarrow M^+$	1011.7	6. $M^{5+} \rightarrow M^{6+}$ 21 268
2. $M^+ \rightarrow M^{2+}$	1903.2	7. $M^{6+} \rightarrow M^{7+}$ 25 397
3. $M^{2+} \rightarrow M^{3+}$	2912	8. $M^{7+} \rightarrow M^{8+}$ 29 854
4. $M^{3+} \rightarrow M^{4+}$	4956	9. $M^{8+} \rightarrow M^{9+}$ 35 867
5. $M^{4+} \rightarrow M^{5+}$	6273	10. $M^{9+} \rightarrow M^{10+}$ 40 958

Principal lines in atomic spectrum

Wavelength/nm	Species	Sensitivity	Application
177.499* (vac)	I		AA but outside range of most equipment
178.287 (vac)	I		AA but outside range of most equipment
178.768 (vac)	I		AA but outside range of most equipment
213.547	I		AA
213.618	I		AA
253.565	I	U2	AE
255.328	I	U3	AE

* Lowest energy transition from ground state to nearest empty orbital

Abundance: Earth's crust 1120 p.p.m.; seawater 0.07 p.p.m. (low at surface)

Biological role: Basis of life (part of DNA); phosphate cycle in nature; P_4 very toxic

Pt

Atomic number: 78

Relative atomic mass (^{12}C=12.0000): 195.08

Chemical properties

Silvery-white metal, lustrous, malleable, ductile. Extracted from Cu–Ni ores. Unaffected by oxygen and water, only dissolves in aqua-regia and fused alkalis. Used in jewellery, drugs, catalysts.

Radii/pm: Pt^{2+} 85; Pt^{4+} 70; atomic 138; covalent 129

Electronegativity: 2.28 (Pauling); 1.44 (Allred-Rochow)

Effective nuclear charge: 4.05 (Slater); 10.75 (Clementi); 15.65 (Froese–Fischer)

Standard reduction potentials $E^\ominus$/V

VI		IV		II		0
PtO_3	2.0	PtO_2	1.045	PtO	0.980	Pt
		PtO_2	0.837	Pt^{2+}	1.188	Pt
		$PtCl_6^{2-}$	0.726	$PtCl_4^{2-}$	0.758	Pt

Oxidation states

Pt^0 (d^{10})	$[Pt(PPh_3)_3]$ $[Pt(PF_3)_4]$
Pt^{II} (d^8)	PtO, $PtCl_2$, $PrBr_3$, PtI_2, $PtCl_4^{2-}$, $[Pt(CN)_4]^{2-}$, complexes
Pt^{IV} (d^6)	PtO_2, $[Pt(OH)_6]^{2-}$ (aq), PtF_4, $PtCl_4$ etc., $PtCl_6^{2-}$ complexes
Pt^V (d^5)	$(PtF_5)_4$, PtF_6^-
Pt^{VI} (d^4)	PtO_3, PtF_6

Physical properties

Melting point/K: 2045

Boiling point/K: 4100±100

ΔH_{fusion}/kJ mol^{-1}: 19.7

ΔH_{vap}/kJ mol^{-1}: 469

Thermodynamic properties (298.15 K, 0.1 MPa)

State	$\Delta_f H^\ominus$/kJ mol^{-1}	$\Delta_f G^\ominus$/kJ mol^{-1}	$S^\ominus$/J K^{-1} mol^{-1}	C_p/J K^{-1} mol^{-1}
Solid	0	0	41.63	25.86
Gas	565.3	520.5	192.406	25.531

Density/kg m^{-3}: 21 450 [293 K]

Thermal conductivity/W m^{-1} K^{-1}: 71.6 [300 K]

Electrical resistivity/Ω m: 10.6×10^{-8} [293 K]

Mass magnetic susceptibility/kg^{-1} m^3: $+1.301 \times 10^{-8}$ (s)

Molar volume/cm^3: 9.10

Coefficient of linear thermal expansion/K^{-1}: 9.0×10^{-6}

Lattice structure (cell dimensions/pm), space group

f.c.c. (a=392.40), Fm3m

X-ray diffraction: mass absorption coefficients (μ/ρ)/cm^2 g^{-1}:
CuK$_\alpha$ 200 MoK$_\alpha$ 113

Known to pre-Columbian South Americans and taken to Europe about 1750

[Spanish, *platina* = silver]

Platinum

Nuclear properties

Atomic number: 78

Thermal neutron capture cross-section/barns: 9 ± 1

Number of isotopes (including nuclear isomers): 32

Isotope mass range: $173 \rightarrow 200$

Key isotopes

Nuclide	Atomic mass	Natural abundance (%)	Half life $T_{1/2}$	Decay mode and energy (MeV)	Nuclear spin I	Nuclear magnetic moment μ	Uses
^{190}Pt	189.960	0.010	6.9×10^{11}y	α (3.18)			
^{192}Pt	191.9614	0.79	*c.* 10^{15}y	α			
^{194}Pt	193.9628	32.9	stable		0		
^{195m}Pt		0	4.1d	IT(0.259); γ		±0.60	tracer
^{195}Pt	194.9648	33.8	stable		1/2	+0.6022	NMR
^{196}Pt	195.9650	25.3	stable		0		
^{197}Pt		0	18h	β^- (0.75); γ	1/2	±0.5	tracer
^{198}Pt	197.9675	7.2	stable				

NMR

	^{195}Pt
Relative sensitivity ($^1H = 1.00$)	9.94×10^{-3}
Absolute sensitivity ($^1H = 1.00$)	3.36×10^{-3}
Receptivity ($^{13}C = 1.00$)	19.1
Magnetogyric ratio/rad T^{-1} s^{-1}	5.7412×10^7
Frequency ($^1H = 100$ MHz; 2.3488 T)/MHz	21.499
Reference: $[Pt(CN)_6]^{2-}$	

Electron shell properties

Ground state electron configuration: $[Xe]4f^{14}5d^96s^1$

Term symbol: $^3D_3$

Electron affinity ($M \rightarrow M^-$)/kJ mol^{-1}: 247

Ionization energies/kJ mol^{-1}

1. $M \rightarrow M^+$	870	6. $M^{5+} \rightarrow M^{6+}$	(7200)
2. $M^+ \rightarrow M^{2+}$	1791	7. $M^{6+} \rightarrow M^{7+}$	(8900)
3. $M^{2+} \rightarrow M^{3+}$	(2800)	8. $M^{7+} \rightarrow M^{8+}$	(10500)
4. $M^{3+} \rightarrow M^{4+}$	(3900)	9. $M^{8+} \rightarrow M^{9+}$	(12300)
5. $M^{4+} \rightarrow M^{5+}$	(5300)	10. $M^{9+} \rightarrow M^{10+}$	(14100)

Principal lines in atomic spectrum

Wavelength/nm	Species	Sensitivity	Application
204.937	I		
265.945	I	U2	AA, AE
283.030	I		AA, AE
299.767	I		AA, AE
306.471	I	U1	AA, AE
331.505*	I		

* Lowest energy transition from ground state to nearest empty orbital

Abundance: Earth's crust 0.01 p.p.m.; seawater n.a. ($<1 \times 10^{-10}$ p.p.m.?)

Biological role: None; non-toxic

Pu

Atomic number: 94

Relative atomic mass (^{12}C=12.0000): (244)

Chemical properties

Radioactive silvery metal obtained in tonne quantities from uranium fuel elements. Attacked by oxygen, steam, and acids, but not alkalis. Used as compact energy source, nuclear fuel, and for nuclear weapons.

Radii/pm: Pu^{3+} 108; Pu^{4+} 93; Pu^{5+} 87; Pu^{6+} 81; atomic 151 (α form)

Electronegativity: 1.28 (Pauling); 1.22 (Allred-Rochow)

Effective nuclear charge: 1.65 (Slater)

Standard reduction potentials $E^{\ominus}$/V

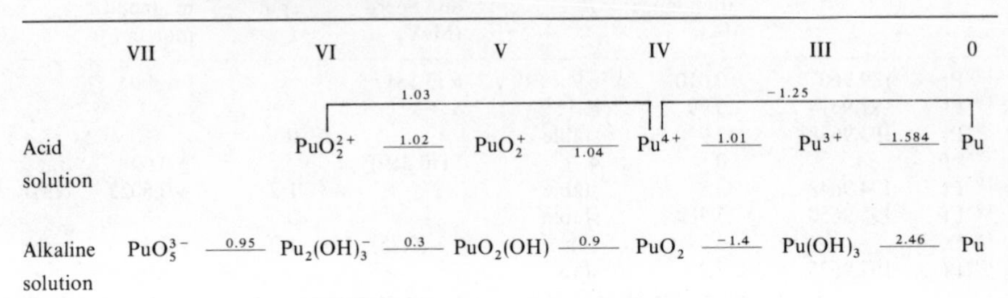

Oxidation states

Pu^{II} (f^6)	PuO, PuH_2
Pu^{III} (f^5)	Pu_2O_3, PuF_3 $PuCl_3$ etc., $[Pu(H_2O)_x]^{3+}$ (aq), Pu^{3+} salts, complexes
Pu^{IV} (f^4)	PuO_2, PuF_4, $[PuCl_6]^{2-}$, $[Pu(H_2O)_x]^{4+}$ (aq) unstable, complexes
Pu^{V} (f^3)	PuO_2^+ (aq) unstable, $CsPuF_6$
Pu^{VI} (f^2)	PuO_2^{2+} (aq), PuF_6
Pu^{VII} (f^1)	Li_5PuO_6, $[PuO_5]^{3-}$ (aq)

Physical properties

Melting point/K: 914 ΔH_{fusion}/kJ mol^{-1}: 2.8

Boiling point/K: 3505 ΔH_{vap}/kJ mol^{-1}: 343.5

Thermodynamic properties (298.15 K, 0.1 MPa)

State	$\Delta_f H^{\ominus}$/kJ mol^{-1}	$\Delta_f G^{\ominus}$/kJ mol^{-1}	$S^{\ominus}$/J K^{-1} mol^{-1}	C_p/J K^{-1} mol^{-1}
Solid	0	0	n.a.	n.a.
Gas	n.a.	n.a.	n.a.	n.a.

Density/kg m^{-3}: 19 840 (α) [298 K]; 16 623 [liquid at m.p.]

Thermal conductivity/W m^{-1} K^{-1}: 6.74 [300 K]

Electrical resistivity/Ω m: 146×10^{-8} [273 K]

Mass magnetic susceptibility/kg^{-1} m^3: $+3.17 \times 10^{-8}$ (s)

Molar volume/cm^3: 12.3

Coefficient of linear thermal expansion/K^{-1}: 55×10^{-6}

Lattice structure (cell dimensions/pm), space group

α-Pu monoclinic (a=618.3; b=482.2; c=1096.3; β=101.79°), $P2_1/m$
β-Pu monoclinic (a=928.4; b=1046.3; c=785.9; β=92.13°), I2/m
γ-Pu orthorhombic (a=315.87; b=576.82; c=1016.2), Fddd
δ-Pu f.c.c. (a=463.71), Fm3m
δ′-Pu tetragonal (a=333.9; c=444.6), I4/mmm
ε-Pu b.c.c. (a=363.48), Im3m
$T(\alpha \rightarrow \beta)$=395 K; $(\beta \rightarrow \gamma)$=473 K; $(\gamma \rightarrow \delta)$=583 K; $(\delta \rightarrow \delta')$=725 K; $(\delta' \rightarrow \varepsilon)$=753 K

X-ray diffraction: mass absorption coefficients (μ/ρ)/cm^2 g^{-1}: n.a.

Discovered in 1940 by G. T. Seaborg, A. C. Wahl, and J. W. Kennedy at Berkeley, California, USA

[Named after the planet Pluto]

Plutonium

Nuclear properties

Atomic number: 94

Thermal neutron capture cross-section/barns: 1.8 ± 0.3 (^{244}Pu)

Number of isotopes (including nuclear isomers): 16

Isotope mass range: 232→246

Key isotopes

Nuclide	Atomic mass	Natural abundance (%)	Half life $T_{1/2}$	Decay mode and energy (MeV)	Nuclear spin I	Nuclear magnetic moment μ	Uses
^{239}Pu	239.0522	0	24400y	α(5.243), SF, γ	1/2	+0.200	NMR
^{242}Pu	242.0587	0	3.79×10^5y	α(4.98)			
^{244}Pu	244.0642	0	8.2×10^7y	α(4.66), SF			

NMR

	^{239}Pu
Relative sensitivity ($^1H = 1.00$)	—
Absolute sensitivity ($^1H = 1.00$)	—
Receptivity ($^{13}C = 1.00$)	—
Magnetogyric ratio/rad T^{-1} s^{-1}	0.972×10^7
Quadrupole moment/m^2	—
Frequency ($^1H = 100$ MHz; 2.3488 T)/MHz	3.63

Electron shell properties

Ground state electron configuration: $[Rn]5f^6 7s^2$

Term symbol: $^7F_0$

Electron affinity ($M \rightarrow M^-$)/kJ mol^{-1}: n.a.

Ionization energies/kJ mol^{-1}

1. $M \rightarrow M^+$	585	6. $M^{5+} \rightarrow M^{6+}$	
2. $M^+ \rightarrow M^{2+}$		7. $M^{6+} \rightarrow M^{7+}$	
3. $M^{2+} \rightarrow M^{3+}$		8. $M^{7+} \rightarrow M^{8+}$	
4. $M^{3+} \rightarrow M^{4+}$		9. $M^{8+} \rightarrow M^{9+}$	
5. $M^{4+} \rightarrow M^{5+}$		10. $M^{9+} \rightarrow M^{10+}$	

Principal lines in atomic spectrum

Wavelength/nm	Species	Sensitivity	Application
321.508	I		
324.416	I		
325.208	I		
327.524	I		
329.256	I		
329.361	I		
329.691	I		

Abundance: Earth's crust: traces in uranium ores; seawater: nil

Biological role: None; toxic due to radioactivity

Po

Atomic number: 84
Relative atomic mass ($^{12}C = 12.0000$): (209)

Chemical properties

Radioactive silver-grey metal produced in gramme amounts from neutron bombardment of bismuth. Soluble in dilute acids. Used as heat source in space equipment and as source of α radiation for research.

Radii/pm: Po^{4+} 65; atomic 167; covalent 153; Po^{2-} 230

Electronegativity: 2.0 (Pauling); 1.76 (Allred-Rochow)

Effective nuclear charge: 6.95 (Slater); 14.22 (Clementi); 18.31 (Froese–Fischer)

Standard reduction potentials $E^\ominus$/V

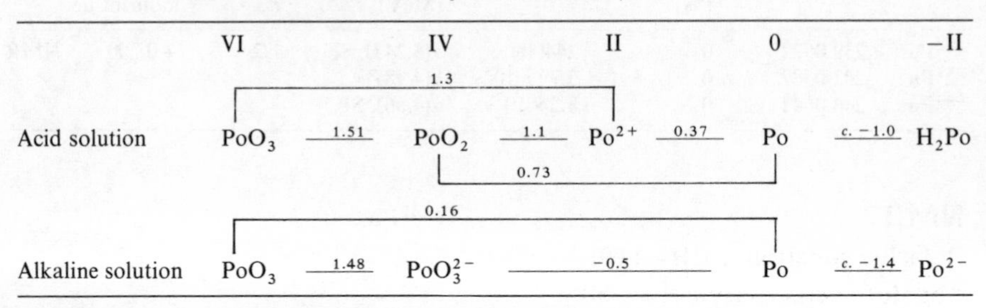

Covalent bonds r/pm E/kJ mol^{-1}

Bond	r/pm	E/kJ mol^{-1}
Po–Cl	238	n.a.
Po–Po	335	n.a.

Oxidation states

Po^{-II}	H_2Po, Na_2Po
Po^{II}	PoO, $PoCl_2$, $PoBr_2$
Po^{IV}	PoO_2, PoO_3^{2-}(aq), $PoCl_4$, $PoBr_4$, PoI_4, PoI_6^{2-}
Po^{VI}	PoO_3 ? PoF_6

Physical properties

Melting point/K: 527

Boiling point/K: 1235

ΔH_{fusion}/kJ mol^{-1}: 10

ΔH_{vap}/kJ mol^{-1}: 100.8

Thermodynamic properties (298.15 K, 0.1 MPa)

State	$\Delta_f H^\ominus$/kJ mol^{-1}	$\Delta_f G^\ominus$/kJ mol^{-1}	$S^\ominus$/J K^{-1} mol^{-1}	C_p/J K^{-1} mol^{-1}
Solid	0	0	n.a.	26.1
Gas	146?	n.a.	n.a.	n.a.

Density/kg m^{-3}: 9320 (α) [293 K]

Thermal conductivity/W m^{-1} K^{-1}: 20 [300 K]

Electrical resistivity/Ω m: 140×10^{-8} [293 K]

Mass magnetic susceptibility/kg^{-1} m^3: n.a.

Molar volume/cm^3: 22.4

Coefficient of linear thermal expansion/K^{-1}: 23.0×10^{-6}

Lattice structure (cell dimensions/pm), space group

α-Po cubic ($a = 335.2$), Pm3m
β-Po rhombohedral ($a = 336.6$, $\alpha = 98°13'$), R$\bar{3}$m

T (α→β) = 309 K

X-ray diffraction: mass absorption coefficients (μ/ρ)/cm^2 g^{-1}: n.a.

Discovered in 1898 by Marie Curie at Paris, France

[Named after Poland]

Polonium

Nuclear properties

Atomic number: 84

Thermal neutron capture cross-section/barns: <0.5 (^{210}Po)

Number of isotopes (including nuclear isomers): 34

Isotope mass range: 192→218

Key isotopes

Nuclide	Atomic mass	Natural abundance (%)	Half life $T_{1/2}$	Decay mode and energy (MeV)	Nuclear spin I	Nuclear magnetic moment μ	Uses
^{209}Po	208.9825	0	103y	α(4.98) 99%; EC; γ	1/2	+0.77	
^{210}Po	209.9829	trace	138.4d	α(5.408); γ	0		tracer, fuel
^{211}Po	210.9866	trace	0.52s	α(7.592); γ			
^{216}Po	216.0019	trace	0.15s	α(6.906); no γ			
^{218}Po	218.0089	trace	3.05m	α(6.111); no γ			

Electron shell properties

Ground state electron configuration: [Xe]$4f^{14}5d^{10}6s^26p^4$

Term symbol: $^3P_2$

Electron affinity (M→M$^-$)/kJ mol^{-1}: 186

Ionization energies/kJ mol^{-1}

1. M →M$^+$	812	6. M^{5+}→M^{6+}	(7000)
2. M$^+$→M^{2+}	(1800)	7. M^{6+}→M^{7+}	(10800)
3. M^{2+}→M^{3+}	(2700)	8. M^{7+}→M^{8+}	(12700)
4. M^{3+}→M^{4+}	(3700)	9. M^{8+}→M^{9+}	(14900)
5. M^{4+}→M^{5+}	(5900)	10. M^{9+}→M^{10+}	(17000)

Principal lines in atomic spectrum

Wavelength/nm	Species	Sensitivity	Application
245.008	I		
255.801*	I		
300.321	I		
417.052	I		

* Lowest energy transition from ground state to nearest empty orbital

Abundance: Earth's crust 3×10^{-10}; seawater nil

Biological role: None; highly toxic due to radioactivity

K

Atomic number: 19
Relative atomic mass ($^{12}C=12.0000$): 39.0983

Chemical properties

Soft white metal, silvery when cut but reacts rapidly with oxygen and vigorously with water. Large deposits of sylvite (KCl) and sylvinite (NaCl, KCl). Metal obtained from Na + KCl at 1100 K. Used in fertilizers, chemicals, and glass.

Radii/pm: K^+ 133; atomic 227; covalent 203; van der Waals 231

Electronegativity: 0.82 (Pauling); 0.91 (Allred-Rochow)

Effective nuclear charge: 2.20 (Slater); 3.50 (Clementi); 4.58 (Froese–Fischer)

Standard reduction potentials $E^{\ominus}$/V

I		0
K^+	$\xrightarrow{-2.924}$	K

Oxidation states

K^{-I} (s^2)	solution in liquid ammonia
K^{I} ([Ar])	K_2O, K_2O_2 (peroxide) KO_2 (superoxide), KO_3 (ozonide), KOH, $[K(H_2O)_4]^+$ (aq), KH, KF, KCl etc., K^+ salts, K_2CO_3, complexes

Physical properties

Melting point/K: 336.80

Boiling point/K: 1047

ΔH_{fusion}/kJ mol^{-1}: 2.40

ΔH_{vap}/kJ mol^{-1}: 79.1

Thermodynamic properties (298.15 K, 0.1 MPa)

State	$\Delta_f H^{\ominus}$/kJ mol^{-1}	$\Delta_f G^{\ominus}$/kJ mol^{-1}	$S^{\ominus}$/J K^{-1} mol^{-1}	C_p/J K^{-1} mol^{-1}
Solid	0	0	64.18	29.58
Gas	89.24	60.59	160.336	20.786

Density/kg m^{-3}: 862 [293 K]; 828 [liquid at m.p.]

Thermal conductivity/W m^{-1} K^{-1}: 102.4 [300 K]

Electrical resistivity/Ω m: 6.15×10^{-8} [273 K]

Mass magnetic susceptibility/kg^{-1} m^3: $+6.7 \times 10^{-9}$ (s)

Molar volume/cm^3: 45.36

Coefficient of linear thermal expansion/K^{-1}: 83×10^{-6}

Lattice structure (cell dimensions/pm), space group

b.c.c. (a = 533.4), Im3m

X-ray diffraction: mass absorption coefficients (μ/ρ)/cm^2 g^{-1}:
CuK$_\alpha$ 143 MoK$_\alpha$ 15.8

Discovered in 1807 by Sir Humphry Davy at London, UK

[English, *potash*; Latin *kalium*]

Potassium

Nuclear properties

Atomic number: 19

Thermal neutron capture cross-section/barns: 2.2±0.2

Number of isotopes (including nuclear isomers): 10

Isotope mass range: 37→45

Key isotopes

Nuclide	Atomic mass	Natural abundance (%)	Half life $T_{1/2}$	Decay mode and energy (MeV)	Nuclear spin I	Nuclear magnetic moment μ	Uses
^{39}K	38.963 71	93.2581	stable		3/2	+0.3914	NMR
^{40}K	39.974	0.0117	1.28×10^{9}y	β^-(1.35); EC; γ	4	−1.2981	
^{41}K	40.962	6.7302	stable		3/2	+0.2149	NMR
^{42}K	41.963	0	12.4h	β^-(0.60); γ	2	−1.1424	tracer, medical
^{43}K	42.964	0	22.4h	β^-(1.82); γ	3/2	+0.163	

NMR

	^{39}K	[^{41}K]
Relative sensitivity (^1H=1.00)	5.08×10^{-4}	8.40×10^{-5}
Absolute sensitivity (^1H=1.00)	4.73×10^{-4}	5.78×10^{-6}
Receptivity (^{13}C=1.00)	2.69	3.28×10^{-2}
Magnetogyric ratio/rad T^{-1} s^{-1}	1.2483×10^{7}	0.6851×10^{7}
Quadrupole moment /m^2	5.5×10^{-30}	6.7×10^{-30}
Frequency (^1H=100 MHz; 2.3488 T)/MHz	4.667	2.561

Reference: K^+ (aq)

Electron shell properties

Ground state electron configuration: [Ar]$4s^1$

Term symbol: $^2S_{1/2}$

Electron affinity ($M\rightarrow M^-$)/kJ mol^{-1}: 48.3

Ionization energies/kJ mol^{-1}

1. $M\rightarrow M^+$	418.8	6. $M^{5+}\rightarrow M^{6+}$	9649
2. $M^+\rightarrow M^{2+}$	3051.4	7. $M^{6+}\rightarrow M^{7+}$	11343
3. $M^{2+}\rightarrow M^{3+}$	4411	8. $M^{7+}\rightarrow M^{8+}$	14942
4. $M^{3+}\rightarrow M^{4+}$	5877	9. $M^{8+}\rightarrow M^{9+}$	16964
5. $M^{4+}\rightarrow M^{5+}$	7975	10. $M^{9+}\rightarrow M^{10+}$	48575

Principal lines in atomic spectrum

Wavelength/nm	Species	Sensitivity	Application
404.414	I	U3	AA, AE
404.720	I	U4	AA, AE
766.491	I	U1	AA, AE
769.898*	I	U2	AA, AE

* Lowest energy transition from ground state to nearest empty orbital

Abundance: Earth's crust 18400 p.p.m.; seawater 380 p.p.m.

Biological role: Essential; non-toxic

Pr

Atomic number: 59

Relative atomic mass (^{12}C=12.0000): **140.90765**

Chemical properties

Soft, malleable, silvery metal of the lanthanide (rare earth) group. Obtained from monazite and bastnaesite. Reacts slowly with oxygen, rapidly with water. Used in alloys, flints, yellow glass for eye protection for welders, etc.

Radii/pm: Pr^{3+} 106; Pr^{4+} 92; atomic 182.8; covalent 165

Electronegativity: 1.13 (Pauling); 1.07 (Allred-Rochow)

Effective nuclear charge: 2.85 (Slater); 7.75 (Clementi); 10.70 (Froese–Fischer)

Standard reduction potentials $E^{\ominus}$/V

	IV		III		0
Acid solution	Pr^{4+}	3.2	Pr^{3+}	−2.35	Pr
Alkaline solution	PrO_2	0.8	$Pr(OH)_3$	−2.79	Pr

Oxidation states

Pr^{III} (f^2)	P_2O_3, $Pr(OH)_3$, $[Pr(H_2O)_x]^{3+}$ (aq), Pr^{3+} salts, PrF_3, $PrCl_3$ etc., complexes
Pr^{IV} (f^1)	PrO_2, PrF_4, Na_2PrF_6

Physical properties

Melting point/K: 1204

Boiling point/K: 3785

ΔH_{fusion}/kJ mol^{-1}: 11.3

ΔH_{vap}/kJ mol^{-1}: 357

Thermodynamic properties (298.15 K, 0.1 MPa)

State	$\Delta_f H^{\ominus}$/kJ mol^{-1}	$\Delta_f G^{\ominus}$/kJ mol^{-1}	$S^{\ominus}$/J K^{-1} mol^{-1}	C_p/J K^{-1} mol^{-1}
Solid	0	0	73.2	27.20
Gas	355.6	320.9	189.808	21.359

Density/kg m^{-3}: 6773 [293 K]

Thermal conductivity/W m^{-1} K^{-1}: 12.5 [300 K]

Electrical resistivity/Ω m: 68×10^{-8} [298 K]

Mass magnetic susceptibility/kg^{-1} m^3: $+4.47 \times 10^{-7}$ (s)

Molar volume/cm^3: 20.80

Coefficient of linear thermal expansion/K^{-1}: 6.79×10^{-6}

Lattice structure (cell dimensions/pm), space group

α-Pr h.c.p. (a=367.25, c=1183.5), $P6_3/mmc$
β-Pr b.c.c. (a=413), Im3m

$T(\alpha \rightarrow \beta)$ = 1065 K

X-ray diffraction: mass absorption coefficients (μ/ρ)/cm^2 g^{-1}:
CuK_α 363 MoK_α 50.7

Separated in 1885 by Baron Auer von Welsbach at Vienna, Austria

[Greek, *prasios didymos* = green twin]

Praseodymium

Nuclear properties

Atomic number: 59

Thermal neutron capture cross-section/barns: 11.5 ± 1.0

Number of isotopes (including nuclear isomers): 15

Isotope mass range: 134→148

Key isotopes

Nuclide	Atomic mass	Natural abundance (%)	Half life $T_{1/2}$	Decay mode and energy (MeV)	Nuclear spin I	Nuclear magnetic moment μ	Uses
^{141}Pr	140.9074	100	stable		5/2	+4.16	NMR
^{142}Pr		0	19.2h	β^-(2.16); γ	2	−0.26	tracer
^{143}Pr		0	13.59d	β^-(0.933); no γ	7/2		tracer

NMR

	^{141}Pr
Relative sensitivity ($^1H = 1.00$)	0.29
Absolute sensitivity ($^1H = 1.00$)	0.29
Receptivity ($^{13}C = 1.00$)	1.62×10^3
Magnetogyric ratio/rad T^{-1} s^{-1}	7.765×10^7
Quadrupole moment /m^2	-5.9×10^{-30}
Frequency ($^1H = 100$ MHz; 2.3488 T)/MHz	29.291

Electron shell properties

Ground state electron configuration: $[Xe]4f^36s^2$

Term symbol: $^4I_{9/2}$

Electron affinity ($M \rightarrow M^-$)/kJ mol^{-1}: $\leqslant 50$

Ionization energies/kJ mol^{-1}

1. $M \rightarrow M^+$	523.1	6. $M^{5+} \rightarrow M^{6+}$	
2. $M^+ \rightarrow M^{2+}$	1018	7. $M^{6+} \rightarrow M^{7+}$	
3. $M^{2+} \rightarrow M^{3+}$	2086	8. $M^{7+} \rightarrow M^{8+}$	
4. $M^{3+} \rightarrow M^{4+}$	3761	9. $M^{8+} \rightarrow M^{9+}$	
5. $M^{4+} \rightarrow M^{5+}$	5543	10. $M^{9+} \rightarrow M^{10+}$	

Principal lines in atomic spectrum

Wavelength/nm	Species	Sensitivity	Application
406.282	II		AE
417.938	II		AE
422.535	II		AE
495.137	I		AA
491.402	I		AA

Abundance: Earth's crust 9.1 p.p.m.; seawater 2×10^{-7} p.p.m.

Biological role: None; toxicity low

Pm

Atomic number: 61

Relative atomic mass (^{12}C=12.0000): (145)

Chemical properties

Radioactive metal of the lanthanide (rare earth) group, obtained in mg quantities from the fission products of nuclear reactors. Used in specialized miniature batteries.

Radii/pm: Pm^{3+} 106; atomic 181.0

Electronegativity: n.a. (Pauling); 1.07 (Allred-Rochow)

Effective nuclear charge: 2.85 (Slater); 9.40 (Clementi); 10.94 (Froese–Fischer)

Standard reduction potentials $E^{\ominus}$/V

	III		0
Acid solution	Pm^{3+}	−2.29	Pm
Alkaline solution	$Pm(OH)_3$	−2.76	Pm

Oxidation states

Pm^{III} (f^4)	Pm_2O_3, $Pm(OH)_3$, $[Pm(H_2O)_x]^{3+}$ (aq), PmF_3, some complexes

Physical properties

Melting point/K: 1441

Boiling point/K: *c.* 3000

ΔH_{fusion}/kJ mol^{-1}: 12.6

ΔH_{vap}/kJ mol^{-1}: n.a.

Thermodynamic properties (298.15 K, 0.1 MPa)

State	$\Delta_f H^{\ominus}$/kJ mol^{-1}	$\Delta_f G^{\ominus}$/kJ mol^{-1}	$S^{\ominus}$/J K^{-1} mol^{-1}	C_p/J K^{-1} mol^{-1}
Solid	0	0	n.a.	26.8
Gas	n.a.	n.a.	187.101	24.255

Density/kg m^{-3}: 7220 [298 K]

Thermal conductivity/W m^{-1} K^{-1}: 17.9 (est.) [300 K]

Electrical resistivity/Ω m: 50×10^{-8} (est.) [273 K]

Mass magnetic susceptibility/kg^{-1} m^3: n.a.

Molar volume/cm^3: 20.1

Coefficient of linear thermal expansion/K^{-1}: n.a.

Lattice structure (cell dimensions/pm), space group

Hexagonal

X-ray diffraction: mass absorption coefficients (μ/ρ)/cm^2 g^{-1}: CuK_α 386 MoK_α 55.9

Produced in 1945 by J. A. Marinsky, L. E. Glendenin, and C. D. Coryell at Oak Ridge, USA

[Greek, *Prometheus*, who stole fire from the gods]

Promethium

Nuclear properties

Atomic number: 61

Thermal neutron capture cross-section/barns: 8400 ± 1680 (^{146}Pm)

Number of isotopes (including nuclear isomers): 14

Isotope mass range: 141→154

Key isotopes

Nuclide	Atomic mass	Natural abundance (%)	Half life $T_{1/2}$	Decay mode and energy (MeV)	Nuclear spin I	Nuclear magnetic moment μ	Uses
^{145}Pm	144.9128	0	17.7y	EC(0.14); γ			
^{146}Pm		0	4.4y	EC(1.72); β^-(1.53); γ			
^{147}Pm	146.9152	0	2.62y	β^-(0.225); no γ	7/2	+2.62	tracer
^{149}Pm		0	53.1h	β^-(1.06); γ	7/2	±3.3	tracer combination
^{151}Pm		0	28h	β^-(1.19); γ	5/2	±1.6	tracer combination

NMR

	^{147}Pm
Relative sensitivity ($^1H = 1.00$)	—
Absolute sensitivity ($^1H = 1.00$)	—
Receptivity ($^{13}C = 1.00$)	—
Magnetogyric ratio/rad T^{-1} s^{-1}	3.613×10^7
Quadrupole moment /m^2	0.67×10^{-28}
Frequency ($^1H = 100$ MHz; 2.3488 T)/MHz	13.51

Electron shell properties

Ground state electron configuration: [Xe]$4f^5 6s^2$

Term symbol: $^6H_{5/2}$

Electron affinity ($M \rightarrow M^-$)/kJ mol^{-1}: ≤50

Ionization energies/kJ mol^{-1}

1. $M \rightarrow M^+$	535.9	6. $M^{5+} \rightarrow M^{6+}$	
2. $M^+ \rightarrow M^{2+}$	1052	7. $M^{6+} \rightarrow M^{7+}$	
3. $M^{2+} \rightarrow M^{3+}$	2150	8. $M^{7+} \rightarrow M^{8+}$	
4. $M^{3+} \rightarrow M^{4+}$	3970	9. $M^{8+} \rightarrow M^{9+}$	
5. $M^{4+} \rightarrow M^{5+}$		10. $M^{9+} \rightarrow M^{10+}$	

Principal lines in atomic spectrum

Wavelength/nm	Species	Sensitivity	Application
371.172	II		
391.026	II		
395.774	II		
398.074	II		
441.796	II		
652.045	II		

Abundance: Earth's crust, traces in uranium ores; seawater nil

Biological role: None; toxic due to radioactivity

Pa

Atomic number: 91

Relative atomic mass ($^{12}C=12.0000$)**:** **231.03588**

Chemical properties

Radioactive, silvery metal found naturally in uranium ores and produced in gramme quantities from uranium fuel elements. Attacked by oxygen, steam, and acids, but not by alkalis. Little used.

Radii/pm: Pa^{3+} 113; Pa^{4+} 98; Pa^{5+} 89; atomic 160.6

Electronegativity: 1.5 (Pauling); 1.14 (Allred-Rochow)

Effective nuclear charge: 1.80 (Slater)

Standard reduction potentials $E^{\ominus}$/V

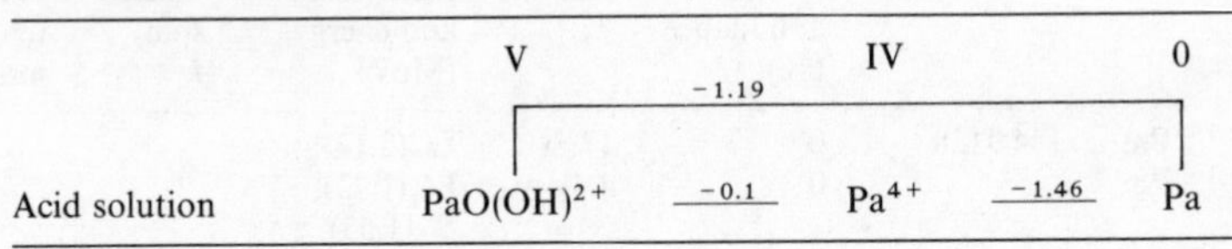

Oxidation states

Pa^{III} (f^2)	PaI_3
Pa^{IV} (f^1)	PaO_2, $[Pa(H_2O)_x]^{4+}$ (aq), PaF_4, $PaCl_4$, etc.
Pa^{V} (f^0, [Rn])	Pa_2O_5, PaO_2^+ (aq), PaF_5, $PaCl_5$, etc., $[PaF_6]^-$, $[PaF_7]^{2-}$, $[PaF_8]^{3-}$

Physical properties

Melting point/K: 2113

Boiling point/K: *c.* 4300

ΔH_{fusion}/kJ mol^{-1}: 16.7

ΔH_{vap}/kJ mol^{-1}: 481

Thermodynamic properties (298.15 K, 0.1 MPa)

State	$\Delta_f H^{\ominus}$/kJ mol^{-1}	$\Delta_f G^{\ominus}$/kJ mol^{-1}	$S^{\ominus}$/J K^{-1} mol^{-1}	C_p/J K^{-1} mol^{-1}
Solid	0	0	51.9	28
Gas	607	563	198.05	22.93

Density/kg m^{-3}: 15 370 (est.)

Thermal conductivity/W m^{-1} K^{-1}: 47 (est.) [300 K]

Electrical resistivity/Ω m: 17.7×10^{-8} [273 K]

Mass magnetic susceptibility/kg^{-1} m^3: n.a.

Molar volume/cm^3: 15.0

Coefficient of linear thermal expansion/K^{-1}: 7.3×10^{-6}

Lattice structure (cell dimensions/pm), space group

Tetragonal ($a=393.2$, $c=323.8$). I4/mmm

X-ray diffraction: mass absorption coefficients (μ/ρ)/cm^2 g^{-1}: n.a.

Discovered in 1917 by Otto Hahn and Lise Meitner at Berlin, by K. Fajans at Karlsruhe, Germany, and by F. Soddy, J. A. Cranston, and A. Fleck at Glasgow, Scotland
[Greek, *protos* = first]

Protactinium

Nuclear properties

Atomic number: 91
Thermal neutron capture cross-section/barns: 200 ± 10 (^{231}Pa)
Number of isotopes (including nuclear isomers): 14
Isotope mass range: $225 \rightarrow 237$

Key isotopes

Nuclide	Atomic mass	Natural abundance (%)	Half life $T_{1/2}$	Decay mode and energy (MeV)	Nuclear spin I	Nuclear magnetic moment μ	Uses
^{231}Pa	231.0359	trace	3.26×10^4y	α(5.148)	3/2	± 1.98	
^{233}Pa	233.040	0	27.0d	β^-(0.571)	3/2	+3.4	tracer
^{234}Pa	234.043	trace	6.75h	β^-(2.23); γ			

Electron shell properties

Ground state electron configuration: [Rn]$5f^2 6d^1 7s^2$
Term symbol: $^4K_{11/2}$
Electron affinity ($M \rightarrow M^-$)/kJ mol^{-1}: n.a.

Ionization energies/kJ mol^{-1}

1. $M \rightarrow M^+$	568	6. $M^{5+} \rightarrow M^{6+}$
2. $M^+ \rightarrow M^{2+}$		7. $M^{6+} \rightarrow M^{7+}$
3. $M^{2+} \rightarrow M^{3+}$		8. $M^{7+} \rightarrow M^{8+}$
4. $M^{3+} \rightarrow M^{4+}$		9. $M^{8+} \rightarrow M^{9+}$
5. $M^{4+} \rightarrow M^{5+}$		10. $M^{9+} \rightarrow M^{10+}$

Principal lines in atomic spectrum

Wavelength/nm	Species	Sensitivity	Application
395.785	II		
398.223	I		
694.572	I		
711.489	I		
736.825	I		
762.679	I		
763.518	I		
766.934	I		

Abundance: Earth's crust, trace in uranium ores; seawater 2×10^{-12} p.p.m.
Biological role: None; toxic due to radioactivity

Ra

Atomic number: 88
Relative atomic mass ($^{12}C = 12.0000$): 226.0254

Chemical properties

Radioactive element found naturally in uranium ores. Silvery, lustrous, soft. Annual production *c.* 100 g. Formerly used in cancer therapy and for luminous paint; both uses now rare. Reacts with oxygen and water.

Radii/pm: Ra^{2+} 152; atomic 223
Electronegativity: 0.89 (Pauling); 0.97 (Allred-Rochow)
Effective nuclear charge: 1.65 (Slater)

Standard reduction potentials $E^{\ominus}$/V

II		0
Ra^{2+}	$\xrightarrow{-2.916}$	Ra
RaO	$\xrightarrow{-1.319}$	Ra

Oxidation states

Ra^{II} ([Rn])	RaO, $Ra(OH)_2$, $[Ra(H_2O)_x]^{2+}$ (aq), Ra^{2+} salts

Physical properties

Melting point/K: 973
Boiling point/K: 1413
ΔH_{fusion}/kJ mol^{-1}: 7.15
ΔH_{vap}/kJ mol^{-1}: 136.7

Thermodynamic properties (298.15 K, 0.1 MPa)

State	$\Delta_f H^{\ominus}$/kJ mol^{-1}	$\Delta_f G^{\ominus}$/kJ mol^{-1}	$S^{\ominus}$/J K^{-1} mol^{-1}	C_p/J K^{-1} mol^{-1}
Solid	0	0	71	27.1
Gas	159	130	176.47	20.79

Density/kg m^{-3}: *c.* 5000 [293 K]
Thermal conductivity/W m^{-1} K^{-1}: 18.6 (est.) [300 K]
Electrical resistivity/Ω m: 100×10^{-8} [273 K]
Mass magnetic susceptibility/kg^{-1} m^3: n.a.
Molar volume/cm^3: 45.2
Coefficient of linear thermal expansion/K^{-1}: 20.2×10^{-6}

Lattice structure (cell dimensions/pm), space group

b.c.c. ($a = 515$)

X-ray diffraction: mass absorption coefficients (μ/ρ)/cm^2 g^{-1}:
CuK_α 304 MoK_α 172

Discovered in 1898 by Pierre and Marie Curie at Paris, France

[Latin, *radius* = ray]

Radium

Nuclear properties

Atomic number: 88

Thermal neutron capture cross-section/barns: 20 ± 3 (^{226}Ra)

Number of isotopes (including nuclear isomers): 16

Isotope mass range: 213→230 (^{218}Ra and ^{219}Ra unknown)

Key isotopes

Nuclide	Atomic mass	Natural abundance (%)	Half life $T_{1/2}$	Decay mode and energy (MeV)	Nuclear spin I	Nuclear magnetic moment μ	Uses
^{223}Ra	226.0186	some	11.43d	α(5.977); γ			
^{224}Ra	224.0202	some	3.64d	α(5.787); γ			
^{226}Ra	226.0254	some	1602y	α(4.781); γ			tracer, medical
^{228}Ra	228.031	some	5.77y	β^-(0.055); no γ			

Electron shell properties

Ground state electron configuration: [Rn]$7s^2$

Term symbol: $^1S_0$

Electron affinity ($M \rightarrow M^-$)/kJ mol^{-1}: n.a.

Ionization energies/kJ mol^{-1}

1. $M \rightarrow M^+$	509.3	6. $M^{5+} \rightarrow M^{6+}$	(7300)
2. $M^+ \rightarrow M^{2+}$	979.0	7. $M^{6+} \rightarrow M^{7+}$	(8600)
3. $M^{2+} \rightarrow M^{3+}$	(3300)	8. $M^{7+} \rightarrow M^{8+}$	(9900)
4. $M^{3+} \rightarrow M^{4+}$	(4400)	9. $M^{8+} \rightarrow M^{9+}$	(13500)
5. $M^{4+} \rightarrow M^{5+}$	(5700)	10. $M^{9+} \rightarrow M^{10+}$	(15100)

Principal lines in atomic spectrum

Wavelength/nm	Species	Sensitivity	Application
381.442	II	V1	AE
468.228	II	V2	AE
482.591	I	U1	AE
714.121*	I		

* Lowest energy transition from ground state to nearest empty orbital

Abundance: Earth's crust 10^{-6} p.p.m.; seawater 3×10^{-11} p.p.m.

Biological role: None; toxic due to radioactivity

Rn

Atomic number: 86

Relative atomic mass (^{12}C=12.0000): **(222)**

Chemical properties

Colourless, odourless gas produced by ^{226}Ra. Little studied because of hazardous radiation which destroys any compounds that are formed. Chemically should be like xenon. May be health hazard in certain localities.

Radii/pm: n.a.

Electronegativity: n.a. (Pauling); n.a. (Allred-Rochow)

Effective nuclear charge: 8.25 (Slater); 16.08 (Clementi); 20.84 (Froese–Fischer)

Solubility in water at 293 K, 230 cm^3 per dm^3

Standard reduction potentials $E^{\ominus}$/V

n.a.

Covalent bonds	r/pm	E/kJ mol^{-1}
Rn–F	n.a.	

Oxidation states	
Rn^{II}	RnF_2

Physical properties

Melting point/K: 202

Boiling point/K: 211.4

ΔH_{fusion}/kJ mol^{-1}: 2.7 (est.)

ΔH_{vap}/kJ mol^{-1}: 18.1

Thermodynamic properties (298.15 K, 0.1 MPa)

State	$\Delta_f H^{\ominus}$/kJ mol^{-1}	$\Delta_f G^{\ominus}$/kJ mol^{-1}	$S^{\ominus}$/J K^{-1} mol^{-1}	C_p/J K^{-1} mol^{-1}
Gas	0	0	176.21	20.786

Density/kg m^{-3}: n.a. [solid]; 4400 [liquid, b.p.]; 9.73 [gas, 273 K]

Thermal conductivity/W m^{-1} K^{-1}: 0.003 64 (est.) [300 K]$_g$

Mass magnetic susceptibility/kg^{-1} m^3: n.a.

Molar volume/cm^3: 50.5 [211 K]

Lattice structure (cell dimensions/pm)

f.c.c.

X-ray diffraction: mass absorption coefficients (μ/ρ)/cm^2 g^{-1}: n.a.

Discovered in 1900 by F. E. Dorn at Halle, Germany

[Named after radium]

Radon

Nuclear properties

Atomic number: 86

Thermal neutron capture cross-section/barns: 0.72±0.07 (^{222}Rn)

Number of isotopes (including nuclear isomers): 20

Isotope mass range: 204→224

Key isotopes

Nuclide	Atomic mass	Natural abundance (%)	Half life $T_{1/2}$	Decay mode and energy (MeV)	Nuclear spin I	Nuclear magnetic moment μ	Uses
^{219}Rn	219.0095	trace	4.0s	α(6.944); γ			
^{220}Rn	220.0114	trace	55s	α(6.405); γ			
^{222}Rn	222.0175	trace	3.82d	α(5.587); γ			tracer, medical

Electron shell properties

Ground state electron configuration: [Xe]$4f^{14}5d^{10}6s^26p^6$

Term symbol: $^1S_0$

Electron affinity ($M \rightarrow M^-$)/kJ mol^{-1}: −41 (calc.)

Ionization energies/kJ mol^{-1}

1. $M \rightarrow M^+$	1037	6. $M^{5+} \rightarrow M^{6+}$	
2. $M^+ \rightarrow M^{2+}$		7. $M^{6+} \rightarrow M^{7+}$	
3. $M^{2+} \rightarrow M^{3+}$		8. $M^{7+} \rightarrow M^{8+}$	
4. $M^{3+} \rightarrow M^{4+}$		9. $M^{8+} \rightarrow M^{9+}$	
5. $M^{4+} \rightarrow M^{5+}$		10. $M^{9+} \rightarrow M^{10+}$	

Principal lines in atomic spectrum

Wavelength/nm	Species	Sensitivity	Application
183.048* (vac)	I		
434.960	I		
705.542	I	U3	AE
745.000	I	U2	AE

* Lowest energy transition from ground state to nearest empty orbital

Abundance: Atmosphere trace; Earth's crust 1.7×10^{-10} p.p.m.; seawater 9×10^{-15} p.p.m.

Biological role: None; toxic due to radioactivity

Re

Atomic number: 75

Relative atomic mass ($^{12}C = 12.0000$): **186.207**

Chemical properties

Silvery metal, usually obtained as grey powder; extracted from flue dusts of Mo smelters. Resists corrosion and oxidation but slowly tarnishes in moist air. Dissolves in nitric and sulfuric acids. Used in filaments, thermistors, and catalysts.

Radii/pm: Re^{4+} 72; Re^{6+} 61; Re^{7+} 60; atomic 137.0; covalent 128

Electronegativity: 1.9 (Pauling); 1.46 (Allred-Rochow)

Effective nuclear charge: 3.60 (Slater); 10.12 (Clementi); 14.62 (Froese–Fischer)

Standard reduction potentials $E^{\ominus}$/V

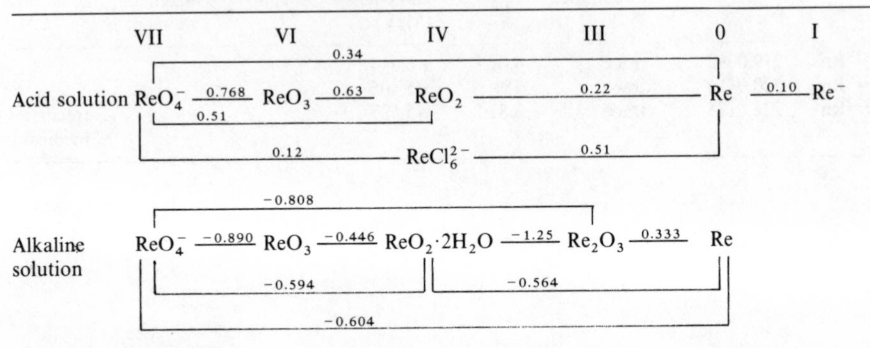

Oxidation states

Re^{-III} (d^{10})	$[Re(CO)_4]^{3-}$	**Re^{IV}** (d^3)	ReO_2, ReF_4, $ReCl_4$, etc., complexes
Re^{-I} (d^8)	$[Re(CO)_5]^-$	**Re^{V}** (d^2)	Re_2O_5, ReF_5, $ReCl_5$, $ReBr_5$, $ReOF_3$, complexes
Re^0 (d^7)	$Re_2(CO)_{10}$	Re^{VI} (d^1)	ReO_3, ReF_6, $ReCl_6$, ReF_8^{2-}, $ReOCl_4$, complexes
Re^I (d^6)	$Re(CO)Cl$, $K_5Re(CN)_6$	Re^{VII} (d^0, f^{14})	Re_2O_7, ReF_7, ReO_3F, $ReOF_5$, ReO_4^- (aq), $[ReH_9]^{2-}$, complexes
Re^{II} (d^5)	ReF_2, $ReCl_2$ etc.		
Re^{III} (d^4)	$Re_2O_3.xH_2O$, Re_3Cl_9, Re_3Br_9, Re_3I_9, $Re_2Cl_8^{2-}$, $[Re(CN)_7]^{4-}$, complexes		

Physical properties

Melting point/K: 3453 ΔH_{fusion}/kJ mol^{-1}: 33.1

Boiling point/K: 5900 ΔH_{vap}/kJ mol^{-1}: 704.25

Thermodynamic properties (298.15 K, 0.1 MPa)

State	$\Delta_f H^{\ominus}$/kJ mol^{-1}	$\Delta_f G^{\ominus}$/kJ mol^{-1}	$S^{\ominus}$/J K^{-1} mol^{-1}	C_p/J K^{-1} mol^{-1}
Solid	0	0	36.86	25.48
Gas	769.9	724.6	188.938	20.786

Density/kg m^{-3}: 21 020 [293 K]; 18 900 [liquid at m.p.]

Thermal conductivity/W m^{-1} K^{-1}: 47.9 [300 K]

Electrical resistivity/Ω m: 19.3×10^{-8} [293 K]

Mass magnetic susceptibility/kg^{-1} m^3: $+4.56 \times 10^{-9}$ (s)

Molar volume/cm^3: 8.86

Coefficient of linear thermal expansion/K^{-1}: 6.63×10^{-6}

Lattice structure (cell dimensions/pm), space group

h.c.p. ($a = 276.09$; $c = 445.76$), $P6_3/mmc$

X-ray diffraction: mass absorption coefficients (μ/ρ)/cm^2 g^{-1}:
CuK_α 179 MoK_α 103

Discovered in 1925 by W. Noddack, Ida Tacke and O. Berg in Berlin, Germany

[Latin, *Rhenus* = Rhine]

Rhenium

Nuclear properties

Atomic number: 75

Thermal neutron capture cross-section/barns: 85±5

Number of isotopes (including nuclear isomers): 20

Isotope mass range: 177→192

Key isotopes

Nuclide	Atomic mass	Natural abundance (%)	Half life $T_{1/2}$	Decay mode and energy (MeV)	Nuclear spin I	Nuclear magnetic moment μ	Uses
^{185}Re	184.9530	37.40	stable		5/2	+3.172	NMR
^{186}Re		0	88.9h	β^-(1.071); EC; γ	1	+1.73	tracer
^{187}Re	186.9560	62.60	4.3×10^{10}y	β^-(<0.01)	5/2	+3.204	NMR
^{188}Re		0	16.7h	β^-(2.116); γ	1	+1.78	tracer

NMR

	[^{185}Re]	^{187}Re
Relative sensitivity (^{1}H = 1.00)	0.13	0.13
Absolute sensitivity (^{1}H = 1.00)	4.93×10^{-2}	8.62×10^{-2}
Receptivity (^{13}C = 1.00)	280	490
Magnetogyric ratio/rad T^{-1} s^{-1}	6.0255×10^{7}	6.0862×10^{7}
Quadrupole moment/m^2	2.8×10^{-28}	2.6×10^{-28}
Frequency (^{1}H = 100 MHz; 2.3488 T)/MHz	22.513	22.744

Reference: $NaReO_4$ (aq)

Electron shell properties

Ground state electron configuration: [Xe]$4f^{14}5d^{2}6s^{2}$

Term symbol: $^6S_{5/2}$

Electron affinity ($M\rightarrow M^-$)/kJ mol^{-1}: 37

Ionization energies/kJ mol^{-1}

1. $M \rightarrow M^{+}$	760	6. $M^{5+}\rightarrow M^{6+}$	(6300)
2. $M^{+}\rightarrow M^{2+}$	1260	7. $M^{6+}\rightarrow M^{7+}$	(7600)
3. $M^{2+}\rightarrow M^{3+}$	2510	8. $M^{7+}\rightarrow M^{8+}$	
4. $M^{3+}\rightarrow M^{4+}$	3640	9. $M^{8+}\rightarrow M^{9+}$	
5. $M^{4+}\rightarrow M^{5+}$	(4900)	10. $M^{9+}\rightarrow M^{10+}$	

Principal lines in atomic spectrum

Wavelength/nm	Species	Sensitivity	Application
345.188	I		AA
346.047	I	U1	AA, AE
346.473	I		AA
488.917	I	U2	AE
527.556*	I		

* Lowest energy transition from ground state to nearest empty orbital

Abundance: Earth's crust 0.0007 p.p.m.; seawater $<1\times10^{-10}$ p.p.m.

Biological role: None; toxicity data n.a.

Rh

Atomic number: 45
Relative atomic mass (^{12}C=12.0000): 102.90550

Chemical properties

Rare; lustrous, silvery, hard metal of the platinum group. Certain Ni–Cu ores contain 0.1% rhodium. Stable in air up to 875 K; inert to all acids; attacked by fused alkalis. Used as a catalyst.

Radii/pm: Rh^{2+} 86; Rh^{3+} 75; Rh^{4+} 67; atomic 134.5; covalent 125
Electronegativity: 2.28 (Pauling); 1.45 (Allred-Rochow)
Effective nuclear charge: 3.90 (Slater); 7.64 (Clementi); 10.85 (Froese–Fischer)

Standard reduction potentials $E^{\ominus}$/V

III		0
Rh^{3+}	0.76	Rh

Oxidation states

Rh^{-I} (d^{10})	$[Rh(CO)_4]^-$
Rh^0 (d^9)	$[Rh_4(CO)_{12}]$
Rh^I (d^8)	$[Rh(PPh_3)_3]^+$
Rh^{II} (d^7)	RhO
Rh^{III} (d^6)	Rh_2O_3 RhF_3, $RhCl_3$ etc., $[RhCl_6]^{3-}$ $[Rh(H_2O)_x]^{3+}$ (aq)
Rh^{IV} (d^5)	RhO_2, RhF_4, $[RhCl_6]^{2-}$
Rh^V (d^4)	$[RhF_5]_4$, $[RhF_6]^-$
Rh^{VI} (d^3)	RhF_6

Physical properties

Melting point/K: 2239
Boiling point/K: 4000
ΔH_{fusion}/kJ mol^{-1}: 21.55
ΔH_{vap}/kJ mol^{-1}: 494.34

Thermodynamic properties (298.15 K, 0.1 MPa)

State	$\Delta_f H^{\ominus}$/kJ mol^{-1}	$\Delta_f G^{\ominus}$/kJ mol^{-1}	$S^{\ominus}$/J K^{-1} mol^{-1}	C_p/J K^{-1} mol^{-1}
Solid	0	0	31.51	24.98
Gas	556.9	510.8	185.808	21.012

Density/kg m^{-3}: 12 410 [293 K]; 10 650 [liquid at m.p.]
Thermal conductivity/W m^{-1} K^{-1}: 150 [300 K]
Electrical resistivity/Ω m: 4.51×10^{-8} [293 K]
Mass magnetic susceptibility/kg^{-1} m^3: $+1.36 \times 10^{-8}$ (s)
Molar volume/cm^3: 8.29
Coefficient of linear thermal expansion/K^{-1}: 8.40×10^{-6}

Lattice structure (cell dimensions/pm), space group

f.c.c. (a=380.36), Fm3m

X-ray diffraction: mass absorption coefficients (μ/ρ)/cm^2 g^{-1}:
CuK$_\alpha$ 194 MoK$_\alpha$ 22.6

Discovered in 1803 by W. H. Wollaston at London, UK

[Greek, *rhodon* = rose]

Rhodium

Nuclear properties

Atomic number: 45

Thermal neutron capture cross-section/barns: 150 ± 5

Number of isotopes (including nuclear isomers): 20

Isotope mass range: $97 \rightarrow 110$

Key isotopes

Nuclide	Atomic mass	Natural abundance (%)	Half life $T_{1/2}$	Decay mode and energy (MeV)	Nuclear spin I	Nuclear magnetic moment μ	Uses
^{103}Rh	102.9048	100	stable		1/2	−0.0883	NMR
^{105}Rh		0	35.88h	β^-(0.565); γ			tracer

NMR

	^{103}Rh
Relative sensitivity ($^1H = 1.00$)	3.11×10^{-5}
Absolute sensitivity ($^1H = 1.00$)	3.11×10^{-5}
Receptivity ($^{13}C = 1.00$)	0.177
Magnetogyric ratio/rad T^{-1} s^{-1}	-0.8520×10^7
Frequency ($^1H = 100$ MHz; 2.3488 T)/MHz	3.172

Reference: mer-$[RhCl_3(SMe_2)_3]$

Electron shell properties

Ground state electron configuration: $[Kr]4d^85s^1$

Term symbol: $^4F_{9/2}$

Electron affinity ($M \rightarrow M^-$)/kJ mol^{-1}: 162

Ionization energies/kJ mol^{-1}

1. $M \rightarrow M^+$	720	6. $M^{5+} \rightarrow M^{6+}$	(8 200)
2. $M^+ \rightarrow M^{2+}$	1744	7. $M^{6+} \rightarrow M^{7+}$	(10 100)
3. $M^{2+} \rightarrow M^{3+}$	2997	8. $M^{7+} \rightarrow M^{8+}$	(12 200)
4. $M^{3+} \rightarrow M^{4+}$	(4400)	9. $M^{8+} \rightarrow M^{9+}$	(14 200)
5. $M^{4+} \rightarrow M^{5+}$	(6500)	10. $M^{9+} \rightarrow M^{10+}$	(22 000)

Principal lines in atomic spectrum

Wavelength/nm	Species	Sensitivity	Application
332.309	I		AE
339.685	I		AA, AE
343.489	I	U1	AA, AE
350.252	I		AA
365.799	I		AA, AE
369.236*	I		AA, AE

* Lowest energy transition from ground state to nearest empty orbital

Abundance: Earth's crust 0.0001 p.p.m.; seawater $< 1 \times 10^{-10}$ p.p.m.

Biological role: None; toxicity n.a.; suspected carcinogenic

Rb

Atomic number: 37
Relative atomic mass (^{12}C=12.0000): 85.4678

Chemical properties

Very soft metal with silvery white lustre when cut. Ignites in air and reacts violently with water. Present in minerals such as lepidolite, pollucite and carnalite. Finds little use outside research.

Radii/pm: Rb^+ 1.49; atomic 247.5; van der Waals 244
Electronegativity: 0.82 (Pauling); 0.89 (Allred-Rochow)
Effective nuclear charge: 2.20 (Slater); 4.98 (Clementi); 6.66 (Froese–Fischer)

Standard reduction potentials $E^\ominus$/V

I		0
Rb^+	$\xrightarrow{-2.924}$	Rb

Oxidation states

Rb^{-I} (s^2)	solution of metal in liquid ammonia
Rb^{I} ([Kr])	Rb_2O, Rb_2O_2 (peroxide), RbO_2 (superoxide), RbOH, RbH, RbF, RbCl etc., $[Rb(H_2O)_x]^+$ (aq), Rb_2CO_3, many salts, some complexes

Physical properties

Melting point/K: 312.2
Boiling point/K: 961
ΔH_{fusion}/kJ mol^{-1}: 2.20
ΔH_{vap}/kJ mol^{-1}: 75.7

Thermodynamic properties (298.15 K, 0.1 MPa)

State	$\Delta_f H^\ominus$/kJ mol^{-1}	$\Delta_f G^\ominus$/kJ mol^{-1}	$S^\ominus$/J K^{-1} mol^{-1}	C_p/J K^{-1} mol^{-1}
Solid	0	0	76.78	31.062
Gas	80.88	53.06	170.089	20.786

Density/kg m^{-3}: 1532 [293 K]; 1475 [liquid at m.p.]
Thermal conductivity/W m^{-1} K^{-1}: 58.2 [300 K]
Electrical resistivity/Ω m: 12.5×10^{-8} [293 K]
Mass magnetic susceptibility/kg^{-1} m^3: $+2.49 \times 10^{-9}$ (s)
Molar volume/cm^3: 55.79
Coefficient of linear thermal expansion/K^{-1}: 90×10^{-6}

Lattice structure (cell dimensions/pm), space group

b.c.c. (a=562), Im3m

X-ray diffraction: mass absorption coefficients (μ/ρ)/cm^2 g^{-1}:
CuK_α 117 MoK_α 90.0

Discovered in 1861 by R. W. Bunsen and G. Kirchoff at University of Heidelberg, Germany

[Latin, *rubidius* = deepest red]

Rubidium

Nuclear properties

Atomic number: 37

Thermal neutron capture cross-section/barns: 0.5 ± 0.1

Number of isotopes (including nuclear isomers): 20

Isotope mass range: $79 \rightarrow 95$

Key isotopes

Nuclide	Atomic mass	Natural abundance (%)	Half life $T_{1/2}$	Decay mode and energy (MeV)	Nuclear spin I	Nuclear magnetic moment μ	Uses
^{83}Rb		0	83d	EC(0.83); γ	5/2	+1.4	tracer
^{85}Rb	84.9117	72.17	stable		5/2	+1.3524	NMR
^{86}Rb	85.911	0	18.66d	β^-(1.78); γ	2	−1.691	tracer
^{87}Rb	86.909	27.83	5×10^{11}y	β^-(0.274); no γ	3/2	+2.750	NMR

NMR

	[^{85}Rb]	^{87}Rb
Relative sensitivity (^{1}H = 1.00)	1.05×10^{-2}	0.17
Absolute sensitivity (^{1}H = 1.00)	7.57×10^{-3}	4.87×10^{-2}
Receptivity (^{13}C = 1.00)	43	277
Magnetogyric ratio/rad T^{-1} s^{-1}	2.5828×10^7	8.7532×10^7
Quadrupole moment/m^2	0.25×10^{-28}	0.12×10^{-28}
Frequency (^{1}H = 100 MHz; 2.3488 T)/MHz	9.655	32.721

Reference: RbCl (aq)

Electron shell properties

Ground state electron configuration: [Kr]$5s^1$

Term symbol: $^2S_{1/2}$

Electron affinity ($M \rightarrow M^-$)/kJ mol^{-1}: 46.9

Ionization energies/kJ mol^{-1}

1. $M \rightarrow M^+$	403.0	6. $M^{5+} \rightarrow M^{6+}$	8 140
2. $M^+ \rightarrow M^{2+}$	2632	7. $M^{6+} \rightarrow M^{7+}$	9 570
3. $M^{2+} \rightarrow M^{3+}$	3900	8. $M^{7+} \rightarrow M^{8+}$	13 100
4. $M^{3+} \rightarrow M^{4+}$	5080	9. $M^{8+} \rightarrow M^{9+}$	14 800
5. $M^{4+} \rightarrow M^{5+}$	6850	10. $M^{9+} \rightarrow M^{10+}$	26 740

Principal lines in atomic spectrum

Wavelength/nm	Species	Sensitivity	Application
420.180	I	U3	AA, AE
421.553	I	U4	AA, AE
780.027	I	U1	AA, AE
794.760*	I	U2	AA, AE

* Lowest energy transition from ground state to nearest empty orbital

Abundance: Earth's crust 78 p.p.m.; seawater 0.12 p.p.m.

Biological role: None; non-toxic, except in large doses; stimulatory

Ru

Atomic number: 44

Relative atomic mass (^{12}C=12.0000): 101.07

Chemical properties

Lustrous, silvery metal of the platinum group. Found in the free state but obtained from the wastes of Ni refining. Unaffected by air, water and acids but dissolves in molten alkali. Used to harden Pt and Pd, and as catalyst.

Radii/pm: Ru^{3+} 77; Ru^{4+} 65; Ru^{8+} 54; atomic 134; covalent 124

Electronegativity: 2.2 (Pauling); 1.42 (Allred-Rochow)

Effective nuclear charge: 3.75 (Slater); 7.45 (Clementi); 10.57 (Froese–Fischer)

Standard reduction potentials $E^\ominus$/V

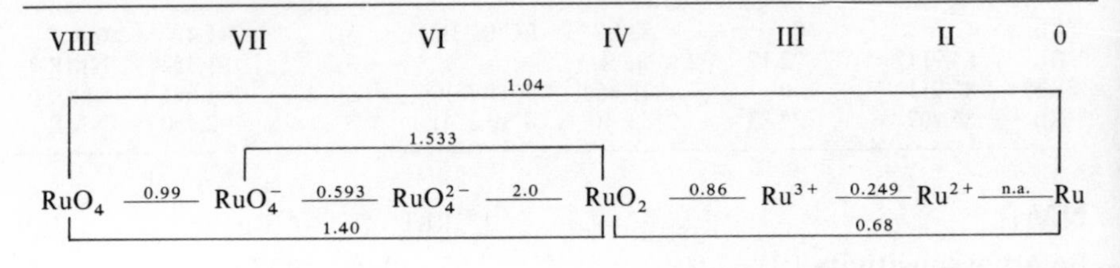

Oxidation states

Ru^{-II}	(d^{10})	rare $[Ru(CO)_4]^{2-}$
Ru	(d^8)	rare $Ru(CO)_5$
Ru^{I}	(d^7)	some complexes
Ru^{II}	(d^6)	$[Ru(H_2O)_6]^{2+}$ (aq), $RuCl_2$, $RuBr_2$, RuI_2, $[Ru(CN)_6]^{2-}$, complexes
Ru^{III}	(d^5)	Ru_2O_3, $[Ru(H_2O)_6]^{3+}$ (aq), RuF_3, $RuCl_3$, etc., $RuCl_6^{3-}$
Ru^{IV}	(d^4)	RuO_2, RuF_4, $RuCl_6^{2-}$
Ru^{V}	(d^3)	RuF_5, RuF_6^-
Ru^{VI}	(d^2)	RuO_3, RuO_4^{2-} (aq), RuF_6
Ru^{VII}	(d^1)	RuO_4^- (aq)
Ru^{VIII}	(d^0, [Kr])	RuO_4

Physical properties

Melting point/K: 2583

Boiling point/K: 4173

ΔH_{fusion}/kJ mol^{-1}: 23.7

ΔH_{vap}/kJ mol^{-1}: 567

Thermodynamic properties (298.15 K, 0.1 MPa)

State	$\Delta_f H^\ominus$/kJ mol^{-1}	$\Delta_f G^\ominus$/kJ mol^{-1}	$S^\ominus$/J K^{-1} mol^{-1}	C_p/J K^{-1} mol^{-1}
Solid	0	0	28.53	24.06
Gas	642.7	595.8	186.507	21.522

Density/kg m^{-3}: 12 370 [293 K]; 10 900 [liquid at m.p.]

Thermal conductivity/W m^{-1} K^{-1}: 117 [300 K]

Electrical resistivity/Ω m: 7.6×10^{-8} [273 K]

Mass magnetic susceptibility/kg^{-1} m^3: $+5.37 \times 10^{-9}$ (s)

Molar volume/cm^3: 8.14

Coefficient of linear thermal expansion/K^{-1}: 9.1×10^{-6}

Lattice structure (cell dimensions/pm), space group

h.c.p. (a=270.58; c=428.11), $P6_3/mmc$

X-ray diffraction: mass absorption coefficients (μ/ρ)/cm^2 g^{-1}:
CuK_α 183 MoK_α 21.1

Discovered in 1808 by J. A. Sniadecki at University of Vilno, Poland; rediscovered in 1828 by G. W. Osann at University of Tartu, Russia
[Latin, *Ruthenia* = Russia]

Ruthenium

Nuclear properties

Atomic number: 44
Thermal neutron capture cross-section/barns: 3.0 ± 0.8
Number of isotopes (including nuclear isomers): 16
Isotope mass range: $93 \rightarrow 108$

Key isotopes

Nuclide	Atomic mass	Natural abundance (%)	Half life $T_{1/2}$	Decay mode and energy (MeV)	Nuclear spin I	Nuclear magnetic moment μ	Uses
^{96}Ru	95.9076	5.52	stable				
^{97}Ru		0	2.88d	EC(*c.* 1.2); γ			tracer
^{98}Ru	97.9055	1.88	stable				
^{99}Ru	98.9061	12.7	stable		5/2	−0.62	NMR
^{100}Ru	99.9030	12.6	stable				
^{101}Ru		17.0	stable		5/2	−0.68	NMR
^{102}Ru	101.9037	31.6	stable				
^{103}Ru		0	39.6d	β^-(0.74); γ			tracer
^{104}Ru	103.9055	18.7	stable				
^{106}Ru		0	367d	β^-(0.0394); no γ			tracer, medical

NMR

	^{99}Ru	^{101}Ru
Relative sensitivity ($^1H = 1.00$)	1.95×10^{-4}	1.41×10^{-3}
Absolute sensitivity ($^1H = 1.00$)	2.48×10^{-5}	2.40×10^{-4}
Receptivity ($^{13}C = 1.00$)	0.83	1.56
Magnetogyric ratio/rad T^{-1} s^{-1}	-1.2343×10^7	-1.3834×10^7
Quadrupole moment/m^2	0.076×10^{-28}	0.44×10^{-28}
Frequency ($^1H = 100$ MHz; 2.3488 T)/MHz	3.389	4.941

Reference: RuO_4

Electron shell properties

Ground state electron configuration: $[Kr]4d^7 5s^1$
Term symbol: $^5F_5$
Electron affinity ($M \rightarrow M^-$)/kJ mol^{-1}: 146

Ionization energies/kJ mol^{-1}

1. $M \rightarrow M^+$	711	6. $M^{5+} \rightarrow M^{6+}$	(7800)
2. $M^+ \rightarrow M^{2+}$	1617	7. $M^{6+} \rightarrow M^{7+}$	(9600)
3. $M^{2+} \rightarrow M^{3+}$	2747	8. $M^{7+} \rightarrow M^{8+}$	(11500)
4. $M^{3+} \rightarrow M^{4+}$	(4500)	9. $M^{8+} \rightarrow M^{9+}$	(18700)
5. $M^{4+} \rightarrow M^{5+}$	(6100)	10. $M^{9+} \rightarrow M^{10+}$	(20900)

Principal lines in atomic spectrum

Wavelength/nm	Species	Sensitivity	Application
343.674	I	U2	AE
349.894	I	U1	AA, AE
359.618	I	U3	AE
372.803	I		AA
396.490*	I		

* Lowest energy transition from ground state to nearest empty orbital

Abundance: Earth's crust 0.0001 p.p.m.; seawater n.a., $<1 \times 10^{-10}$ p.p.m.
Biological role: None; toxicity n.a.

Sm

Atomic number: 62

Relative atomic mass (^{12}C=12.0000): **150.36**

Chemical properties

Silvery white metal of the lanthanide (rare earth) group. Obtained from monazite. Relatively stable in dry air but in moist air an oxide coating forms. Used in special glass, catalysts, ceramics, and electronics.

Radii/pm: Sm^{2+} 111; Sm^{3+} 100; atomic 180.2; covalent 166

Electronegativity: 1.17 (Pauling); 1.07 (Allred-Rochow)

Effective nuclear charge: 2.85 (Slater); 8.01 (Clementi); 11.06 (Froese–Fischer)

Standard reduction potentials $E^{\ominus}$/V

	III		II		0
		(Sm³⁺ → Sm: −2.30)			
Acid solution	Sm^{3+}	−1.55	Sm^{2+}	−2.67	Sm
Alkaline solution	$Sm(OH)_3$		−2.80		Sm

Oxidation states

Sm^{II} (f^6)	SmO, SmS, SmF_2, $SmCl_2$, etc.
Sm^{III} (f^5)	Sm_2O_3, $Sm(OH)_3$, SmF_3, $SmCl_3$ etc., $[Sm(H_2O)_x]^{3+}$ (aq), Sm^{3+} salts, complexes

Physical properties

Melting point/K: 1350

Boiling point/K: 2064

ΔH_{fusion}/kJ mol^{-1}: 10.9

ΔH_{vap}/kJ mol^{-1}: 164.8

Thermodynamic properties (298.15 K, 0.1 MPa)

State	$\Delta_f H^{\ominus}$/kJ mol^{-1}	$\Delta_f G^{\ominus}$/kJ mol^{-1}	$S^{\ominus}$/J K^{-1} mol^{-1}	C_p/J K^{-1} mol^{-1}
Solid	0	0	69.58	29.54
Gas	206.7	172.8	183.042	30.355

Density/kg m^{-3}: 7520 [293 K]

Thermal conductivity/W m^{-1} K^{-1}: 13.3 [300 K]

Electrical resistivity/Ω m: 88.0×10^{-8} [298 K]

Mass magnetic susceptibility/kg^{-1} m^3: $+1.52 \times 10^{-7}$ (s)

Molar volume/cm^3: 20.00

Coefficient of linear thermal expansion/K^{-1}: 10.4×10^{-6}

Lattice structure (cell dimensions/pm), space group

α-Sm rhombohedral (a=899.6, α=23°13′), R$\bar{3}$m
β-Sm cubic (a=407), Im3m

$T(\alpha \rightarrow \beta)$=1190 K

High pressure form: h.c.p. (a=361.8, c=1166), $P6_3$/mmc

X-ray diffraction: mass absorption coefficients (μ/ρ)/cm^2 g^{-1}:
CuK_α 397 MoK_α 58.6

Discovered by P.-E. Lecoq de Boisbaudran in 1879 at Paris, France

[Named after the mineral Samarskite]

Samarium

Nuclear properties

Atomic number: 62

Thermal neutron capture cross-section/barns: 5820±100

Number of isotopes (including nuclear isomers): 17

Isotope mass range: 142→157

Key isotopes

Nuclide	Atomic mass	Natural abundance (%)	Half life $T_{1/2}$	Decay mode and energy (MeV)	Nuclear spin I	Nuclear magnetic moment μ	Uses
^{144}Sm	143.9115	3.1	stable				
^{146}Sm	145.9129	0	7×10^7y	α(2.46)			
^{147}Sm	146.9147	15.1	1.05×10^{11}y	α(2.314)	7/2	−0.813	NMR
^{148}Sm	147.9146	11.3	$>12\times10^{14}$y	α(2.001)			
^{149}Sm	148.9169	13.9	$>1\times10^{15}$y	α	7/2	−0.670	NMR
^{150}Sm	149.9170	7.4	stable				
^{152}Sm	151.9195	26.6	stable				
^{153}Sm		0	46.8h	β^-(0.801); γ	3/2	−0.0217	tracer
^{154}Sm	153.9220	22.6	stable				

NMR

	^{147}Sm	^{149}Sm
Relative sensitivity ($^1H=1.00$)	1.48×10^{-3}	7.47×10^{-4}
Absolute sensitivity ($^1H=1.00$)	2.21×10^{-4}	1.03×10^{-4}
Receptivity ($^{13}C=1.00$)	1.28	0.665
Magnetogyric ratio/rad T^{-1} s^{-1}	-1.1124×10^7	-0.9175×10^7
Quadrupole moment/m^2	-0.208×10^{-28}	6×10^{-30}
Frequency ($^1H=100$ MHz; 2.3488 T)/MHz	4.128	3.289

Electron shell properties

Ground state electron configuration: [Xe]$4f^6 6s^2$

Term symbol: $^7F_0$

Electron affinity ($M\rightarrow M^-$)/kJ mol^{-1}: $\leqslant 50$

Ionization energies/kJ mol^{-1}

1. $M \rightarrow M^+$	543.3	6. $M^{5+}\rightarrow M^{6+}$	
2. $M^+\rightarrow M^{2+}$	1068	7. $M^{6+}\rightarrow M^{7+}$	
3. $M^{2+}\rightarrow M^{3+}$	2260	8. $M^{7+}\rightarrow M^{8+}$	
4. $M^{3+}\rightarrow M^{4+}$	3990	9. $M^{8+}\rightarrow M^{9+}$	
5. $M^{4+}\rightarrow M^{5+}$		10. $M^{9+}\rightarrow M^{10+}$	

Principal lines in atomic spectrum

Wavelength/nm	Species	Sensitivity	Application
356.827	II		
359.260	II		
429.674	I		AA
442.434	II	V1	AE
443.432	II	V2	AE
476.027	I		AA

Abundance: Earth's crust 7.0 p.p.m.; seawater 2×10^{-7} p.p.m.

Biological role: None; toxicity n.a., but probably low

Sc

Atomic number: 21
Relative atomic mass (^{12}C=12.0000): 44.955910

Chemical properties

Soft, silvery-white metal. Tarnishes in air and burns easily. Reacts with water to form hydrogen gas. Forms salts with acids. Little used.

Radii/pm: Sc^{3+} 83; atomic 160.6; covalent 144
Electronegativity: 1.36 (Pauling); 1.20 (Allred-Rochow)
Effective nuclear charge: 3.00 (Slater); 4.63 (Clementi); 6.06 (Froese–Fischer)

Standard reduction potentials $E^{\ominus}$/V

III		0
Sc^{3+}	$\xrightarrow{-2.03}$	Sc

Oxidation states

Sc^{II} (d^1)	$CsScCl_3$
Sc^{III} ([Ar])	Sc^2O_3, ScO . OH, '$Sc(OH)_3$', $Sc(OH)_6^{3-}$, ScF_3, $ScCl_3$, etc., ScF_6^{3-}, complexes

ScH_2 is probably $Sc^{III}H^-$ with complex bonding

Physical properties

Melting point/K: 1814
Boiling point/K: 3104
ΔH_{fusion}/kJ mol^{-1}: 15.9
ΔH_{vap}/kJ mol^{-1}: 376.1

Thermodynamic properties (298.15 K, 0.1 MPa)

State	$\Delta_f H^{\ominus}$/kJ mol^{-1}	$\Delta_f G^{\ominus}$/kJ mol^{-1}	$S^{\ominus}$/J K^{-1} mol^{-1}	C_p/J K^{-1} mol^{-1}
Solid	0	0	34.64	25.52
Gas	377.8	336.03	174.79	22.09

Density/kg m^{-3}: 2989 [273 K]
Thermal conductivity/W m^{-1} K^{-1}: 15.8 [300 K]
Electrical resistivity/Ω m: 61.0×10^{-8} [295 K]
Mass magnetic susceptibility/kg^{-1} m^3: $+8.8 \times 10^{-8}$ (s)
Molar volume/cm^3: 15.04
Coefficient of linear thermal expansion/K^{-1}: 10.0×10^{-6}

Lattice structure (cell dimensions/pm), space group

α-Sc h.c.p. (a=330.90, c=527.3), $P6_3/mmc$
β-Sc cubic, Im3m

$T(\alpha \rightarrow \beta)$=1223 K

X-ray diffraction: mass absorption coefficients $(\mu/\rho)/cm^2\ g^{-1}$:
CuK_α 184 MoK_α 21.1

Discovered in 1879 by L. F. Nilson at Uppsala, Sweden

[Latin, *Scandia* = Scandinavia]

Scandium

Nuclear properties

Atomic number: 21

Thermal neutron capture cross-section/barns: 25±2

Number of isotopes (including nuclear isomers): 15

Isotope mass range: 40→50

Key isotopes

Nuclide	Atomic mass	Natural abundance (%)	Half life $T_{1/2}$	Decay mode and energy (MeV)	Nuclear spin I	Nuclear magnetic moment μ	Uses
^{44}Sc		0	3.92h	β^-(3.647); γ	2	+2.56	tracer
^{45}Sc	44.955 92	100	stable		7/2	+4.7559	NMR
^{46}Sc	45.955	0	83.80d	β^-(2.367); γ	4	+3.03	tracer
^{47}Sc		0	3.43d	β^-(0.60); γ	7/2	+5.34	tracer

NMR

	^{45}Sc
Relative sensitivity (^1H = 1.00)	0.30
Absolute sensitivity (^1H = 1.00)	0.30
Receptivity (^{13}C = 1.00)	1710
Magnetogyric ratio/rad T^{-1} s^{-1}	6.4982×10^7
Quadrupole momentm^2	-0.22×10^{-28}
Frequency (^1H = 100 MHz; 2.3488 T)/MHz	24.290

Reference: $Sc(ClO_4)_3$ (aq)

Electron shell properties

Ground state electron configuration: $[Ar]3d^14s^2$

Term symbol: $^2D_{3/2}$

Electron affinity ($M \to M^-$)/kJ mol^{-1}: −70

Ionization energies/kJ mol^{-1}

1. $M \to M^+$	631	6. $M^{5+} \to M^{6+}$	10 720
2. $M^+ \to M^{2+}$	1235	7. $M^{6+} \to M^{7+}$	13 320
3. $M^{2+} \to M^{3+}$	2389	8. $M^{7+} \to M^{8+}$	15 310
4. $M^{3+} \to M^{4+}$	7089	9. $M^{8+} \to M^{9+}$	17 369
5. $M^{4+} \to M^{5+}$	8844	10. $M^{9+} \to M^{10+}$	21 740

Principal lines in atomic spectrum

Wavelength/nm	Species	Sensitivity	Application
361.384	II	V2	AE
390.749	I	U2	AA, AE
391.181	I	U1	AA, AE
402.040	I	U4	AA, AE
402.369	I	U3	AA, AE
637.882*	I		

* Lowest energy transition from ground state to nearest empty orbital

Abundance: Earth's crust 25 p.p.m.; seawater 4×10^{-5} p.p.m.

Biological role: None; toxicity n.a.; suspected carcinogen

Se

Atomic number: 34

Relative atomic mass (^{12}C=12.0000): 78.96

Chemical properties

Obtained as silvery metallic allotrope or red amorphous powder, which is less stable. Found in sulfide ores. Burns in air, unaffected by water, dissolves in concentrated nitric acid and alkalis. Used in photoelectric cells, photocopiers, solar cells, and semiconductors.

Radii/pm: Se^{4+} 69; atomic 215.2 (grey); covalent 117; van der Waals 200; Se^{2-} 191

Electronegativity: 2.55 (Pauling); 2.48 (Allred-Rochow)

Effective nuclear charge: 6.95 (Slater); 8.29 (Clementi); 9.96 (Froese–Fischer)

Standard reduction potentials $E^\ominus$/V

	VI		IV		0		−II
Acid solution	SeO_4^{2-}	1.1	H_2SeO_3	0.74	Se	−0.11	H_2Se
Alkaline solution	SeO_4^{2-}	0.03	SeO_3^{2-}	−0.36	Se	−0.67	Se^{2-}

Covalent bonds

	r/pm	E/kJ mol^{-1}
Se–H	146	305
Se–C	198	245
Se–O	161	343
Se–F	170	285
Se–Cl	220	245
Se–Se (Se_8)	232	330

Oxidation states

Se^{-II}	H_2Se
Se^{0-I}	Se cluster cations, e.g. Se_4^{2+}, Se_8^{2+}
Se^{I}	Se_2Cl_2, Se_2Br_2
Se^{II}	?
Se^{IV}	SeO_2, H_2SeO_3, SeO_3^{2-} (aq), $SeOF_2$, $SeOCl_4$, $SeOBr_2$, SeF_4, $SeCl_4$, $SeBr_4$, $SeBr_6^{2-}$
Se^{VI}	SeO_3 ? H_2SeO_4, SeO_4^{2-} (aq), SeO_2F_2, SeO_2Cl_2, SeF_6

Physical properties

Melting point/K: 490

Boiling point/K: 958.1

ΔH_{fusion}/kJ mol^{-1}: 5.1

ΔH_{vap}/kJ mol^{-1}: 90

Thermodynamic properties (298.15 K, 0.1 MPa)

State	$\Delta_f H^\ominus$/kJ mol^{-1}	$\Delta_f G^\ominus$/kJ mol^{-1}	$S^\ominus$/J K^{-1} mol^{-1}	C_p/J K^{-1} mol^{-1}
Solid (α)	0	0	42.442	25.363
Gas	227.07	187.03	176.72	20.820

Density/kg m^{-3}: 4790 (grey) [293 K]; 3987 [liquid at m.p.]

Thermal conductivity/W m^{-1} K^{-1}: 2.04 [300 K]

Electrical resistivity/Ω m: 0.01 [293 K]

Mass magnetic susceptibility/kg^{-1} m^3: -4.0×10^{-9} (s)

Molar volume/cm^3: 16.48

Coefficient of linear thermal expansion/K^{-1}: 36.9×10^{-6}

Lattice structure (cell dimensions/pm), space group

Grey hexagonal (a=436.56, c=495.90), $P3_121$, metallic form
α-Se monoclinic (a=906.4, b=907.2, c=115.6, β=90°52′), $P2_1/a$, Se_8
β-Se monoclinic (a=1285, b=807, c=931, β=93°8′), $P2_1/a$, Se_8
α′-Se cubic (a=297.0), Pm3m
β′-Se cubic (a=604), Fd3m

X-ray diffraction: mass absorption coefficients (μ/ρ)/cm^2 g^{-1}: CuK_α 91.4 MoK_α 74.7

Discovered by J. J. Berzelius in 1817 at Stockholm, Sweden

[Greek, *selene* = moon]

Selenium

Nuclear properties

Atomic number: 34

Thermal neutron capture cross-section/barns: 12.2 ± 0.6

Number of isotopes (including nuclear isomers): 20

Isotope mass range: 70→85

Key isotopes

Nuclide	Atomic mass	Natural abundance (%)	Half life $T_{1/2}$	Decay mode and energy (MeV)	Nuclear spin I	Nuclear magnetic moment μ	Uses
^{74}Se	73.9225	0.9	stable		0		
^{75}Se	74.923	0	120.4d	β^-(0.865); γ	5/2		tracer, medical
^{76}Se	75.9192	9.0	stable		0		
^{77}Se	76.9199	7.6	stable		1/2	+0.534	NMR
^{78}Se	77.9173	23.5	stable		0		
^{80}Se	79.9165	49.6	stable		0		
^{82}Se	81.9167	9.4	stable		0		

NMR

	^{77}Se
Relative sensitivity (^1H = 1.00)	6.93×10^{-3}
Absolute sensitivity (^1H = 1.00)	5.25×10^{-4}
Receptivity (^{13}C = 1.00)	2.98
Magnetogyric ratio/rad T^{-1} s^{-1}	5.1018×10^7
Frequency (^1H = 100 MHz; 2.3488 T)/MHz	19.092

Reference: $S(CH_3)_2$

Electron shell properties

Ground state electron configuration: $[Ar]3d^{10}4s^24p^4$

Term symbol: $^3P_2$

Electron affinity ($M \rightarrow M^-$)/kJ mol^{-1}: 195.0

Ionization energies/kJ mol^{-1}

1. $M \rightarrow M^+$	940.9	6. $M^{5+} \rightarrow M^{6+}$	7883
2. $M^+ \rightarrow M^{2+}$	2044	7. $M^{6+} \rightarrow M^{7+}$	14990
3. $M^{2+} \rightarrow M^{3+}$	2974	8. $M^{7+} \rightarrow M^{8+}$	(19500)
4. $M^{3+} \rightarrow M^{4+}$	4144	9. $M^{8+} \rightarrow M^{9+}$	(23300)
5. $M^{4+} \rightarrow M^{5+}$	6590	10. $M^{9+} \rightarrow M^{10+}$	(27200)

Principal lines in atomic spectrum

Wavelength/nm	Species	Sensitivity	Application
196.026	I		AA
203.985	I	U2	AA, AE
206.279	I	U3	AA, AE
207.479*	I		
1032.726			

* Lowest energy transition from ground state to nearest empty orbital

Abundance: Earth's crust 0.05 p.p.m.; seawater 0.004 p.p.m.

Biological role: Essential; toxic in excess of dietary requirement; stimulatory; carcinogenic; teratogenic

Si

Atomic number: 14
Relative atomic mass (^{12}C=12.0000): 28.0855

Chemical properties

Black amorphous Si obtained by reduction of sand (SiO_2) with carbon; ultrapure semiconductor grade crystals are blue–grey metallic. Bulk Si unreactive towards oxygen, water, acids (except HF), but dissolves in hot alkali. Used in semiconductors, alloys, polymers.

Radii/pm: Si^{4+} 26; atomic 117; covalent 117; van der Waals 200; Si^{4-} 271

Electronegativity: 1.90 (Pauling); 1.74 (Allred-Rochow)

Effective nuclear charge: 4.15 (Slater); 4.29 (Clementi); 4.48 (Froese–Fischer)

Standard reduction potentials $E^{\ominus}$/V

	IV		II		0		−IV
Acid solution	SiO_2	$\xrightarrow{-0.967}$	'SiO'	$\xrightarrow{-0.808}$	Si	$\xrightarrow{-0.143}$	SiH_4

[Alkaline solutions contain many different forms]

Covalent bonds

	r/pm	E/kJ mol^{-1}
Si–H	148.0	326
Si–C	187	301
Si–O	151	486
Si–F	155	582
Si–Cl	202	391
Si–Si	232	226

Oxidation states

Si^{II}	SiF_2 (gas)
Si^{IV}	SiO_2, 'H_4SiO_4', silicates, zeolites etc., SiH_4 etc., SiF_4, $SiCl_4$ etc., SiF_6^{2-}, metal silicides, e.g. Ca_2Si, CaSi, organosilicon compounds

Physical properties

Melting point/K: 1683
Boiling point/K: 2628
ΔH_{fusion}/kJ mol^{-1}: 39.6
ΔH_{vap}/kJ mol^{-1}: 383.3

Thermodynamic properties (298.15 K, 0.1 MPa)

State	$\Delta_f H^{\ominus}$/kJ mol^{-1}	$\Delta_f G^{\ominus}$/kJ mol^{-1}	$S^{\ominus}$/J K^{-1} mol^{-1}	C_p/J K^{-1} mol^{-1}
Solid	0	0	18.83	20.00
Gas	455.6	411.3	167.97	22.251

Density/kg m^{-3}: 2329 [293 K]; 2525 [liquid at m.p.]
Thermal conductivity/W m^{-1} K^{-1}: 148 [300 K]
Electrical resistivity/Ω m: 0.001 [273 K]
Mass magnetic susceptibility/kg^{-1} m^3: -1.8×10^{-9} (s)
Molar volume/cm^3: 12.06
Coefficient of linear thermal expansion/K^{-1}: 4.2×10^{-6}

Lattice structure (cell dimensions/pm), space group

Cubic (a=543.07), Fd3m, diamond structure
High pressure forms: (a=468.6, c=258.5), $I4_1$/amd
(a=664), Ia3
(a=380, c=628) $P6_3mc$

X-ray diffraction: mass absorption coefficients (μ/ρ)/cm^2 g^{-1}:
CuK_α 60.6 MoK_α 6.44

Discovered in 1824 by J. J. Berzelius at Stockholm, Sweden

[Latin, *silicis* = flint]

Silicon

Nuclear properties

Atomic number: 14

Thermal neutron capture cross-section/barns: 0.160 ± 0.020

Number of isotopes (including nuclear isomers): 8

Isotope mass range: 25→31

Key isotopes

Nuclide	Atomic mass	Natural abundance (%)	Half life $T_{1/2}$	Decay mode and energy (MeV)	Nuclear spin I	Nuclear magnetic moment μ	Uses
^{28}Si	27.97693	92.23	stable		0		
^{29}Si	28.97649	4.67	stable		1/2	−0.55526	NMR
^{30}Si	29.97376	3.10	stable		0		
^{32}Si	31.9740	0	*c.* 650y	β^-(0.21); no γ			tracer

NMR

	^{29}Si
Relative sensitivity ($^1H = 1.00$)	7.84×10^{-3}
Absolute sensitivity ($^1H = 1.00$)	3.69×10^{-4}
Receptivity ($^{13}C = 1.00$)	2.09
Magnetogyric ratio/rad T^{-1} s^{-1}	-5.3146×10^7
Frequency ($^1H = 100$ MHz; 2.3488 T)/MHz	19.865

Reference: $Si(CH_3)_4$

Electron shell properties

Ground state electron configuration: $[Ne]3s^23p^2$

Term symbol: $^3P_0$

Electron affinity ($M \rightarrow M^-$)/kJ mol^{-1}: 133.6

Ionization energies/kJ mol^{-1}

1. $M \rightarrow M^+$	786.5	6. $M^{5+} \rightarrow M^{6+}$	19784
2. $M^+ \rightarrow M^{2+}$	1577.1	7. $M^{6+} \rightarrow M^{7+}$	23786
3. $M^{2+} \rightarrow M^{3+}$	3231.4	8. $M^{7+} \rightarrow M^{8+}$	29252
4. $M^{3+} \rightarrow M^{4+}$	4355.5	9. $M^{8+} \rightarrow M^{9+}$	33876
5. $M^{4+} \rightarrow M^{5+}$	16091	10. $M^{9+} \rightarrow M^{10+}$	38732

Principal lines in atomic spectrum

Wavelength/nm	Species	Sensitivity	Application
250.690	I	U4	AA, AE
251.431*	I		AA
251.611	I	U3	AA, AE
252.851	I	U2	AA, AE
288.151	I	U1	AE
other lines of equal intensity at higher wavelengths			

* Lowest energy transition from ground state to nearest empty orbital

Abundance: Earth's crust 272000 p.p.m.; seawater 3.0 p.p.m.

Biological role: Essential; non-toxic but some silicates are carcinogenic (e.g. asbestos)

Ag

Atomic number: 47
Relative atomic mass (^{12}C=12.0000): 107.8682

Chemical properties

Soft, malleable metal with characteristic silver sheen. Stable to water and oxygen but attacked by sulfur compounds in air to form black sulfide layer. Dissolves in sulfuric and nitric acid. Used in photography, silverware, jewellery, electrical industry, and glass (mirrors).

Radii/pm: Ag^+ 113; Ag^{2+} 89; atomic 144.4; covalent 134
Electronegativity: 1.93 (Pauling); 1.42 (Allred-Rochow)
Effective nuclear charge: 4.20 (Slater); 8.03 (Clementi); 11.35 (Froese–Fischer)

Standard reduction potentials $E^{\ominus}$/V

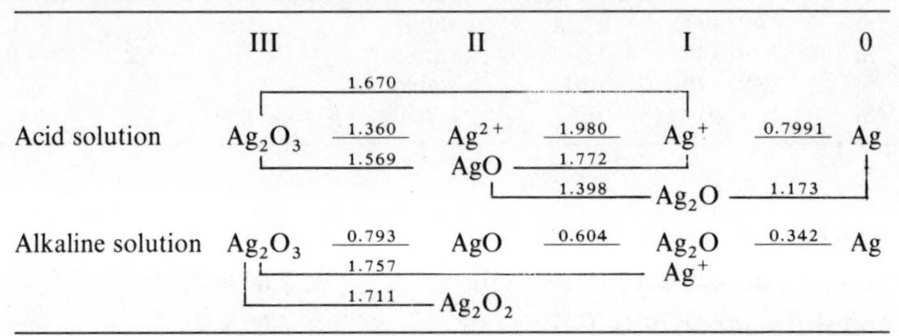

Oxidation states

Ag^0 ($d^{10}s^1$)	rare $Ag(CO)_3$ at 10 K
Ag (d^{10})	Ag_2O, $Ag(OH)_2^-$ (aq), $Ag(H_2O)_4^+$ (aq), AgF, AgCl etc., Ag^+ salts, e.g. $AgNO_3$, Ag_2S, $Ag(CN)_2^-$, and other complexes
Ag^{II} (d^9)	AgF_2, $[Ag(C_5H_5N)_2]^+$, AgO is not Ag^{II} but $Ag^IAg^{III}O_2$
Ag^{III} (d^8)	rare AgF_4^-, AgF_6^{3-}

Physical properties

Melting point/K: 1235.08
Boiling point/K: 2485
ΔH_{fusion}/kJ mol^{-1}: 11.3
ΔH_{vap}/kJ mol^{-1}: 257.7

Thermodynamic properties (298.15 K, 0.1 MPa)

State	$\Delta_f H^{\ominus}$/kJ mol^{-1}	$\Delta_f G^{\ominus}$/kJ mol^{-1}	$S^{\ominus}$/J K^{-1} mol^{-1}	C_p/J K^{-1} mol^{-1}
Solid	0	0	42.55	25.351
Gas	284.55	245.65	172.997	20.786

Density/kg m^{-3}: 10 500 [293 K]; 9345 [liquid at m.p.]
Thermal conductivity/W m^{-1} K^{-1}: 429 [300 K]
Electrical resistivity/Ω m: 1.59×10^{-8} [293 K]
Mass magnetic susceptibility/kg^{-1} m^3: -2.27×10^{-9} (s)
Molar volume/cm^3: 10.27
Coefficient of linear thermal expansion/K^{-1}: 19.2×10^{-6}

Lattice structure (cell dimensions/pm), space group

f.c.c. (a=408.626), Fm3m)

X-ray diffraction: mass absorption coefficients (μ/ρ)/cm^2 g^{-1}:
CuK_α 218 MoK_α 25.8

Known to ancient civilizations

[Anglo-Saxon, *siolfur* = silver; Latin, *argentum*]

Silver

Nuclear properties

Atomic number: 47
Thermal neutron capture cross-section/barns: 63.8 ± 0.6
Number of isotopes (including nuclear isomers): 27
Isotope mass range: $102 \rightarrow 117$

Key isotopes

Nuclide	Atomic mass	Natural abundance (%)	Half life $T_{1/2}$	Decay mode and energy (MeV)	Nuclear spin I	Nuclear magnetic moment μ	Uses
^{107}Ag	106.905 09	51.83	stable		1/2	−0.1135	NMR
^{109}Ag	108.9047	48.17	stable		1/2	−0.1305	NMR
^{110m}Ag		0	253d	β^-(*c.* 2.9); γ	6	+3.604	tracer
^{111}Ag		0	7.5d	β^-(1.05); γ	1/2	−0.145	tracer

NMR

	^{107}Ag	^{109}Ag
Relative sensitivity ($^1H = 1.00$)	6.62×10^{-5}	1.01×10^{-4}
Absolute sensitivity ($^1H = 1.00$)	3.43×10^{-5}	4.86×10^{-5}
Receptivity ($^{13}C = 1.00$)	0.195	0.276
Magnetogyric ratio/rad T^{-1} s^{-1}	-1.0828×10^7	-1.2448×10^7
Frequency ($^1H = 100$ MHz; 2.3488 T)/MHz	4.046	4.652

Reference: Ag^+ (aq)

Electron shell properties

Ground state electron configuration: $[Kr]4d^{10}5s^1$
Term symbol: $^2S_{1/2}$
Electron affinity ($M \rightarrow M^-$)/kJ mol^{-1}: 125.7

Ionization energies/kJ mol^{-1}

1. $M \rightarrow M^+$	731.0		6. $M^{5+} \rightarrow M^{6+}$	(8 600)
2. $M^+ \rightarrow M^{2+}$	2073		7. $M^{6+} \rightarrow M^{7+}$	(11 200)
3. $M^{2+} \rightarrow M^{3+}$	3361		8. $M^{7+} \rightarrow M^{8+}$	(13 400)
4. $M^{3+} \rightarrow M^{4+}$	(5000)		9. $M^{8+} \rightarrow M^{9+}$	(15 600)
5. $M^{4+} \rightarrow M^{5+}$	(6700)		10. $M^{9+} \rightarrow M^{10+}$	(18 000)

Principal lines in atomic spectrum

Wavelength/nm	Species	Sensitivity	Application
328.068	I	U1	AA, AE
338.289*	I	U2	AA, AE
520.907	I	U3	AE
546.549	I	U4	AE

* Lowest energy transition from ground state to nearest empty orbital

Abundance: Earth's crust 0.08 p.p.m.; seawater 0.0003 p.p.m.
Biological role: None; toxic to lower organisms but not animals; suspected carcinogen

Na

Atomic number: 11
Relative atomic mass ($^{12}C=12.0000$): **22.989 768**

Chemical properties

Soft, silvery white metal which oxidizes rapidly when cut. Reacts vigorously with water. Produced in large quantities and used as metal in heat exchangers in atomic reactors. Large deposits of NaCl and trona. NaCl is key industrial chemical, used to make Cl_2, NaOH, etc.

Radii/pm: Na^+ 98; atomic 153.7; van der Waals 231
Electronegativity: 0.93 (Pauling); 1.01 (Allred-Rochow)
Effective nuclear charge: 2.20 (Slater); 2.51 (Clementi); 3.21 (Froese–Fischer)

Standard reduction potentials $E^\ominus$/V

I		0
Na^+	$\xrightarrow{-2.713}$	Na

Oxidation states

Na^{-I} (s^2)	Na metal solutions in liquid NH_3
Na^I ([Ne])	Na_2O, Na_2O_2 (peroxide), NaOH, NaH, NaF, NaCl etc., $[Na(H_2O)_4]^+$ (aq), $NaHCO_3$, Na_2CO_3, Na^+ salts, some complexes

Physical properties

Melting point/K: 370.96
Boiling point/K: 1156.1
ΔH_{fusion}/kJ mol^{-1}: 2.64
ΔH_{vap}/kJ mol^{-1}: 99.2

Thermodynamic properties (298.15 K, 0.1 MPa)

State	$\Delta_f H^\ominus$/kJ mol^{-1}	$\Delta_f G^\ominus$/kJ mol^{-1}	$S^\ominus$/J K^{-1} mol^{-1}	C_p/J K^{-1} mol^{-1}
Solid	0	0	51.21	28.24
Gas	107.32	76.761	153.712	20.786

Density/kg m^{-3}: 971 [293 K]; 928 [liquid at m.p.]
Thermal conductivity/W m^{-1} K^{-1}: 141 [300 K]
Electrical resistivity/Ω m: 4.2×10^{-8} [273 K]
Mass magnetic susceptibility/kg^{-1} m^3: $+8.8\times10^{-9}$ (s)
Molar volume/cm^3: 23.68
Coefficient of linear thermal expansion/K^{-1}: 70.6×10^{-6}

Lattice structure (cell dimensions/pm), space group

α-Na hexagonal ($a=376.7$, $c=615.4$), $P6_3/mmc$
β-Na b.c.c. ($a=429.06$), Im3m
T(b.c.c.→hexagonal)=5 K

X-ray diffraction: mass absorption coefficients (μ/ρ)/cm^2 g^{-1}:
CuK_α 30.1 MoK_α 3.21

Isolated by Sir Humphry Davy in 1807 at the Royal Institution, London, UK

[English, soda; Latin, *natrium*]

Sodium

Nuclear properties

Atomic number: 11

Thermal neutron capture cross-section/barns: 0.534 ± 0.005

Number of isotopes (including nuclear isomers): 7

Isotope mass range: 20→26

Key isotopes

Nuclide	Atomic mass	Natural abundance (%)	Half life $T_{1/2}$	Decay mode and energy (MeV)	Nuclear spin I	Nuclear magnetic moment μ	Uses
^{22}Na	21.9944	0	2.602y	β^+(1.82); EC; γ	3	+1.746	tracer
^{23}Na	22.9898	100	stable		3/2	+2.217 40	NMR
^{24}Na	23.990 96	0	15.0h	β^-(4.17); γ	4	+1.690	tracer, medical

NMR

	^{23}Na
Relative sensitivity (^{1}H = 1.00)	9.25×10^{-2}
Absolute sensitivity (^{1}H = 1.00)	9.25×10^{-2}
Receptivity (^{13}C = 1.00)	525
Magnetogyric ratio/rad T^{-1} s^{-1}	7.0761×10^7
Quadrupole moment/m^2	0.12×10^{-28}
Frequency (^{1}H = 100 MHz; 2.3488 T)/MHz	26.451
Reference: NaCl (aq)	

Electron shell properties

Ground state electron configuration: [Ne]3s^1

Term symbol: $^2S_{1/2}$

Electron affinity ($M \rightarrow M^-$)/kJ mol^{-1}: 52.9

Ionization energies/kJ mol^{-1}

1. $M \rightarrow M^+$	495.8		6. $M^{5+} \rightarrow M^{6+}$	16 610
2. $M^+ \rightarrow M^{2+}$	4 562.4		7. $M^{6+} \rightarrow M^{7+}$	20 114
3. $M^{2+} \rightarrow M^{3+}$	6 912		8. $M^{7+} \rightarrow M^{8+}$	25 490
4. $M^{3+} \rightarrow M^{4+}$	9 543		9. $M^{8+} \rightarrow M^{9+}$	28 933
5. $M^{4+} \rightarrow M^{5+}$	13 353		10. $M^{9+} \rightarrow M^{10+}$	141 360

Principal lines in atomic spectrum

Wavelength/nm	Species	Sensitivity	Application
330.232	I	U3	AA, AE
330.299	I	U4	AA, AE
568.824	I		AE
588.995	I	U1	AA, AE
589.592*	I	U2	AE

* Lowest energy transition from ground state to nearest empty orbital

Abundance: Earth's crust 22 700 p.p.m.; seawater 10 600 p.p.m.

Biological role: Essential; non-toxic

Sr

Atomic number: 38

Relative atomic mass (^{12}C=12.0000): 87.62

Chemical properties

Silvery white, soft metal obtained by high temperature reduction of SrO with Al. Protected as bulk metal by oxide film but will burn in air and reacts with water. Used in fireworks and flares to give red colour. Found as celestite, $SrSO_4$.

Radii/pm: Sr^{2+} 127; atomic 215.1 (α form); covalent 192

Electronegativity: 0.95 (Pauling); 0.99 (Allred-Rochow)

Effective nuclear charge: 2.85 (Slater); 6.07 (Clementi); 8.09 (Froese–Fischer)

Standard reduction potentials $E^{\ominus}$/V

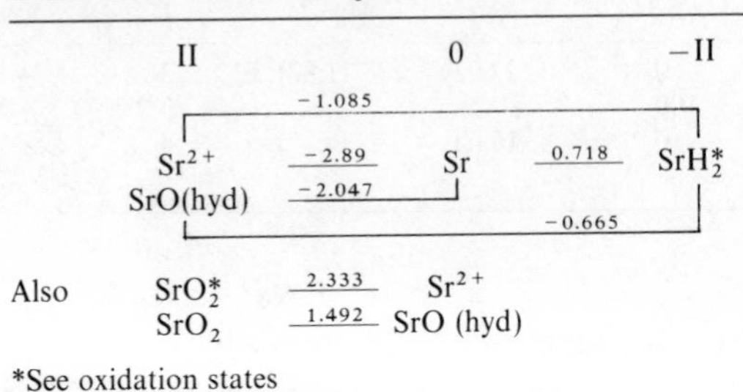

Also SrO_2^* —2.333— Sr^{2+}
SrO_2 —1.492— SrO (hyd)

*See oxidation states

Oxidation states

Sr^{II} ([Kr])	SrO, SrO_2 (peroxide), $Sr(OH)_2$, $[Sr(H_2O)_x]^{2+}$ (aq), Sr^{2+} salts, SrF_2, $SrCl_2$ etc., $SrCO_3$, some complexes SrH_2 is $Sr^{2+}2H^-$

Physical properties

Melting point/K: 1042

Boiling point/K: 1657

ΔH_{fusion}/kJ mol^{-1}: 9.16

ΔH_{vap}/kJ mol^{-1}: 154.4

Thermodynamic properties (298.15 K, 0.1 MPa)

State	$\Delta_f H^{\ominus}$/kJ mol^{-1}	$\Delta_f G^{\ominus}$/kJ mol^{-1}	$S^{\ominus}$/J K^{-1} mol^{-1}	C_p/J K^{-1} mol^{-1}
Solid	0	0	52.3	26.4
Gas	164.4	130.9	164.62	20.786

Density/kg m^{-3}: 2540 [293 K]; 2375 [liquid at m.p.]

Thermal conductivity/W m^{-1} K^{-1}: 35.3 [300 K]

Electrical resistivity/Ω m: 23.0×10^{-8} [293 K]

Mass magnetic susceptibility/kg^{-1} m^3: $+1.32 \times 10^{-8}$ (s)

Molar volume/cm^3: 34.50

Coefficient of linear thermal expansion/K^{-1}: 23×10^{-6}

Lattice structure (cell dimensions/pm), space group

α-Sr f.c.c. (a=608.49), Fm3m
β-Sr h.c.p. (a=432, c=706), $P6_3/mmc$
γ-Sr b.c.c. (a=485), Im3m

$T(\alpha \rightarrow \beta)$=506 K
$T(\beta \rightarrow \gamma)$=813 K

X-ray diffraction: mass absorption coefficients (μ/ρ)/cm^2 g^{-1}:
CuK_α 125 MoK_α 95.0

Recognized as an element in 1790 by A. Crawford at Edinburgh, Scotland. Isolated in 1808 by Sir Humphry Davy at London, UK

[Named after Strontian, Scotland]

Strontium

Nuclear properties

Atomic number: 38

Thermal neutron capture cross-section/barns: 1.21 ± 0.06

Number of isotopes (including nuclear isomers): 18

Isotope mass range: 80→95

Key isotopes

Nuclide	Atomic mass	Natural abundance (%)	Half life $T_{1/2}$	Decay mode and energy (MeV)	Nuclear spin I	Nuclear magnetic moment μ	Uses
^{82}Sr		0	25d	EC(*c.* 0.4); γ			tracer
^{84}Sr	83.9134	0.56	stable				
^{85}Sr	84.913	0	64d	EC(1.11); γ			tracer medical
^{86}Sr	85.9094	9.86	stable		0		
^{87}Sr	86.9089	7.00	stable		9/2	−1.093	NMR
^{88}Sr	87.9056	82.58	stable		0		
^{89}Sr	88.907	0	52.7d	β^-(1.463); γ			
^{90}Sr	89.907	0	28.1y	β^-(0.546); no γ			tracer, medical

NMR

	^{87}Sr
Relative sensitivity (^1H = 1.00)	2.69×10^{-3}
Absolute sensitivity (^1H = 1.00)	1.88×10^{-4}
Receptivity (^{13}C = 1.00)	1.07
Magnetogyric ratio/rad T^{-1} s^{-1}	-1.1593×10^7
Quadrupole moment/m^2	0.36×10^{-28}
Frequency (^1H = 100 MHz; 2.3488 T)/MHz	4.333

Reference: Sr^{2+} (aq)

Electron shell properties

Ground state electron configuration: [Kr]$5s^2$

Term symbol: $^1S_0$

Electron affinity ($M \rightarrow M^-$)/kJ mol^{-1}: −146

Ionization energies/kJ mol^{-1}

1. $M \rightarrow M^+$	549.5	6. $M^{5+} \rightarrow M^{6+}$	8 760
2. $M^+ \rightarrow M^{2+}$	1064.2	7. $M^{6+} \rightarrow M^{7+}$	10 200
3. $M^{2+} \rightarrow M^{3+}$	4210	8. $M^{7+} \rightarrow M^{8+}$	11 800
4. $M^{3+} \rightarrow M^{4+}$	5500	9. $M^{8+} \rightarrow M^{9+}$	15 600
5. $M^{4+} \rightarrow M^{5+}$	6910	10. $M^{9+} \rightarrow M^{10+}$	17 100

Principal lines in atomic spectrum

Wavelength/nm	Species	Sensitivity	Application
407.771	II	V1	AE
460.733	I	U1	AA, AE
483.208	I	U2	AE
487.249	I	U3	AE
689.258*	I		

* Lowest energy transition from ground state to nearest empty orbital

Abundance: Earth's crust 384 p.p.m.; seawater 8 p.p.m.

Biological role: None; non-toxic; mimics calcium

S

Atomic number: 16
Relative atomic mass (^{12}C=12.0000): 32.066

Chemical properties

Several allotropes, of which orthorhombic S_8 is most stable. Stable to air and water but burns if ignited. Attacked by oxidizing acids. Found as native sulfur and as metal sulfide ores. Recovered from H_2S in natural gas. Key industrial chemical.

Radii/pm: S^{4+} 37; S^{6+} 29; atomic 104 (S_8); covalent 104: van der Waals 185

Electronegativity: 2.58 (Pauling); 2.44 (Allred-Rochow)

Effective nuclear charge: 5.45 (Slater); 5.48 (Clementi); 6.04 (Froese–Fischer)

Standard reduction potentials $E^{\ominus}$/V

	VI	V	IV	III	II*	0	−II
Acid solution	SO_4^{2-}	$\xrightarrow{-0.07}$ $S_2O_6^{2-}$	$\xrightarrow{0.57}$ H_2SO_3	$\xrightarrow{-0.07}$ $HS_2O_4^-$	$\xrightarrow{0.87}$ $S_2O_3^{2-}$	$\xrightarrow{0.60}$ S	$\xrightarrow{0.14}$ H_2S

(Acid solution: SO_4^{2-} → H_2SO_3 0.16; H_2SO_3 → $S_2O_3^{2-}$ 0.40; H_2SO_3 → S 0.50)

Also: $S_2O_8^{2-}$ $\xrightarrow{0.25}$ SO_4^-

Alkaline solution: SO_4^{2-} $\xrightarrow{-0.94}$ SO_3^{2-} $\xrightarrow{-0.58}$ $S_2O_3^{2-}$ $\xrightarrow{-0.74}$ S $\xrightarrow{-0.45}$ S^{2-} (SO_3^{2-} → S −0.66)

*Average oxidation state

Covalent bonds

Bond	r/pm	E/kJ mol^{-1}
S—H	133.5	347
S—C	182	272
S=C	155	476
S—O	150	265
S=O	144	c. 525
S—F	156	328
S—Cl	199	255
S—S	208	226

Oxidation states

S^{-II}	H_2S, S^{2-} (polysulfides S_n^{2-})
S^{-I}	H_2S_2, etc. (polysulfides S_n^{2-})
S^0	S_6, S_8, etc. (polysulfides S_n^{2-})
S^I	S_2O?, S_2F_2, S_2Cl_2
S^{II}	SF_2, SCl_2
S^{III}	$Na_2S_2O_4$
S^{IV}	SO_2, SO_3^{2-} (aq), SF_4, SCl_4, $SOCl_2$
S^V	$Na_2S_2O_6$, S_2F_{10}
S^{VI}	SO_3, H_2SO_4, SO_4^{2-} (aq), etc., SF_6, HSO_3F, SO_2Cl_2

Physical properties

Melting point/K: 386.0 (α) 392.2 (β) 380.0 (γ)

Boiling point/K: 717.824

ΔH_{fusion}/kJ mol^{-1}: 1.23 **ΔH_{vap}**/kJ mol^{-1}: 9.62

Thermodynamic properties (298.15 K, 0.1 MPa)

State	$\Delta_f H^{\ominus}$/kJ mol^{-1}	$\Delta_f G^{\ominus}$/kJ mol^{-1}	$S^{\ominus}$/J K^{-1} mol^{-1}	C_p/J K^{-1} mol^{-1}
Solid (α)	0	0	31.80	22.64
Solid (β)	0.33	n.a.	n.a.	n.a.
Gas	278.805	238.250	167.821	23.673

Density/kg m^{-3}: 2070 (α), 1957 (β) [293 K]; 1819 [liquid, 393 K]

Thermal conductivity/W m^{-1} K^{-1}: 0.269 (α) [300 K]

Electrical resistivity/Ω m: 2×10^{15} [293 K]

Mass magnetic susceptibility/kg^{-1} m^3: -6.09×10^{-9} (α); -5.83×10^{-9} (β)

Molar volume/cm^3: 15.49

Coefficient of linear thermal expansion/K^{-1}: 74.33×10^{-6}

Lattice structure (cell dimensions/pm), space group

α-S_8 orthorhombic (a=1046.46, b=1286.60, c=2448.60), Fddd
β-S_8 monoclinic (a=1102, b=1096, c=1090, α=96.7°), $P2_1/c$
γ-S_8 monoclinic (a=857, b=1305, c=823, α=112°54′), P2/c
ε-S_6 rhombohedral (a=646, α=115°18′), $R\bar{3}$
In addition to the above ring forms there are also S_7, S_{9-12}, S_{18} and S_{20} rings. Plastic sulfur is long chains of Sn also known in several forms χ, ψ, φ, μ and ω.
$T(\alpha\rightarrow\beta)$=366.7 K

X-ray diffraction: mass absorption coefficients (μ/ρ)/cm^2 g^{-1}:
CuK_α 89.1 MoK_α 9.55

Known to ancient civilizations

[Sanskrit, *sulvere* = sulfur: Latin *sulphurium*]

Sulfur

Nuclear properties

Atomic number: 16
Thermal neutron capture cross-section/barns: 0.51
Number of isotopes (including nuclear isomers): 10
Isotope mass range: 29→38

Key isotopes

Nuclide	Atomic mass	Natural abundance (%)	Half life $T_{1/2}$	Decay mode and energy (MeV)	Nuclear spin I	Nuclear magnetic moment μ	Uses
^{32}S	31.972 07	95.02	stable		0		
^{33}S	32.971 46	0.75	stable		3/2	+0.6435	NMR
^{34}S	33.967 86	4.21	stable		0		
^{35}S	34.969 0	0	87.9d	β^-(0.1674); no γ	3/2	±1.00	tracer
^{36}S	35.967 09	0.02	stable		0		

NMR

	^{33}S
Relative sensitivity (^1H=1.00)	2.26×10^{-3}
Absolute sensitivity (^1H=1.00)	1.72×10^{-5}
Receptivity (^{13}C=1.00)	0.0973
Magnetogyric ratio/rad T^{-1} s^{-1}	2.0534×10^7
Quadrupole moment/m^2	-0.05×10^{-28}
Frequency (^1H=100 MHz; 2.3488 T)/MHz	7.670

Reference: CS_2

Electron shell properties

Ground state electron configuration: $[Ne]3s^23p^4$
Term symbol: $^3P_2$
Electron affinity ($M \rightarrow M^-$)/kJ mol^{-1}: 200.4

Ionization energies/kJ mol^{-1}

1.	$M \rightarrow M^+$	999.6	6. $M^{5+} \rightarrow M^{6+}$	8 495
2.	$M^+ \rightarrow M^{2+}$	2251	7. $M^{6+} \rightarrow M^{7+}$	27 106
3.	$M^{2+} \rightarrow M^{3+}$	3361	8. $M^{7+} \rightarrow M^{8+}$	31 669
4.	$M^{3+} \rightarrow M^{4+}$	4564	9. $M^{8+} \rightarrow M^{9+}$	36 578
5.	$M^{4+} \rightarrow M^{5+}$	7013	10. $M^{9+} \rightarrow M^{10+}$	43 138

Principal lines in atomic spectrum

Wavelength/nm	Species	Sensitivity	Application
191.468* (vac)	I		
469.413	I	U7	AE
921.291	I	U4	AE
922.811	I	U5	AE
923.749	I	U6	AE

* Lowest energy transition from ground state to nearest empty orbital

Abundance: Atmosphere SO_2 up to 1 p.p.m., H_2S 0.0001 p.p.m. (volume); Earth's crust 340 p.p.m.; seawater 900 p.p.m.
Biological role: Essential; non-toxic as element or sulfate

Ta

Atomic number: 73
Relative atomic mass (^{12}C=12.0000): **180.9479**

Chemical properties

Shiny, silvery metal, soft when pure. Found as tantalite (Fe,Mn)Ta_2O_6 but mostly obtained as by-product of tin extraction. Very corrosion resistant due to oxide film; attacked by HF and fused alkalis. Used in chemical plants and surgery.

Radii/pm: Ta^{3+} 72; Ta^{4+} 68; Ta^{5+} 64; atomic 143; covalent 134

Electronegativity: 1.5 (Pauling); 1.33 (Allred-Rochow)

Effective nuclear charge: 3.30 (Slater); 9.53 (Clementi); 13.78 (Froese–Fischer)

Standard reduction potentials $E^{\ominus}$/V

V		0
Ta_2O_5	$\xrightarrow{-0.81}$	Ta

Oxidation states

Ta^{-III}	(d^8)	$[Ta(CO)_5]^{3-}$
Ta^{-I}	(d^6)	$[Ta(CO)_5]^-$
Ta^{I}	(d^4)	$[(C_5H_5)Ta(CO)_4]$
Ta^{II}	(d^3)	TaO?
Ta^{III}	(d^2)	TaF_3, $TaCl_3$, $TaBr_3$
Ta^{IV}	(d^1)	TaO_2, $TaCl_4$, $TaBr_4$, TaI_4
Ta^{V}	(d^0, f^{14})	Ta_2O_5, $[Ta_6O_{19}]^{8-}$ (aq), TaF_5, $TaCl_5$, TaF_6^-, TaF_7^{2-}, $TaOF_3$, TaO_2F

Physical properties

Melting point/K: 3269
Boiling point/K: 5698±100
ΔH_{fusion}/kJ mol^{-1}: 31.4
ΔH_{vap}/kJ mol^{-1}: 758.22

Thermodynamic properties (298.15 K, 0.1 MPa)

State	$\Delta_f H^{\ominus}$/kJ mol^{-1}	$\Delta_f G^{\ominus}$/kJ mol^{-1}	$S^{\ominus}$/J K^{-1} mol^{-1}	C_p/J K^{-1} mol^{-1}
Solid	0	0	41.51	25.36
Gas	782.0	739.3	185.214	20.857

Density/kg m^{-3}: 16 654 [293 K]; 15 000 [liquid at m.p.]
Thermal conductivity/W m^{-1} K^{-1}: 57.5 [300 K]
Electrical resistivity/Ω m: 12.45×10^{-8} [298 K]
Mass magnetic susceptibility/kg^{-1} m^3: $+1.07 \times 10^{-8}$ (s)
Molar volume/cm^3: 10.87
Coefficient of linear thermal expansion/K^{-1}: 6.6×10^{-6}

Lattice structure (cell dimensions/pm), space group

b.c.c. (a=330.29), Im3m

X-ray diffraction: mass absorption coefficients (μ/ρ)/cm^2 g^{-1}:
CuK_α 166 MoK_α 95.4

Discovered in 1802 by A. G. Ekeberg at Uppsala, Sweden

[Greek, *Tantalos* = father of Niobe]

Tantalum

Nuclear properties

Atomic number: 73

Thermal neutron capture cross-section/barns: 22 ± 1

Number of isotopes (including nuclear isomers): 18

Isotope mass range: 172→186

Key isotopes

Nuclide	Atomic mass	Natural abundance (%)	Half life $T_{1/2}$	Decay mode and energy (MeV)	Nuclear spin I	Nuclear magnetic moment μ	Uses
^{180}Ta	179.9415	0.012	$>1 \times 10^{12}$y	EC, β^-			
^{181}Ta	180.9480	99.988	stable		7/2	+2.35	NMR
^{182}Ta		0	115.1d	β^-(1.811); γ			tracer

NMR

	^{181}Ta
Relative sensitivity (^{1}H = 1.00)	3.60×10^{-2}
Absolute sensitivity (^{1}H = 1.00)	3.60×10^{-2}
Receptivity (^{13}C = 1.00)	204
Magnetogyric ratio/rad T^{-1} s^{-1}	3.2073×10^7
Quadrupole moment /m^2	3×10^{-28}
Frequency (^{1}H = 100 MHz; 2.3488 T)/MHz	11.970

Reference: TaF_6^- (aq)

Electron shell properties

Ground state electron configuration: [Xe]$4f^{14}5d^36s^2$

Term symbol: $^4F_{3/2}$

Electron affinity ($M \rightarrow M^-$)/kJ mol^{-1}: 14

Ionization energies/kJ mol^{-1}

1. $M \rightarrow M^+$ 761		6. $M^{5+} \rightarrow M^{6+}$
2. $M^+ \rightarrow M^{2+}$ (1500)		7. $M^{6+} \rightarrow M^{7+}$
3. $M^{2+} \rightarrow M^{3+}$ (2100)		8. $M^{7+} \rightarrow M^{8+}$
4. $M^{3+} \rightarrow M^{4+}$ (3200)		9. $M^{8+} \rightarrow M^{9+}$
5. $M^{4+} \rightarrow M^{5+}$ (4300)		10. $M^{9+} \rightarrow M^{10+}$

Principal lines in atomic spectrum

Wavelength/nm	Species	Sensitivity	Application
265.327	I		
260.863	I		AE
271.467	I		AA
331.116	I	U1	AE
540.251*	I		

* Lowest energy transition from ground state to nearest empty orbital

Abundance: Earth's crust 1.7 p.p.m.; seawater <0.0025 p.p.m.

Biological role: None; non-toxic

Tc

Atomic number: 43
Relative atomic mass (^{12}C=12.0000): 98.9062

Chemical properties

Radioactive metal produced in tonne quantities from fission products of uranium nuclear fuel. Silvery as metal but usually obtained as grey powder. Resists oxidation, slowly tarnishes in moist air, burns in oxygen, dissolves in nitric and sulfuric acids.

Radii/pm: Tc^{2+} 95; Tc^{4+} 72; Tc^{7+} 56; atomic 135.8
Electronegativity: 1.9 (Pauling); 1.36 (Allred-Rochow)
Effective nuclear charge: 3.60 (Slater); 7.23 (Clementi); 10.28 (Froese–Fischer)

Standard reduction potentials E°/V

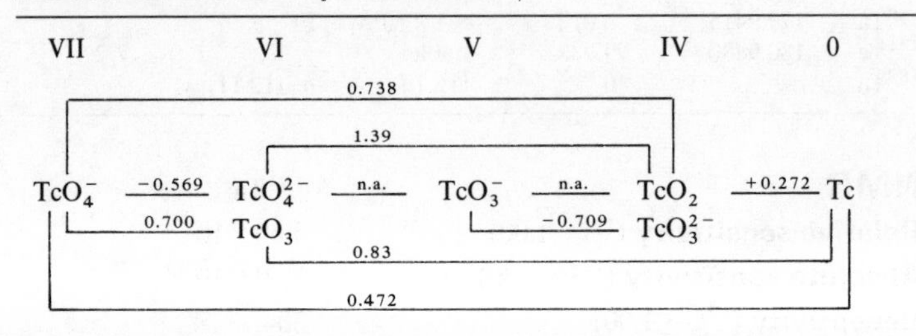

Oxidation states

Tc^{-I}	(d^8)	$[Tc(CO)_5]^-$
Tc^0	(d^7)	$Tc_2(CO)_{10}$
Tc^{IV}	(d^3)	TcO_2, TcO_3^{2-} (aq), $TcCl_4$, complexes
Tc^V	(d^2)	TcO_3^- (aq), $TcCl_5$, complexes
Tc^{VI}	(s^1)	TcO_3 ?, TcF_6, $TcOCl_4$, complexes
Tc^{VII}	(d^0, [Kr])	Tc_2O_7, TcO_4^- (aq), TcO_3Cl, complexes

Physical properties

Melting point/K: 2445
Boiling point/K: 5150
ΔH_{fusion}/kJ mol^{-1}: 23.81
ΔH_{vap}/kJ mol^{-1}: 585.22

Thermodynamic properties (298.15 K, 0.1 MPa)

State	$\Delta_f H^\circ$/kJ mol^{-1}	$\Delta_f G^\circ$/kJ mol^{-1}	S°/J K^{-1} mol^{-1}	C_p/J K^{-1} mol^{-1}
Solid	0	0	n.a.	25
Gas	678	n.a.	181.07	20.79

Density/kg m^{-3}: 11 500 (est.) [293 K]
Thermal conductivity/W m^{-1} K^{-1}: 50.6 [300 K]
Electrical resistivity/Ω m: 22.6×10^{-8} [393 K]
Mass magnetic susceptibility/kg^{-1} m^3: $+3.1 \times 10^{-8}$ (s)
Molar volume/cm^3: 8.6 (est.)
Coefficient of linear thermal expansion/K^{-1}: 8.06×10^{-6}

Lattice structure (cell dimensions/pm), space group

h.c.p. (a=274.3, c=440.0), $P6_3/mmc$

X-ray diffraction: mass absorption coefficients (μ/ρ)/cm^2 g^{-1}:
CuK_α 172 MoK_α 19.7

Discovered in 1937 by C. Perrier and E. G. Segrè at Palermo, Italy

[Greek, *technikos* = artificial]

Technetium

Nuclear properties

Atomic number: 43

Thermal neutron capture cross-section/barns: 22 ± 3 (^{99}Tc)

Number of isotopes (including nuclear isomers): 23

Isotope mass range: 92→107

Key isotopes

Nuclide	Atomic mass	Natural abundance (%)	Half life $T_{1/2}$	Decay mode and energy (MeV)	Nuclear spin I	Nuclear magnetic moment μ	Uses
^{97}Tc		0	2.6×10^6y	EC(*c.* 0.3)			
^{98}Tc	97.9072	0	1.5×10^6y	β^-(1.7); γ			
^{99}Tc	98.903	0	2.12×10^5y	β^-(0.292); no γ	9/2	+5.681	NMR
^{99m}Tc		0	6.049h	IT(0.1427); γ			tracer, medical

NMR

	^{99}Tc
Relative sensitivity (^{1}H = 1.00)	—
Absolute sensitivity (^{1}H = 1.00)	—
Receptivity (^{13}C = 1.00)	—
Magnetogyric ratio/rad T^{-1} s^{-1}	6.0503×10^7
Quadrupole moment /m^2	-0.13×10^{-28}
Frequency (^{1}H = 100 MHz; 2.3488 T)/MHz	22.508
Reference: TcO_4^- (aq)	

Electron shell properties

Ground state electron configuration: [Kr]$4d^5 5s^2$

Term symbol: $^6S_{5/2}$

Electron affinity (M→M^-)/kJ mol^{-1}: 96

Ionization energies/kJ mol^{-1}

1. M → M^+	702	6. M^{5+} → M^{6+}	(7300)
2. M^+ → M^{2+}	1472	7. M^{6+} → M^{7+}	(9100)
3. M^{2+} → M^{3+}	2850	8. M^{7+} → M^{8+}	(15600)
4. M^{3+} → M^{4+}	(4100)	9. M^{8+} → M^{9+}	(17800)
5. M^{4+} → M^{5+}	(5700)	10. M^{9+} → M^{10+}	(19900)

Principal lines in atomic spectrum

Wavelength/nm	Species	Sensitivity	Application
403.163	I		
426.227	I		
429.706	I		
485.359	I		
608.523*	I		

* Lowest energy transition from ground state to nearest empty orbital

Abundance: Earth's crust 0.0007 p.p.m.; seawater nil

Biological role: None; toxicity n.a., but radioactive

Te

Atomic number: 52
Relative atomic mass (^{12}C=12.0000): 127.60

Chemical properties

Silvery white, metallic looking in bulk but usually obtained as dark grey powder. Semi-metal. Obtained from anode slime in Cu refining. Burns in air or oxygen. Unaffected by water or HCl but dissolves in HNO_3. Used in alloys to improve machinability.

Radii/pm: Te^{4+} 97; Te^{6+} 56; atomic 143.2; covalent 137; van der Waals 220; Te^{2-} 211

Electronegativity: 2.1 (Pauling); 2.01 (Allred-Rochow)

Effective nuclear charge: 6.95 (Slater); 10.81 (Clementi); 13.51 (Froese–Fischer)

Standard reduction potentials E°/V

	VI		IV		0		−I		−II
Acid solution	H_2TeO_4	0.93	Te^{4+}	0.57	Te	−0.74	Te_2^{2-}	−0.64	H_2Te
	(H_2TeO_4 → TeO_2)	1.00	TeO_2	0.53 (→ Te_2^{2-})					
Alkaline solution	TeO_4^{2-}	0.07	TeO_3^{2-}	−0.42	Te			−1.14	Te^{2-}

Covalent bonds r/pm E/kJ mol^{-1}

Bond	r/pm	E/kJ mol^{-1}
Te–H	c. 170	c. 240
Te–C	205	
Te–O	200	268
Te–F	185	335
Te–Cl	231	251
Te–Te	286	235

Oxidation states

Te^{-II}	H_2Te, Te^{2-}
Te^{-I}	Te_2^{2-}
Te^{0-I}	cluster cations Te_4^{2+}, Te_6^{4+}
Te^{II}	TeO, $TeCl_2$, $TeBr_2$
Te^{IV}	TeO_2, H_2TeO_3, TeO_3^{2-} (aq), TeF_4, $TeCl_4$, TeF_5^-
Te^{V}	Te_2F_{10}
Te^{VI}	TeO_3, H_2TeO_4, TeO_4^{2-} (aq), H_6TeO_6, TeF_6, TeF_8^{2-}

Physical properties

Melting point/K: 722.7

Boiling point/K: 1263.0

ΔH_{fusion}/kJ mol^{-1}: 13.5 ΔH_{vap}/kJ mol^{-1}: 104.6

Thermodynamic properties (298.15 K, 0.1 MPa)

State	$\Delta_f H^\circ$/kJ mol^{-1}	$\Delta_f G^\circ$/kJ mol^{-1}	S°/J K^{-1} mol^{-1}	C_p/J K^{-1} mol^{-1}
Solid	0	0	49.71	25.73
Gas	196.73	157.08	182.74	20.786

Density/kg m^{-3}: 6240 [293 K]; 5797 [liquid at m.p.]

Thermal conductivity/W m^{-1} K^{-1}: 2.35 [300 K]

Electrical resistivity/Ω m: 4.36×10^{-3} [298 K]

Mass magnetic susceptibility/kg^{-1} m^3: -3.9×10^{-9} (s)

Molar volume/cm^3: 20.45

Coefficient of linear thermal expansion/K^{-1}: 16.75×10^{-6}

Lattice structure (cell dimensions/pm), space group

Hexagonal (a=445.65, c=592.68), $P3_121$ or $P3_221$
High pressure forms: (a=420.8, c=1203.6), $R\bar{3}m$
(a=460.3, c=382.2), $R\bar{3}m$

X-ray diffraction: mass absorption coefficients (μ/ρ)/cm^2 g^{-1}:
CuK_α 282 MoK_α 35.0

Discovered in 1783 by Baron Franz Joseph Müller von Reichenstein at Sibiu, Romania

[Latin, *tellus* = earth]

Tellurium

Nuclear properties

Atomic number: 52

Thermal neutron capture cross-section/barns: 4.7 ± 0.1

Number of isotopes (including nuclear isomers): 29

Isotope mass range: 115→135

Key isotopes

Nuclide	Atomic mass	Natural abundance (%)	Half life $T_{1/2}$	Decay mode and energy (MeV)	Nuclear spin I	Nuclear magnetic moment μ	Uses
^{120}Te	119.9045	0.096	stable				
^{122}Te	121.9030	2.60	stable				
^{123}Te	122.9042	0.908	1.2×10^{13}y	EC(*c.* 0.06); no γ	1/2	−0.7359	NMR
^{124}Te	123.9028	4.816	stable				
^{125m}Te		0	58d	IT; γ	3/2	+0.60	tracer
^{125}Te	124.9044	7.18	stable		1/2	−0.8872	NMR
^{126}Te	125.9032	18.95	stable		0		
^{127}Te		0	9.4h	β^-(0.69); γ			tracer
^{128}Te	127.9047	31.69	stable		0		
^{130}Te	129.9067	33.80	stable		0		

NMR

	[^{123}Te]	^{125}Te
Relative sensitivity ($^1H = 1.00$)	1.80×10^{-2}	3.15×10^{-2}
Absolute sensitivity ($^1H = 1.00$)	1.56×10^{-4}	2.20×10^{-3}
Receptivity ($^{13}C = 1.00$)	0.89	12.5
Magnetogyric ratio/rad $T^{-1} s^{-1}$	-7.0006×10^7	-8.4398×10^7
Frequency ($^1H = 100$ MHz; 2.3488 T)/MHz	26.207	31.596

Reference: $Te(CH_3)_2$

Electron shell properties

Ground state electron configuration: $[Kr]4d^{10}5s^25p^4$

Term symbol: $^3P_2$

Electron affinity ($M \rightarrow M^-$)/kJ mol^{-1}: 190.2

Ionization energies/kJ mol^{-1}

1.	$M \rightarrow M^+$	869.2	6.	$M^{5+} \rightarrow M^{6+}$	6822
2.	$M^+ \rightarrow M^{2+}$	1795	7.	$M^{6+} \rightarrow M^{7+}$	13200
3.	$M^{2+} \rightarrow M^{3+}$	2698	8.	$M^{7+} \rightarrow M^{8+}$	(15800)
4.	$M^{3+} \rightarrow M^{4+}$	3610	9.	$M^{8+} \rightarrow M^{9+}$	(18500)
5.	$M^{4+} \rightarrow M^{5+}$	5668	10.	$M^{9+} \rightarrow M^{10+}$	(21200)

Principal lines in atomic spectrum

Wavelength/nm	Species	Sensitivity	Application
200.202	I		
214.281	I		AE
225.902*	I		
238.326	I	U3	AE
238.578	I	U2	AE

* Lowest energy transition from ground state to nearest empty orbital

Abundance: Earth's crust 0.001 p.p.m.; seawater n.a. $< 1 \times 10^{-10}$ p.p.m.

Biological role: None; very toxic; teratogenic

Tb

Atomic number: 65

Relative atomic mass ($^{12}C = 12.0000$): **158.92534**

Chemical properties

Silvery white metal, rare member of the lanthanide (rare earth) group. Obtained from monazite. Slowly oxidized by air, reacts with cold water. Used in solid state devices and lasers.

Radii/pm: Tb^{3+} 93; Tb^{4+} 81; atomic 178.2; covalent 159

Electronegativity: n.a. (Pauling); 1.10 (Allred-Rochow)

Effective nuclear charge: 2.85 (Slater); 8.30 (Clementi); 11.39 (Froese–Fischer)

Standard reduction potentials $E^{\ominus}$/V

	IV		III		0
Acid solution	Tb^{4+}	—3.1—	Tb^{3+}	—−2.31—	Tb
Alkaline solution	TbO_2	—0.9—	$Tb(OH)_3$	—−2.82—	Tb

Oxidation states

Tb^{III} (f^8)	Tb_2O_3, $Tb(OH)_3$, $[Tb(H_2O)_x]^{3+}$ (aq), TbF_3, $TbCl_3$, etc., Tb^{3+} salts, complexes
Tb^{IV} (f^7)	TbO_2, TbF_4

Physical properties

Melting point/K: 1629

Boiling point/K: 3396

ΔH_{fusion}/kJ mol^{-1}: 16.3

ΔH_{vap}/kJ mol^{-1}: 391

Thermodynamic properties (298.15 K, 0.1 MPa)

State	$\Delta_f H^{\ominus}$/kJ mol^{-1}	$\Delta_f G^{\ominus}$/kJ mol^{-1}	$S^{\ominus}$/J K^{-1} mol^{-1}	C_p/J K^{-1} mol^{-1}
Solid	0	0	73.22	28.91
Gas	388.7	349.7	203.58	24.56

Density/kg m^{-3}: 8229 [293 K]

Thermal conductivity/W m^{-1} K^{-1}: 11.1 [300 K]

Electrical resistivity/Ω m: 114×10^{-8} [298 K]

Mass magnetic susceptibility/kg^{-1} m^3: $+1.15 \times 10^{-5}$ (s)

Molar volume/cm^3: 19.31

Coefficient of linear thermal expansion/K^{-1}: 7.0×10^{-6}

Lattice structure (cell dimensions/pm), space group

Tb rhombic ($a = 359.0$, $b = 626.0$, $c = 571.5$), Cmcm
α-Tb h.c.p. ($a = 360.10$, $c = 569.36$), $P6_3/mmc$
β-Tb b.c.c. ($a = 402$), Im3m

$T(\alpha \rightarrow \text{rhombic}) = 220$ K
$T(\alpha \rightarrow \beta) = 1590$ K

X-ray diffraction: mass absorption coefficients (μ/ρ)/cm^2 g^{-1}:
CuK_α 273 MoK_α 67.5

Discovered in 1843 by C. G. Mosander at Stockholm, Sweden

[Named after Ytterby, Sweden]

Terbium

Nuclear properties

Atomic number: 65

Thermal neutron capture cross-section/barns: 30 ± 10

Number of isotopes (including nuclear isomers): 24

Isotope mass range: $147 \rightarrow 164$

Key isotopes

Nuclide	Atomic mass	Natural abundance (%)	Half life $T_{1/2}$	Decay mode and energy (MeV)	Nuclear spin I	Nuclear magnetic moment μ	Uses
^{159}Tb	159.9250	100	stable		3/2	+2.008	NMR
^{160}Tb		0	72.1d	$\beta^-(1.72)$; γ	3	± 1.70	tracer

NMR

	^{159}Tb
Relative sensitivity (^{1}H = 1.00)	5.83×10^{-2}
Absolute sensitivity (^{1}H = 1.00)	5.83×10^{-2}
Receptivity (^{13}C = 1.00)	3.94×10^{2}
Magnetogyric ratio/rad T^{-1} s^{-1}	6.4306×10^{7}
Quadrupole moment/m^2	1.3×10^{-28}
Frequency (^{1}H = 100 MHz; 2.3488 T)/MHz	22.678

Electron shell properties

Ground state electron configuration: [Xe]$4f^9 6s^2$

Term symbol: $^6H_{15/2}$

Electron affinity ($M \rightarrow M^-$)/kJ mol^{-1}: $\leqslant 50$

Ionization energies/kJ mol^{-1}

1. $M \rightarrow M^+$	564.6	6. $M^{5+} \rightarrow M^{6+}$	
2. $M^+ \rightarrow M^{2+}$	1112	7. $M^{6+} \rightarrow M^{7+}$	
3. $M^{2+} \rightarrow M^{3+}$	2114	8. $M^{7+} \rightarrow M^{8+}$	
4. $M^{3+} \rightarrow M^{4+}$	3839	9. $M^{8+} \rightarrow M^{9+}$	
5. $M^{4+} \rightarrow M^{5+}$		10. $M^{9+} \rightarrow M^{10+}$	

Principal lines in atomic spectrum

Wavelength/nm	Species	Sensitivity	Application
350.917	II		AE
356.174	II		AE
387.418	II		AE
431.883	I		AA
432.643	I		AA

Abundance: Earth's crust 1.2 p.p.m.; seawater 1.4×10^{-6} p.p.m.

Biological role: None; toxicity n.a., but probably low

Tl

Atomic number: 81

Relative atomic mass ($^{12}C = 12.0000$): 204.3833

Chemical properties

Soft, silvery-grey metal. Tarnishes readily in moist air and with steam reacts to form TlOH. Attacked by acids, rapidly by HNO_3. Little used because of its toxicity, but still employed in special glass.

Radii/pm: Tl^+ 149; Tl^{3+} 105; atomic 170.4 (α form); covalent 155

Electronegativity: 1.62 (Tl^I) 2.04 (Tl^{III}) (Pauling); 1.44 (Allred-Rochow)

Effective nuclear charge: 5.00 (Slater); 12.25 (Clementi); 13.50 (Froese–Fischer)

Standard reduction potentials $E^\ominus$/V

	III		I		0
Acid solution	Tl^{3+}	1.25	Tl^+	−0.3363	Tl
			0.72 (Tl^{3+} → Tl)		

Covalent bonds

	r/pm	E/kJ mol^{-1}
Tl^I–H	187.0	185
Tl^{III}–C	230	125
Tl^{III}–O	226	375
Tl^{III}–F	195	460
Tl^{III}–Cl	248	368
Tl–Tl	340.8	*c.* 63

Oxidation states

Tl^{-I} (s^2)	Tl_2O, TlOH, Tl_2CO_3, $[Tl(H_2O)_6]^+$ (aq), Tl^+ salts, TlF, TlCl, etc.
Tl^{III} (d^{10})	Tl_2O_3, $[Tl(H_2O)_6]^{3+}$ (aq), TlF_3, $TlCl_3$, $TlBr_3$, $(CH_3)_2Tl^+$ (aq), $TlCl_6^{3-}$

Physical properties

Melting point/K: 576.7

Boiling point/K: 1730

ΔH_{fusion}/kJ mol^{-1}: 4.31

ΔH_{vap}/kJ mol^{-1}: 166.1

Thermodynamic properties (298.15 K, 0.1 MPa)

State	$\Delta_f H^\ominus$/kJ mol^{-1}	$\Delta_f G^\ominus$/kJ mol^{-1}	$S^\ominus$/J K^{-1} mol^{-1}	C_p/J K^{-1} mol^{-1}
Solid	0	0	64.18	26.32
Gas	182.21	147.41	180.963	20.786

Density/kg m^{-3}: 11 850 [293 K]; 11 290 [liquid at m.p.]

Thermal conductivity/W m^{-1} K^{-1}: 46.1 [300 K]

Electrical resistivity/Ω m: 18.0×10^{-8} [273 K]

Mass magnetic susceptibility/kg^{-1} m^3: -3.13×10^{-9} (s)

Molar volume/cm^3: 17.24

Coefficient of linear thermal expansion/K^{-1}: 28×10^{-6}

Lattice structure (cell dimensions/pm), space group

α-Tl hexagonal ($a = 345.6$, $c = 552.5$), $P6_3/mmc$
β-Tl cubic ($a = 388.2$), Im3m
γ-Tl f.c.c. ($a = 485.1$), Fm3m
$T(\alpha \rightarrow \beta) = 503$ K

X-ray diffraction: mass absorption coefficients ((μ/ρ)/cm^2 g^{-1}):
CuK_α 224 MoK_α 119

Discovered in 1861 by W. Crookes at London, UK; isolated in 1862 by C.-A. Lamy at Paris, France

[Greek, *thallos* = green twig]

Thallium

Nuclear properties

Atomic number: 81

Thermal neutron capture cross-section/barns: 3.4 ± 0.5

Number of isotopes (including nuclear isomers): 28

Isotope mass range: 191→210

Key isotopes

Nuclide	Atomic mass	Natural abundance (%)	Half life $T_{1/2}$	Decay mode and energy (MeV)	Nuclear spin I	Nuclear magnetic moment μ	Uses
^{203}Tl	202.9723	29.524	stable		1/2	+1.6115	NMR
^{204}Tl		0	3.81y	β^-(0.763); EC	2	±0.089	tracer
^{205}Tl	204.9745	70.476	stable		1/2	+1.6274	NMR
^{208}Tl		trace	3.10m	β^-(4.994); γ			

NMR

	[^{203}Tl]	^{205}Tl
Relative sensitivity ($^1H = 1.00$)	0.18	0.19
Absolute sensitivity ($^1H = 1.00$)	5.51×10^{-2}	0.13
Receptivity ($^{13}C = 1.00$)	289	769
Magnetogyric ratio/rad $T^{-1} s^{-1}$	15.3078×10^7	15.4584×10^7
Frequency ($^1H = 100$ MHz; 2.3488 T)/MHz	57.149	57.708

Reference: $TlNO_3$ (aq)

Electron shell properties

Ground state electron configuration: [Xe]$4f^{14}5d^{10}6s^26p^1$

Term symbol: $^2P_{1/2}$

Electron affinity ($M \rightarrow M^-$)/kJ mol^{-1}: 30

Ionization energies/kJ mol^{-1}

1. $M \rightarrow M^+$	589.3	6. $M^{5+} \rightarrow M^{6+}$	(8 300)
2. $M^+ \rightarrow M^{2+}$	1971.0	7. $M^{6+} \rightarrow M^{7+}$	(9 500)
3. $M^{2+} \rightarrow M^{3+}$	2878	8. $M^{7+} \rightarrow M^{8+}$	(11 300)
4. $M^{3+} \rightarrow M^{4+}$	(4900)	9. $M^{8+} \rightarrow M^{9+}$	(14 000)
5. $M^{4+} \rightarrow M^{5+}$	(6100)	10. $M^{9+} \rightarrow M^{10+}$	(16 000)

Principal lines in atomic spectrum

Wavelength/nm	Species	Sensitivity	Application
237.969	I		AA
276.787	I		AA
351.924	I	U3	AE
377.572*	I	U2	AA, AE
535.046	I	U1	AE

* Lowest energy transition from ground state to nearest empty orbital

Abundance: Earth's crust 0.7 p.p.m.; seawater 1×10^{-5} p.p.m.

Biological role: None; very toxic; teratogenic

Th

Atomic number: 90
Relative atomic mass (^{12}C=12.0000): 232.0381

Chemical properties

Radioactive silvery metal, found in large deposits. Metal protected by oxide coating. Attacked by steam and slowly by acids. Metal itself is soft and ductile but alloys can be very strong. Used in incandescent gas mantles.

Radii/pm: Th^{3+} 101; Th^{4+} 99; atomic 179.8

Electronegativity: 1.3 (Pauling); 1.11 (Allred-Rochow)

Effective nuclear charge: 1.95 (Slater)

Standard reduction potentials $E^\ominus$/V

	IV		0
Acid solution	Th^{4+}	$\xrightarrow{-1.83}$	Th
Alkaline solution	ThO_2	$\xrightarrow{-2.56}$	Th

Oxidation states

Th^{II} (d^2)	ThO, ThH_2
Th^{III} (d^1)	ThI_3
Th^{IV} ([Rn])	ThO_2, $[Th(H_2O)_x]^{4+}$(aq), ThF_4, $ThCl_4$etc., ThF_7^{3-}, Th^{4+} salts, complexes

Physical properties

Melting point/K: 2023

Boiling point/K: *c.* 5060

ΔH_{fusion}/kJ mol^{-1}: <19.2

ΔH_{vap}/kJ mol^{-1}: 513.67

Thermodynamic properties (298.15 K, 0.1 MPa)

State	$\Delta_f H^\ominus$/kJ mol^{-1}	$\Delta_f G^\ominus$/kJ mol^{-1}	$S^\ominus$/J K^{-1} mol^{-1}	C_p/J K^{-1} mol^{-1}
Solid	0	0	53.39	27.32
Gas	598.3	557.53	190.15	20.79

Density/kg m^{-3}: 11 720 [293 K]

Thermal conductivity/W m^{-1} K^{-1}: 54.0 [300 K]

Electrical resistivity/Ω m: 13.0×10^{-8} [273 K]

Mass magnetic susceptibility/kg^{-1} m^3: $+7.2 \times 10^{-9}$ (s)

Molar volume/cm^3: 19.80

Coefficient of linear thermal expansion/K^{-1}: 12.5×10^{-6}

Lattice structure (cell dimensions/pm), space group

α-Th f.c.c. (a=508.42). Fm3m
β-Th b.c.c. (a=411), Im3m
$T(\alpha \rightarrow \beta)$=1673 K

X-ray diffraction: mass absorption coefficients (μ/ρ)/cm^2 g^{-1}:
CuK$_\alpha$ 327 MoK$_\alpha$ 143

Discovered in 1815 by J. J. Berzelius at Stockholm, Sweden

[Called after Thor, Scandanavian god of war]

Thorium

Nuclear properties

Atomic number: 90

Thermal neutron capture cross-section/barns: 7.4±0.1

Number of isotopes (including nuclear isomers): 12

Isotope mass range: 223→234

Key isotopes

Nuclide	Atomic mass	Natural abundance (%)	Half life $T_{1/2}$	Decay mode and energy (MeV)	Nuclear spin I	Nuclear magnetic moment μ	Uses
^{228}Th	228.0287	trace	1.913y	α(5.521); γ			
^{229}Th	229.0316	0	7340y	α(5.167); γ			NMR
^{230}Th	230.0331	trace	8.0×10^4y	α(4.767); γ			
^{231}Th		trace	25.5h	β^-(0.318); γ			
^{232}Th	232.0382	100	1.41×10^{10}y	α(4.08); γ			tracer
^{234}Th	234.0436	trace	24.1d	β^-(0.263); γ			

NMR

	^{229}Th
Relative sensitivity (^{1}H=1.00)	—
Absolute sensitivity (^{1}H=1.00)	—
Receptivity (^{13}C=1.00)	—
Magnetogyric ratio/rad T^{-1} s^{-1}	0.40×10^7
Quadrupole moment/m^2	4.4×10^{-28}
Frequency (^{1}H=100 MHz; 2.3488 T)/MHz	1.5

Electron shell properties

Ground state electron configuration: [Rn]$6d^27s^2$

Term symbol: $^3F_2$

Electron affinity ($M\rightarrow M^-$)/kJ mol^{-1}: n.a.

Ionization energies/kJ mol^{-1}

1. $M \rightarrow M^+$	587	6. $M^{5+}\rightarrow M^{6+}$	
2. $M^+\rightarrow M^{2+}$	1110	7. $M^{6+}\rightarrow M^{7+}$	
3. $M^{2+}\rightarrow M^{3+}$	1978	8. $M^{7+}\rightarrow M^{8+}$	
4. $M^{3+}\rightarrow M^{4+}$	2780	9. $M^{8+}\rightarrow M^{9+}$	
5. $M^{4+}\rightarrow M^{5+}$		10. $M^{9+}\rightarrow M^{10+}$	

Principal lines in atomic spectrum

Wavelength/nm	Species	Sensitivity	Application
324.445	I		AA
324.576	II		AA
360.103	II		AE
401.914	II		AE
576.055	I		AA

Abundance: Earth's crust 8.1 p.p.m.; seawater 0.0007 p.p.m.

Biological role: None; low toxicity, but dangerous due to radioactivity

Tm

Atomic number: 69

Relative atomic mass (^{12}C=12.0000): 168.93421

Chemical properties

Silvery metal, rarest member of the lanthanide group (rare earths). Obtained from euxenite and monazite. Tarnishes in air and reacts with water. Few uses but some employed as radiation source in portable X-ray equipment.

Radii/pm: Tm^{3+} 87; Tm^{4+} 94; atomic 174.6; covalent 156

Electronegativity: 1.25 (Pauling); 1.11 (Allred-Rochow)

Effective nuclear charge: 2.85 (Slater); 8.58 (Clementi); 11.80 (Froese–Fischer)

Standard reduction potentials $E^\ominus$/V

	III		II		0
Acid solution	Tm^{3+}	−2.3	Tm^{2+}	−2.3	Tm
	(III → 0: −2.32)				
Alkaline solution	$Tm(OH)_3$	−2.83			Tm

Oxidation states

Tm^{II} (f^{13})	$TmCl_2$, $TmBr_2$, TmI_2
Tm^{III} (f^{12})	Tm_2O_3, $Tm(OH)_3$, $[Tm(H_2O)_x]^{3+}$ (aq), Tm^{3+} salts, TmF_3, $TmCl_3$ etc., complexes

Physical properties

Melting point/K: 1818

Boiling point/K: 2220

ΔH_{fusion}/kJ mol^{-1}: 18.4

ΔH_{vap}/kJ mol^{-1}: 247

Thermodynamic properties (298.15 K, 0.1 MPa)

State	$\Delta_f H^\ominus$/kJ mol^{-1}	$\Delta_f G^\ominus$/kJ mol^{-1}	$S^\ominus$/J K^{-1} mol^{-1}	C_p/J K^{-1} mol^{-1}
Solid	0	0	74.01	27.03
Gas	232.2	197.5	190.113	20.786

Density/kg m^{-3}: 9321 [293 K]

Thermal conductivity/W m^{-1} K^{-1}: 16.8 [300 K]

Electrical resistivity/Ω m: 79.0×10^{-8} [298 K]

Mass magnetic susceptibility/kg^{-1} m^3: $+1.90 \times 10^{-6}$ (s)

Molar volume/cm^3: 18.12

Coefficient of linear thermal expansion/K^{-1}: 13.3×10^{-6}

Lattice structure (cell dimensions/pm), space group

h.c.p. (a=353.75; c=555.46), $P6_3/mmc$

X-ray diffraction: mass absorption coefficients (μ/ρ)/cm^2 g^{-1}:
CuK_α 140 MoK_α 80.8

Discovered in 1879 by P. T. Cleve at Uppsala, Sweden

[Called after *Thule*, ancient Scandinavia]

Thulium

Nuclear properties

Atomic number: 69
Thermal neutron capture cross-section/barns: 115±15
Number of isotopes (including nuclear isomers): 18
Isotope mass range: 161→176

Key isotopes

Nuclide	Atomic mass	Natural abundance (%)	Half life $T_{1/2}$	Decay mode and energy (MeV)	Nuclear spin I	Nuclear magnetic moment μ	Uses
^{169}Tm	168.9344	100	stable		1/2	−0.231	NMR
^{170}Tm		0	134d	β^-(0.96); γ	1	±0.246	tracer

NMR

	^{169}Tm
Relative sensitivity (^{1}H=1.00)	5.66×10^{-4}
Absolute sensitivity (^{1}H=1.00)	5.66×10^{-4}
Receptivity (^{13}C=1.00)	3.21
Magnetogyric ratio/rad T^{-1} s^{-1}	-2.21×10^7
Frequency (^{1}H=100 MHz; 2.3488 T)/MHz	8.271

Electron shell properties

Ground state electron configuration: [Xe]4f^{13}6s^2
Term symbol: $^2F_{7/2}$
Electron affinity ($M \rightarrow M^-$)/kJ mol^{-1}: ≤50

Ionization energies/kJ mol^{-1}

1. $M \rightarrow M^+$	596.7	6. $M^{5+} \rightarrow M^{6+}$	
2. $M^+ \rightarrow M^{2+}$	1163	7. $M^{6+} \rightarrow M^{7+}$	
3. $M^{2+} \rightarrow M^{3+}$	2285	8. $M^{7+} \rightarrow M^{8+}$	
4. $M^{3+} \rightarrow M^{4+}$	4119	9. $M^{8+} \rightarrow M^{9+}$	
5. $M^{4+} \rightarrow M^{5+}$		10. $M^{9+} \rightarrow M^{10+}$	

Principal lines in atomic spectrum

Wavelength/nm	Species	Sensitivity	Application
371.791	I		AA
376.133	II		AE
376.192	II		AE
409.419	I		AA
410.584	I		AA

Abundance: Earth's crust 0.5 p.p.m.; seawater 4×10^{-8} p.p.m.
Biological role: None; non-toxic; stimulatory

Sn

Atomic number: 50
Relative atomic mass (^{12}C=12.0000): 118.710

Chemical properties

Soft, pliable, silvery-white metal. Chief ore is cassiteite (SnO_2). Unreactive to oxygen (protected by oxide film) and water but dissolves in acids and bases. Used in solders, alloys, tinplate, polymer additives and antifouling paints.

Radii/pm: Sn^{2+} 93; Sn^{4+} 74; atomic 140.5; covalent 140; Sn^{4+} 294

Electronegativity: 1.96 (Pauling); 1.72 (Allred-Rochow)

Effective nuclear charge: 5.65 (Slater); 9.10 (Clementi); 11.11 (Froese–Fischer)

Standard reduction potentials $E^\ominus$/V

	IV		II		0		−IV
Acid solution	SnO_2	−0.088	SnO	−0.104	Sn	−1.071	SnH_4
	Sn^{4+}	0.15	Sn^{2+}	−0.137			

[Alkaline solutions contain many different forms]

Covalent bonds

	r/pm	E/kJ mol^{-1}
Sn–H	170.1	<314
Sn–C	217	225
Sn^{II}–O	195	557
Sn^{IV}–F	188	322
Sn^{IV}–Cl	231	315
Sn–Sn (α)	281	195

Oxidation states

Sn^{II}	SnO, SnF_2, $SnCl_2$, etc., $[Sn(OH)]^+$ (aq), $[Sn_3(OH)_4]^{2+}$ (aq), Sn^{2+} salts
Sn^{IV}	SnO_2, SnH_4, SnF_4, $SnCl_4$, etc., $[SnCl_6]^{2-}$ (aq HCl), $[Sn(OH)_6]^{2-}$ (aqbase), organotin compounds

Physical properties

Melting point/K: 505.118

Boiling point/K: 2543

ΔH_{fusion}/kJ mol^{-1}: 7.20

ΔH_{vap}/kJ mol^{-1}: 296.2

Thermodynamic properties (298.15 K, 0.1 MPa)

State	$\Delta_f H^\ominus$/kJ mol^{-1}	$\Delta_f G^\ominus$/kJ mol^{-1}	$S^\ominus$/J K^{-1} mol^{-1}	C_p/J K^{-1} mol^{-1}
Solid (α)	−2.09	0.13	44.14	25.77
Solid (β)	0	0	51.55	26.99
Gas	302.1	267.3	168.486	21.259

Density/kg m^{-3}: 5750 (α); 7310 (β) [293 K]; 6973 [liquid at m.p.]

Thermal conductivity/W m^{-1} K^{-1}: 66.6 (α) [300 K]

Electrical resistivity/Ω m: 11.0×10^{-8} (α) [273 K]

Mass magnetic susceptibility/kg^{-1} m^3: -4.0×10^{-9} (α); $+3.3 \times 10^{-10}$ (β)

Molar volume/cm^3: 16.24 (β)

Coefficient of linear thermal expansion/K^{-1}: 5.3×10^{-6} (α); 21.2×10^{-6} (β)

Lattice structure (cell dimensions/pm), space group

α-Sn (grey) cubic (a=648.92), Fd3m
β–Sn (white) tetragonal (a=583.16, c=318.13), $I4_2$/amd
T(α→β)=286.4 K [β form at room temperatures]

X-ray diffraction: mass absorption coefficients (μ/ρ)/cm^2 g^{-1}:
CuK_α 256 MoK_α 31.1

Known to ancient civilizations

[Anglo-Saxon, *tin*; Latin, *stannum*]

Tin

Nuclear properties

Atomic number: 50
Thermal neutron capture cross-section/barns: 0.63 ± 0.01
Number of isotopes (including nuclear isomers): 28
Isotope mass range: 108→128

Key isotopes

Nuclide	Atomic mass	Natural abundance (%)	Half life $T_{1/2}$	Decay mode and energy (MeV)	Nuclear spin I	Nuclear magnetic moment μ	Uses
^{112}Sn	111.9040	1.0	stable				
^{113}Sn		0	115d	EC(1.02); γ	1/2	± 0.88	tracer
^{114}Sn	113.9030	0.7	stable				
^{115}Sn	114.9035	0.4	stable		1/2	-0.9178	NMR
^{116}Sn	115.9021	14.7	stable		0		
^{117}Sn	116.9031	7.7	stable		1/2	-1.000	NMR
^{118}Sn	117.9018	24.3	stable		0		
^{119}Sn	118.9034	8.6	stable		1/2	-1.0461	NMR
^{120}Sn		32.4	stable		0		
^{121}Sn		0	27.5h	β^-(0.383); no γ	3/2	± 0.699	tracer
^{122}Sn	121.9034	4.6	stable				
^{124}Sn	123.9052	5.6	stable				

NMR

	[^{115}Sn]	[^{117}Sn]	^{119}Sn
Relative sensitivity ($^1H = 1.00$)	3.5×10^{-2}	4.52×10^{-2}	5.18×10^{-2}
Absolute sensitivity ($^1H = 1.00$)	1.22×10^{-4}	3.44×10^{-3}	4.44×10^{-3}
Receptivity ($^{13}C = 1.00$)	0.693	19.54	25.2
Magnetogyric ratio/rad T^{-1} s^{-1}	-8.7475×10^7	-9.5319×10^7	-9.9756×10^7
Frequency ($^1H = 100$ MHz; 2.3488 T)/MHz	32.699	35.625	37.272

Reference: $Sn(CH_3)_4$

Electron shell properties

Ground state electron configuration: $[Kr]4d^{10}5s^25p^2$
Term symbol: $^3P_0$
Electron affinity ($M \rightarrow M^-$)/kJ mol^{-1}: 121

Ionization energies/kJ mol^{-1}

1. $M \rightarrow M^+$	708.6	6. $M^{5+} \rightarrow M^{6+}$	(9900)
2. $M^+ \rightarrow M^{2+}$	1411.8	7. $M^{6+} \rightarrow M^{7+}$	(12200)
3. $M^{2+} \rightarrow M^{3+}$	2943.0	8. $M^{7+} \rightarrow M^{8+}$	(14600)
4. $M^{3+} \rightarrow M^{4+}$	3930.2	9. $M^{8+} \rightarrow M^{9+}$	(17000)
5. $M^{4+} \rightarrow M^{5+}$	6974	10. $M^{9+} \rightarrow M^{10+}$	(20600)

Principal lines in atomic spectrum

Wavelength/nm	Species	Sensitivity	Application
224.605	I		AA
283.999	I	U1	AE
286.333*	I	U2	AA, AE
303.412	I		AA, AE
326.233	I	U3	AE

* Lowest energy transition from ground state to nearest empty orbital

Abundance: Earth's crust 2.1 p.p.m.; seawater 0.003 p.p.m.
Biological role: Non-toxic but organotins used as biocides

Ti

Atomic number: 22

Relative atomic mass (^{12}C=12.0000): 47.88

Chemical properties

Hard, lustrous, silvery metal. Main ores are ilmenite, $FeTiO_3$, and rutile, TiO_2. Resists corrosion due to oxide layer, but powdered metal burns in air. Unaffected both by acids, (except HF and concentrated H_2SO_4), and alkalis. Used in paints and lightweight alloys.

Radii/pm: Ti^{2+} 80; Ti^{3+} 69; atomic 144.8; covalent 132

Electronegativity: 1.54 (Pauling); 1.32 (Allred-Rochow)

Effective nuclear charge: 3.15 (Slater); 4.82 (Clementi); 6.37 (Froese–Fischer)

Standard reduction potentials $E^\ominus$/V

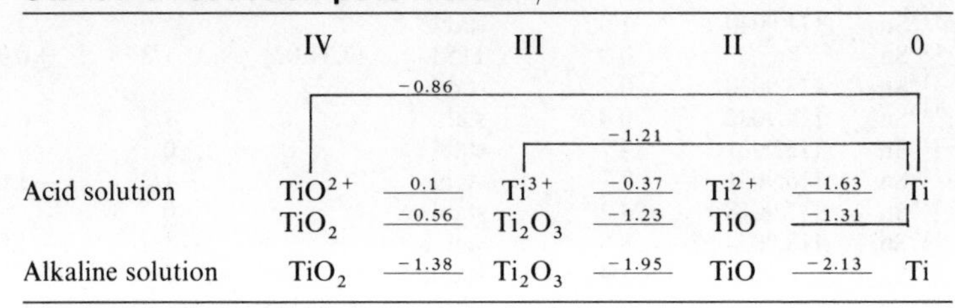

Oxidation states

Ti^{-I}	(d^5)	rare $[Ti(bipyridyl)_3]^-$
Ti^0	(d^4)	rare $[Ti(bipyridyl)_3]$
Ti^{II}	(d^2)	TiO, $TiCl_2$, $TiBr_2$, TiI_2, no solution chemistry (reduces H_2O); complexes
Ti^{III}	(d^1)	Ti_2O_3, $[Ti(H_2O)_6]^{3+}$ (aq), TiF_3, $TiCl_3$ etc., complexes
Ti^{IV}	([Ar])	TiO_2, TiO^{2+} (aq), $[Ti(OH)_3]^{2+}$ (aq), TiF_4, $TiCl_4$ etc., titanates (TiO_4^{4-}, TiO_3^{2-}), complexes

Physical properties

Melting point/K: 1933

Boiling point/K: 3560

ΔH_{fusion}/kJ mol^{-1}: 20.9

ΔH_{vap}/kJ mol^{-1}: 425.5

Thermodynamic properties (298.15 K, 0.1 MPa)

State	$\Delta_f H^\ominus$/kJ mol^{-1}	$\Delta_f G^\ominus$/kJ mol^{-1}	$S^\ominus$/J K^{-1} mol^{-1}	C_p/J K^{-1} mol^{-1}
Solid	0	0	30.63	25.02
Gas	469.9	425.1	180.298	24.430

Density/kg m^{-3}: 4540 [293 K]; 4110 [liquid at m.p.]

Thermal conductivity/W m^{-1} K^{-1}: 21.9 [300 K]

Electrical resistivity/Ω m: 42.0×10^{-8} [293 K]

Mass magnetic susceptibility/kg^{-1} m^3: $+4.01 \times 10^{-8}$ (s)

Molar volume/cm^3: 10.55

Coefficient of linear thermal expansion/K^{-1}: 8.35×10^{-6}

Lattice structure (cell dimensions/pm), space group

α-Ti h.c.p. (a=295.11, c=468.43), $P6_3/mmc$
β-Ti b.c.c. (a=330.65), Im3m

$T(\alpha \rightarrow \beta)$ = 1155 K

High pressure form: (a=462.5; c=281.3), $P\bar{3}m1$

X-ray diffraction: mass absorption coefficients (μ/ρ)/cm^2 g^{-1}: CuK_α 208 MoK_α 24.2

Discovered in 1791 by Rev. W. Gregor at Creed, Cornwall, UK, and independently by M. H. Klaproth in 1795 at Berlin, Germany

[Called after the Titans, sons of the Earth goddess]

Titanium

Nuclear properties

Atomic number: 22

Thermal neutron capture cross-section/barns: 6.1 ± 0.2

Number of isotopes (including nuclear isomers): 9

Isotope mass range: 43→51

Key isotopes

Nuclide	Atomic mass	Natural abundance (%)	Half life $T_{1/2}$	Decay mode and energy (MeV)	Nuclear spin I	Nuclear magnetic moment μ	Uses
^{44}Ti		0	48y	EC(0.16); γ			tracer
^{46}Ti	45.952 63	8.2	stable				
^{47}Ti	45.951 8	7.4	stable		5/2	−0.788 46	NMR
^{48}Ti	45.948	73.8	stable				
^{49}Ti	45.947 87	5.4	stable		7/2	−1.04 14	NMR
^{50}Ti	49.944 8	5.2	stable				

NMR

	^{47}Ti	^{49}Ti
Relative sensitivity ($^1H = 1.00$)	2.09×10^{-3}	3.76×10^{-3}
Absolute sensitivity ($^1H = 1.00$)	1.52×10^{-4}	2.07×10^{-4}
Receptivity ($^{13}C = 1.00$)	0.864	1.18
Magnetogyric ratio/rad T^{-1} s^{-1}	1.5084×10^7	1.5080×10^7
Quadrupole moment/m^2	$+0.29 \times 10^{-28}$	$+0.24 \times 10^{-28}$
Frequency ($^1H = 100$ MHz; 2.3488 T)/MHz	5.637	5.638

Reference: TiF_6^{2-} (conc. HF)

Electron shell properties

Ground state electron configuration: $[Ar]3d^24s^2$

Term symbol: $^3F_2$

Electron affinity ($M \rightarrow M^-$)/kJ mol^{-1}: −2

Ionization energies/kJ mol^{-1}

1. $M \rightarrow M^+$	658	6. $M^{5+} \rightarrow M^{6+}$	11 516
2. $M^+ \rightarrow M^{2+}$	1310	7. $M^{6+} \rightarrow M^{7+}$	13 590
3. $M^{2+} \rightarrow M^{3+}$	2652	8. $M^{7+} \rightarrow M^{8+}$	16 260
4. $M^{3+} \rightarrow M^{4+}$	4175	9. $M^{8+} \rightarrow M^{9+}$	18 640
5. $M^{4+} \rightarrow M^{5+}$	9573	10. $M^{9+} \rightarrow M^{10+}$	20 830

Principal lines in atomic spectrum

Wavelength/nm	Species	Sensitivity	Application
319.992	I		AA
334.941	II		
364.268	I		AA, AE
365.350	I	U2	AA, AE
498.173	I	U1	

Abundance: Earth's crust 6320 p.p.m.; seawater 0.001 p.p.m.

Biological role: None; non-toxic; suspected carcinogen; stimulatory

W

Atomic number: 74

Relative atomic mass ($^{12}C=12.0000$): 183.85

Chemical properties

Obtained as dull grey powder, difficult to melt. Metal is lustrous and silvery white. Ores are scheelite $CaWO_4$ and wolframite $(Fe,Mn)WO_4$. Resists attack by oxygen, acids, and alkalis. Used in alloys, light bulb filaments, and cutting tools.

Radii/pm: W^{4+} 68; W^{6+} 62; atomic 137.0; covalent 130

Electronegativity: 2.36 (Pauling); 1.40 (Allred-Rochow)

Effective nuclear charge: 4.35 (Slater); 9.85 (Clementi); 14.22 (Froese–Fischer)

Standard reduction potentials $E^{\ominus}$/V

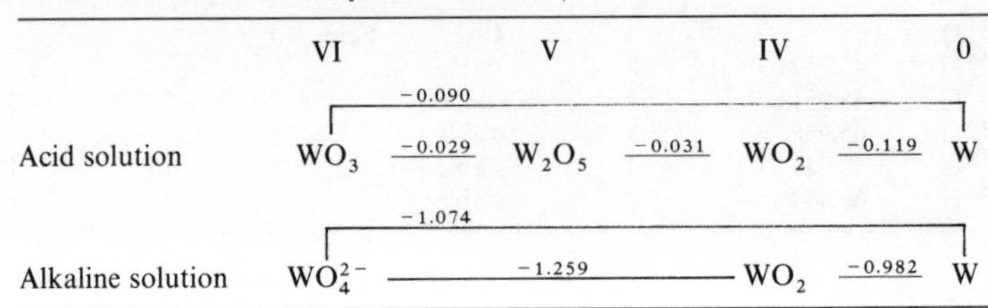

Oxidation states

W^{-IV}	(d^{10})	$[W(CO)_4]^{4-}$
W^{-II}	(d^8)	$[W(CO)_5]^{2-}$
W^{-I}	(d^7)	$[W_2(CO)_{10}]^{2-}$
W^0	(d^6)	$W(CO)_6$
W^{II}	(d^4)	WCl_2, WBr_2, WI_2, complexes
W^{III}	(d^3)	WCl_3, WBr_3, WI_3, complexes
W^{IV}	(d^2)	WO_2, WF_4, WCl_4, etc. WS_2, complexes
W^{V}	(d^1)	W_2O_5, WF_5, WCl_5, WF_6^-, complexes
W^{VI}	(d^0, f^{14})	WO_3, WO_4^{2-}, WF_6, WCl_6, $WOCl_4$, polytungstates, complexes

NB. There are no aqua ions of W in any oxidation state.

Physical properties

Melting point/K: 3680±20

Boiling point/K: 5930

ΔH_{fusion}/kJ mol^{-1}: 35.2

ΔH_{vap}/kJ mol^{-1}: 824.2

Thermodynamic properties (298.15 K, 0.1 MPa)

State	$\Delta_f H^{\ominus}$/kJ mol^{-1}	$\Delta_f G^{\ominus}$/kJ mol^{-1}	$S^{\ominus}$/J K^{-1} mol^{-1}	C_p/J K^{-1} mol^{-1}
Solid	0	0	32.64	24.27
Gas	849.4	807.1	173.950	21.309

Density/kg m^{-3}: 19 300 [293 K]; 17 700 [liquid at m.p.]

Thermal conductivity/W m^{-1} K^{-1}: 174 [300 K]

Electrical resistivity/Ω m: 5.65×10^{-8} [300 K]

Mass magnetic susceptibility/kg^{-1} m^3: $+4.0\times10^{-9}$ (s)

Molar volume/cm^3: 9.53

Coefficient of linear thermal expansion/K^{-1}: 4.59×10^{-6}

Lattice structure (cell dimensions/pm), space group

b.c.c. ($a=316.522$), Im3m

X-ray diffraction: mass absorption coefficients (μ/ρ)/cm^2 g^{-1}:
CuK_α 172 MoK_α 99.1

Isolated in 1783 by J. J. and F. Elhuijar at Vergara, Spain

[Swedish, *tung sten* = heavy stone]

Tungsten (Wolfram)

Nuclear properties

Atomic number: 74

Thermal neutron capture cross-section/barns: 18.5 ± 0.5

Number of isotopes (including nuclear isomers): 22

Isotope mass range: $173 \rightarrow 189$

Key isotopes

Nuclide	Atomic mass	Natural abundance (%)	Half life $T_{1/2}$	Decay mode and energy (MeV)	Nuclear spin I	Nuclear magnetic moment μ	Uses
^{180}W	179.9470	0.10	stable				
^{182}W	181.9483	26.3	stable		0		
^{183}W	182.9503	14.3	stable		1/2	+0.1169	NMR
^{184}W	183.9510	30.7	stable		0		
^{185}W		0	75d	$\beta^-(0.432)$; no γ	3/2		tracer
^{186}W	185.9543	28.6	stable		0		
^{187}W		0	23.9h	$\beta^-(1.315)$; γ	3/2		tracer

NMR

	^{183}W
Relative sensitivity ($^1H = 1.00$)	7.20×10^{-4}
Absolute sensitivity ($^1H = 1.00$)	1.03×10^{-5}
Receptivity ($^{13}C = 1.00$)	0.0589
Magnetogyric ratio/rad T^{-1} s^{-1}	1.1145×10^7
Frequency ($^1H = 100$ MHz; 2.3488 T)/MHz	4.161

Reference: WF_6

Electron shell properties

Ground state electron configuration: $[Xe]4f^{14}5d^46s^2$

Term symbol: $^5D_0$

Electron affinity ($M \rightarrow M^-$)/kJ mol^{-1}: 119

Ionization energies/kJ mol^{-1}

1. $M \rightarrow M^+$	770	6. $M^{5+} \rightarrow M^{6+}$	(5900)
2. $M^+ \rightarrow M^{2+}$	(1700)	7. $M^{6+} \rightarrow M^{7+}$	
3. $M^{2+} \rightarrow M^{3+}$	(2300)	8. $M^{7+} \rightarrow M^{8+}$	
4. $M^{3+} \rightarrow M^{4+}$	(3400)	9. $M^{8+} \rightarrow M^{9+}$	
5. $M^{4+} \rightarrow M^{5+}$	(4600)	10. $M^{9+} \rightarrow M^{10+}$	

Principal lines in atomic spectrum

Wavelength/nm	Species	Sensitivity	Application
202.998	II		
255.135	I		AA
268.142	I		AA
400.875	I	U3	AE
429.461	I	U2	AE
430.211	I	U1	AE
498.259*	I		

* Lowest energy transition from ground state to nearest empty orbital

Abundance: Earth's crust 1.2 p.p.m.; seawater $<1 \times 10^{-6}$ p.p.m.

Biological role: None; probably of low toxicity

U

Atomic number: 92

Relative atomic mass ($^{12}C=12.0000$): 238.0289

Chemical properties

Radioactive silvery metal; large deposits. Malleable, ductile, and tarnishes in air. Attacked by steam and acids but not by alkalis. Used a nuclear fuel and in nuclear weapons.

Radii/pm: U^{3+} 103; U^{4+} 97; U^{5+} 89; U^{6+} 80; atomic 138.5

Electronegativity: 1.38 (Pauling); 1.22 (Allred-Rochow)

Effective nuclear charge: 1.80 (Slater)

Standard reduction potentials $E^{\ominus}$/V

	VI		V		IV		III		0
Acid solution	UO_2^{2+}	0.16	UO_2^+	−0.38	U^{4+}	−0.52	U^{3+}	−1.66	U
Alkaline solution	$UO_2(OH)_2$		−0.3		UO_2	−2.6	$U(OH)_3$	−2.10	U

(Acid solution: VI→IV −0.027; IV→0 −1.38)

Oxidation states

U^{II} (f^3 d^1)	UO?
U^{III} (f^3)	$[U(H_2O)_x]^{3+}$ (aq) unstable, UF_3, UCl_3 etc., $[U(C_5H_5)_3]$
U^{IV} (f^2)	UO_2, $[U(H_2O)_x]^{4+}$ (aq), salts, UF_4, UCl_4 etc., $[UCl_6]^{2-}$
U^{V} (f^1)	U_2O_5, UO_2^+ (aq) unstable, UF_5, UCl_5, UBr_5, UF_6^-, UF_7^{2-}, UF_8^{3-}
U^{VI} (f^0, [Rn])	UO_3 (U_3O_8), UO_2^{2+} (aq), salts, UF_6, UCl_6, complexes

Physical properties

Melting point/K: 1405.5

Boiling point/K: 4018

ΔH_{fusion}/kJ mol^{-1}: 15.5

ΔH_{vap}/kJ mol^{-1}: 417.1

Thermodynamic properties (298.15 K, 0.1 MPa)

State	$\Delta_f H^{\ominus}$/kJ mol^{-1}	$\Delta_f G^{\ominus}$/kJ mol^{-1}	$S^{\ominus}$/J K^{-1} mol^{-1}	C_p/J K^{-1} mol^{-1}
Solid	0	0	50.21	27.665
Gas	535.6	491.2	199.77	23.694

Density/kg m^{-3}: 18 950 [293 K]; 17 907 [liquid at m.p.]

Thermal conductivity/W m^{-1} K^{-1}: 27.6 [300 K]

Electrical resistivity/Ω m: 30.8×10^{-8} [273 K]

Mass magnetic susceptibility/kg^{-1} m^3: $+2.16 \times 10^{-8}$ (s)

Molar volume/cm^3: 12.56

Coefficient of linear thermal expansion/K^{-1}: 12.6×10^{-6}

Lattice structure (cell dimensions/pm), space group

α-U orthorhombic ($a=284.785$, $b=585.801$, $c=494.553$), Cmcm

β-U tetragonal ($a=1076.0$, $c=565.2$), $P4_2/mnm$ or $P4_2nm$

γ-U b.c.c. ($a=352.4$), Im3m

$T(\alpha \rightarrow \beta)=941$ K

$T(\beta \rightarrow \gamma)=1047$ K

X-ray diffraction: mass absorption coefficients (μ/ρ)/cm^2 g^{-1}:
CuK_α 352 MoK_α 153

Discovered in 1789 by M. H. Klaproth at Berlin, Germany; isolated in 1841 by E.-M. Peligot at Paris, France

[Named after the planet Uranus]

Uranium

Nuclear properties

Atomic number: 92

Thermal neutron capture cross-section/barns: 7.60 ± 0.07

Number of isotopes (including nuclear isomers): 15

Isotope mass range: $227 \rightarrow 240$

Key isotopes

Nuclide	Atomic mass	Natural abundance (%)	Half life $T_{1/2}$	Decay mode and energy (MeV)	Nuclear spin I	Nuclear magnetic moment μ	Uses
^{234}U	234.0409	0.005	2.47×10^5y	$\alpha(4.856)$; γ			
^{235}U	235.0439	0.720	7.00×10^8y	$\alpha(4.681)$; SF; γ	7/2	-0.43	NMR
^{236}U	236.0457	0	2.39×10^7y	$\alpha(4.573)$; SF; γ			
^{238}U	238.0508	99.275	4.51×10^9y	$\alpha(4.268)$; SF: γ			

NMR

	^{235}U
Relative sensitivity ($^1H = 1.00$)	1.21×10^{-4}
Absolute sensitivity ($^1H = 1.00$)	8.71×10^{-7}
Receptivity ($^{13}C = 1.00$)	5.4×10^{-3}
Magnetogyric ratio/rad T^{-1} s^{-1}	-0.4926×10^7
Quadrupole moment/m^2	4.55×10^{-28}
Frequency ($^1H = 100$ MHz; 2.3488 T)/MHz	1.790

Reference: UF_6

Electron shell properties

Ground state electron configuration: $[Rn]5f^3 6d^1 7s^2$

Term symbol: $^5L_6$

Electron affinity ($M \rightarrow M^-$)/kJ mol^{-1}: n.a.

Ionization energies/kJ mol^{-1}

1. $M \rightarrow M^+$	584	6. $M^{5+} \rightarrow M^{6+}$
2. $M^+ \rightarrow M^{2+}$	1420	7. $M^{6+} \rightarrow M^{7+}$
3. $M^{2+} \rightarrow M^{3+}$		8. $M^{7+} \rightarrow M^{8+}$
4. $M^{3+} \rightarrow M^{4+}$		9. $M^{8+} \rightarrow M^{9+}$
5. $M^{4+} \rightarrow M^{5+}$		10. $M^{9+} \rightarrow M^{10+}$

Principal lines in atomic spectrum

Wavelength/nm	Species	Sensitivity	Application
356.659	I		AA
358.488	I		AA
367.007	II		AE
385.958	II		
591.540	I		AE

Abundance: Earth's crust 2.3 p.p.m.; seawater 0.003 p.p.m.

Biological role: None; toxic and dangerous due to radioactivity

V

Atomic number: 23

Relative atomic mass (^{12}C=12.0000): 50.9415

Chemical properties

Shiny, silvery metal, soft when pure. Found as patronite VS_4 but obtained as by-product from other ores and Venezuelan oils. Resists corrosion due to protective oxide film. Attacked by concentrated acids but not by fused alkalis. Used mainly as alloys and in steel.

Radii/pm: V^{2+} 72; V^{3+} 65; V^{4+} 61; V^{5+} 59; atomic 132.1

Electronegativity: 1.63 (Pauling); 1.45 (Allred-Rochow)

Effective nuclear charge: 3.30 (Slater); 4.98 (Clementi); 6.65 (Froese–Fischer)

Standard reduction potentials $E^{\ominus}$/V

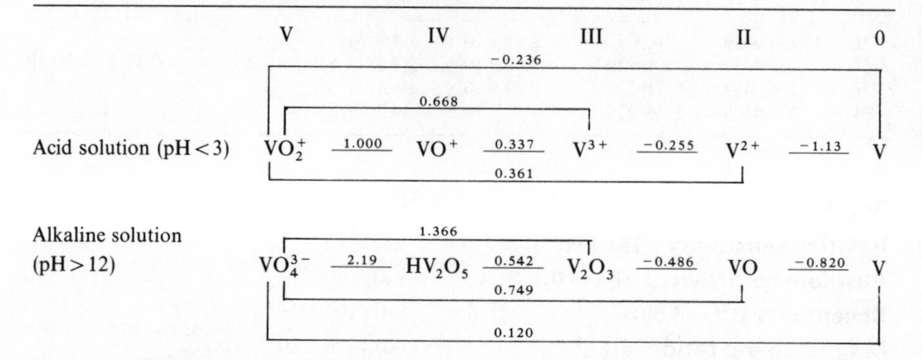

Oxidation states

V^{-III}	(d^8)	rare $[V(CO)_5]^{3-}$
V^{-I}	(d^6)	$[V(CO)_6]^-$
V^0	(d^5)	$[V(CO)_6]$
V^I	(d^4)	$[V(bipyridyl)_3]^+$
V^{II}	(d^3)	VO, $[V(H_2O)_6]^{2+}$ (aq), VF_2, VCl_2, complexes
V^{III}	(d^2)	V_2O_3, $[V(H_2O)_6]^{3+}$ (aq), VF_3, VCl_3, $[VCl_4]^-$
V^{IV}	(d^1)	VO_2, VO^{2+} (aq), VF_4, VCl_4, complexes
V^V	(d^0, [Ar])	V_2O_5, VO_2^+ (aq), VO_4^{3-} (aq, alkali), VF_5, VF_6^-, complexes

Physical properties

Melting point/K: 2160

Boiling point/K: 3650

ΔH_{fusion}/kJ mol^{-1}: 17.6

ΔH_{vap}/kJ mol^{-1}: 459.70

Thermodynamic properties (298.15 K, 0.1 MPa)

State	$\Delta_f H^{\ominus}$/kJ mol^{-1}	$\Delta_f G^{\ominus}$/kJ mol^{-1}	$S^{\ominus}$/J K^{-1} mol^{-1}	C_p/J K^{-1} mol^{-1}
Solid	0	0	28.91	24.89
Gas	514.21	754.43	182.298	26.012

Density/kg m^{-3}: 6110 [292 K]; 5550 [liquid at m.p.]

Thermal conductivity/W m^{-1} K^{-1}: 30.7 [300 K]

Electrical resistivity/K: 24.8×10^{-8} [293 K]

Mass magnetic susceptibility/kg^{-1} m^3: $+6.28 \times 10^{-8}$ (s)

Molar volume/cm^3: 8.34

Coefficient of linear thermal expansion/K^{-1}: 8.3×10^{-6}

Lattice structure (cell dimensions/pm), space group

b.c.c. (a=302.40), Im3m

X-ray diffraction: mass absorption coefficients (μ/ρ)/cm^2 g^{-1}:
CuK_α 233 MoK_α 27.5

Discovered in 1801 by A. M. del Rio at Mexico City, Mexico; rediscovered in 1831 by N. G. Selfström at Falun, Sweden

[Named after *Vanadis*, Scandinavian goddess]

Vanadium

Nuclear properties

Atomic number: 23

Thermal neutron capture cross-section/barns: 5.06 ± 0.06

Number of isotopes (including nuclear isomers): 9

Isotope mass range: 46→54

Key isotopes

Nuclide	Atomic mass	Natural abundance (%)	Half life $T_{1/2}$	Decay mode and energy (MeV)	Nuclear spin I	Nuclear magnetic moment μ	Uses
^{48}V	47.952	0	16.0d	β^+(4.013), EC; γ	4	±1.6	tracer
^{49}V	48.948	0	330d	EC(0.601); no γ	7/2	±4.5	tracer
^{50}V	49.9472	0.250	6×10^{15}y	EC	6	+3.3470	NMR
^{51}V	50.9440	99.750	stable		7/2	+5.1485	NMR

NMR

	[^{50}V]	^{51}V
Relative sensitivity (^1H=1.00)	5.55×10^{-2}	0.38
Absolute sensitivity (^1H=1.00)	1.33×10^{-4}	0.38
Receptivity (^{13}C=1.00)	0.755	2150
Magnetogyric ratio/rad T^{-1} s^{-1}	2.6491×10^7	7.0362×10^7
Quadrupole moment/m^2	$+0.21 \times 10^{-28}$	-0.052×10^{-28}
Frequency (^1H=100 MHz; 2.3488 T)/MHz	9.970	26.289

Reference: $VOCl_3$

Electron shell properties

Ground state electron configuration: $[Ar]3d^34s^2$

Term symbol: $^4F_{3/2}$

Electron affinity ($M \rightarrow M^-$)/kJ mol^{-1}: 61

Ionization energies/kJ mol^{-1}

1. $M \rightarrow M^+$	650	6. $M^{5+} \rightarrow M^{6+}$	12 362
2. $M^+ \rightarrow M^{2+}$	1414	7. $M^{6+} \rightarrow M^{7+}$	14 489
3. $M^{2+} \rightarrow M^{3+}$	2828	8. $M^{7+} \rightarrow M^{8+}$	16 760
4. $M^{3+} \rightarrow M^{4+}$	4507	9. $M^{8+} \rightarrow M^{9+}$	19 860
5. $M^{4+} \rightarrow M^{5+}$	6294	10. $M^{9+} \rightarrow M^{10+}$	22 240

Principal lines in atomic spectrum

Wavelength/nm	Species	Sensitivity	Application
309.311	II	V1	AE
318.398	I		AA, AE
318.540	I	U2	AA, AE
437.924	I	U1	AA, AE
485.148*	I		

* Lowest energy transition from ground state to nearest empty orbital

Abundance: Earth's crust 136 p.p.m.; seawater 0.002 p.p.m.

Biological role: Essential trace element; some compounds quite toxic; stimulatory

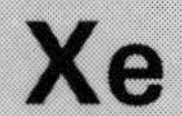

Atomic number: 54
Relative atomic mass (^{12}C = 12.0000): 131.29

Chemical properties

Colourless, odourless gas obtained from liquid air. Inert towards most other chemicals but reacts with fluorine gas to form xenon fluorides. Oxides, acids and salts known. Little used outside research.

Radii/pm: Xe^{+} 190; atomic 218; covalent 209

Electronegativity: 2.6 (Pauling); n.a. (Allred-Rochow)

Effective nuclear charge: 8.25 (Slater); 12.42 (Clementi); 15.61 (Froese–Fischer)

Standard reduction potentials $E^{\ominus}$/V

	VIII		IV		II		I		0
Acid solution	H_4XeO_6	2.42	XeO_3	2.12					Xe
					XeF_2	0.9	XeF	3.4	Xe
Alkaline solution	$HXeO_6^{3-}$	0.99	$HXeO_4^-$	1.24					Xe

Acid solution: H_4XeO_6 → Xe 2.18; XeF_2 → Xe 2.32. Alkaline solution: $HXeO_6^{3-}$ → Xe 1.18.

Covalent bonds

	r/pm	E/kJ mol^{-1}
Xe–O (XeO_3)	176	84
Xe–F	194	133

Oxidation states

Xe^0	clathrates: $Xe_8(H_2O)_{46}$, Xe (quinol)$_3$
Xe^{II}	XeF_2, $[XeF]^+[AsF_6]^-$
Xe^{IV}	XeF_4
Xe^{VI}	XeO_3, $XeOF_4$, XeO_2F_2, XeF_6, XeF_7^-, XeF_8^{2-}, $[XeF_5]^+[AsF_6]^-$
Xe^{VIII}	XeO_4, XeO_3F_2, Ba_2XeO_6, XeO_6^{4-} (aq)

Physical properties

Melting point/K: 161.3

Boiling point/K: 166.1

ΔH_{fusion}/kJ mol^{-1}: 3.10

ΔH_{vap}/kJ mol^{-1}: 12.65

Thermodynamic properties (298.15 K, 0.1 MPa)

State	$\Delta_f H^{\ominus}$/kJ mol^{-1}	$\Delta_f G^{\ominus}$/kJ mol^{-1}	$S^{\ominus}$/J K^{-1} mol^{-1}	C_p/J K^{-1} mol^{-1}
Gas	0	0	169.683	20.786

Density/kg m^{-3}: 3540 [solid, m.p.]; 2939 [liquid, b.p.]; 5.8971 [gas, 273 K]

Thermal conductivity/W m^{-1} K^{-1}: 0.005 69 [300 K]$_g$

Mass magnetic susceptibility/kg^{-1} m^3: -4.20×10^{-9} (g)

Molar volume/cm^3: 37.09 [161 K]

Lattice structure (cell dimensions/pm), space group

f.c.c. (88 K) (a = 619.7), Fm3m

X-ray diffraction: mass absorption coefficients (μ/ρ)/cm^2 g^{-1}:
CuK$_\alpha$ 306 MoK$_\alpha$ 39.2

Discovered in 1898 by Sir William Ramsay and M. W. Travers at London, UK

[Greek, *xenos* = stranger]

Xenon

Nuclear properties

Atomic number: 54

Thermal neutron capture cross-section/barns: 24.5

Number of isotopes (including nuclear isomers): 31

Isotope mass range: 118→142

Key isotopes

Nuclide	Atomic mass	Natural abundance (%)	Half life $T_{1/2}$	Decay mode and energy (MeV)	Nuclear spin I	Nuclear magnetic moment μ	Uses
^{127}Xe		0	36.4d	EC(0.44); γ			tracer
^{129}Xe	128.9048	26.4	stable		1/2	−0.7768	NMR
^{130}Xe	129.9035	4.1	stable				
^{131}Xe	130.9051	21.2	stable		3/2	+0.6908	NMR
^{132}Xe	131.9042	26.9	stable		0		
^{133}Xe		0	5.270d	β^-(0.427); γ			tracer, medical
^{134}Xe	133.9054	10.4	stable		0		
^{136}Xe	135.9072	8.9	stable		0		

NMR

	^{129}Xe	[^{131}Xe]
Relative sensitivity (^1H=1.00)	2.12×10^{-2}	2.76×10^{-3}
Absolute sensitivity (^1H=1.00)	5.60×10^{-3}	5.84×10^{-4}
Receptivity (^{13}C=1.00)	31.8	3.31
Magnetogyric ratio/rad T^{-1} s^{-1}	-7.4003×10^{7}	2.1939×10^{7}
Quadrupole moment/m^2	—	-0.12×10^{-28}
Frequency (^1H = 100 MHz; 2.3488 T)/MHz	27.660	8.199

Reference: $XeOF_4$

Electron shell properties

Ground state electron configuration: $[Kr]4d^{10}5s^25p^6$

Term symbol: $^1S_0$

Electron affinity ($M\rightarrow M^-$)/kJ mol^{-1}: −41 (calc.)

Ionization energies/kJ mol^{-1}

1. $M \rightarrow M^+$	1170.4	6. $M^{5+}\rightarrow M^{6+}$	(6600)
2. $M^+\rightarrow M^{2+}$	2046	7. $M^{6+}\rightarrow M^{7+}$	(9300)
3. $M^{2+}\rightarrow M^{3+}$	3097	8. $M^{7+}\rightarrow M^{8+}$	(10600)
4. $M^{3+}\rightarrow M^{4+}$	(4300)	9. $M^{8+}\rightarrow M^{9+}$	(19800)
5. $M^{4+}\rightarrow M^{5+}$	(5500)	10. $M^{9+}\rightarrow M^{10+}$	(23000)

Principal lines in atomic spectrum

Wavelength/nm	Species	Sensitivity	Application
146.961* (vac)	I		
450.098	I	U4	AE
462.428	I	U3	AE
467.123	I	U2	AE
823.164	I		

* Lowest energy transition from ground state to nearest empty orbital

Abundance: Atmosphere 0.086 p.p.m. (by volume); Earth's crust nil; seawater 4.7×10^{-5} p.p.m.

Biological role: None; non-toxic as element

Yb

Atomic number: 70
Relative atomic mass ($^{12}C=12.0000$): 173.04

Chemical properties

Soft, silvery-white metal of the lanthanide (rare earth) group. Obtained from euxenite. Slowly oxidized by air, reacts with water. Little used.

Radii/pm: Yb^{2+} 113; Yb^{3+} 86; atomic 194; covalent 170

Electronegativity: n.a. (Pauling); 1.06 (Allred-Rochow)

Effective nuclear charge: 2.85 (Slater); 8.59 (Clementi); 11.90 (Froese–Fischer)

Standard reduction potentials $E^{\ominus}/V$

	III		II		0
		−2.22 (III→0)			
Acid solution	Yb^{3+}	−1.05	Yb^{2+}	−2.8	Yb
Alkaline solution	$Yb(OH)_3$		−2.74		Yb

Oxidation states

Yb^{II} (f^{14})	YbO, YbS, YbF_2, $YbCl_2$ etc.
Yb^{III} (f^{13})	Yb_2O_3, $Yb(OH)_3$, $[Yb(H_2O)_x]^{3+}$ (aq), Yb^{3+} salts, YbF_3, $YbCl_3$, etc. $YbCl_6^{3-}$, complexes

Physical properties

Melting point/K: 1097
Boiling point/K: 1466
ΔH_{fusion}/kJ mol^{-1}: 9.20
ΔH_{vap}/kJ mol^{-1}: 159

Thermodynamic properties (298.15 K, 0.1 MPa)

State	$\Delta_f H^{\ominus}$/kJ mol^{-1}	$\Delta_f G^{\ominus}$/kJ mol^{-1}	$S^{\ominus}$/J K^{-1} mol^{-1}	C_p/J K^{-1} mol^{-1}
Solid	0	0	59.87	26.74
Gas	152.3	118.4	173.126	20.786

Density/kg m^{-3}: 6965 [293 K]
Thermal conductivity/W m^{-1} K^{-1}: 34.9 [300 K]
Electrical resistivity/Ω m: 29.0×10^{-8} [293 K]
Mass magnetic susceptibility/kg^{-1} m^3: $+1.81\times10^{-8}$ (s)
Molar volume/cm^3: 24.84
Coefficient of linear thermal expansion/K^{-1}: 25.0×10^{-6}

Lattice structure (cell dimensions/pm), space group

α-Yb f.c.c. ($a=548.62$), Fm3m
β-Yb b.c.c. ($a=444$), Im3m
$T(\alpha\rightarrow\beta)=1073$ K

X-ray diffraction: mass absorption coefficients $(\mu/\rho)/cm^2\ g^{-1}$:
CuK_α 146 MoK_α 84.5

Discovered in 1878 by J. C. G. de Marignac at Geneva, Switzerland

[Named after *Ytterby*, Sweden]

Ytterbium

Nuclear properties

Atomic number: 70

Thermal neutron capture cross-section/barns: 37 ± 3

Number of isotopes (including nuclear isomers): 16

Isotope mass range: $164 \rightarrow 177$

Key isotopes

Nuclide	Atomic mass	Natural abundance (%)	Half life $T_{1/2}$	Decay mode and energy (MeV)	Nuclear spin I	Nuclear magnetic moment μ	Uses
^{168}Yb	167.9339	0.14	stable				
^{169}Yb		0	31.8d	EC(*c.* 1.2); γ			tracer, medical
^{170}Yb	169.9349	3.06	stable				
^{171}Yb	170.9365	14.4	stable		1/2	+0.4919	NMR
^{172}Yb	171.9366	21.9	stable				
^{173}Yb	172.9383	16.1	stable		5/2	−0.6776	NMR
^{174}Yb	173.9390	31.8	stable				
^{175}Yb		0	101h	β^-(0.467); γ	(7/2)	±0.3	tracer
^{176}Yb	175.9427	12.7	stable				

NMR (rarely studied)

	^{171}Yb	^{173}Yb
Relative sensitivity ($^1H = 1.00$)	5.46×10^{-3}	1.33×10^{-3}
Absolute sensitivity ($^1H = 1.00$)	7.81×10^{-4}	2.14×10^{-4}
Receptivity ($^{13}C = 1.00$)	4.05	1.14
Magnetogyric ratio/rad T^{-1} s^{-1}	4.718×10^7	1.310×10^7
Quadrupole moment/m^2	—	$0.4\text{–}3.9 \times 10^{-28}$
Frequency ($^1H = 100$ MHz; 2.3488 T)/MHz	17.613	4.852

Electron shell properties

Ground state electron configuration: $[Xe]4f^{14}6s^2$

Term symbol: $^1S_0$

Electron affinity ($M \rightarrow M^-$)/kJ mol^{-1}: $\leqslant 50$

Ionization energies/kJ mol^{-1}

1. $M \rightarrow M^+$ 603.4
2. $M^+ \rightarrow M^{2+}$ 1176
3. $M^{2+} \rightarrow M^{3+}$ 2415
4. $M^{3+} \rightarrow M^{4+}$ 4220
5. $M^{4+} \rightarrow M^{5+}$
6. $M^{5+} \rightarrow M^{6+}$
7. $M^{6+} \rightarrow M^{7+}$
8. $M^{7+} \rightarrow M^{8+}$
9. $M^{8+} \rightarrow M^{9+}$
10. $M^{9+} \rightarrow M^{10+}$

Principal lines in atomic spectrum

Wavelength/nm	Species	Sensitivity	Application
246.450	I		AA
328.937	II		AE
346.437	I		AA
369.420	II		AE
398.799	I		AA, AE

Abundance: Earth's crust 3.1 p.p.m.; seawater 2×10^{-7} p.p.m.

Biological role: None; non-toxic

Y

Atomic number: 39
Relative atomic mass ($^{12}C = 12.0000$): **88.90585**

Chemical properties

Soft, silvery-white metal. Stable in air due to formation of oxide film. Burns easily. Reacts with water to form hydrogen, H_2. Used as yttrium phosphors to give red colours in television screens.

Radii/pm: Y^{3+} 106; atomic 181; covalent 162

Electronegativity: 1.22 (Pauling); 1.11 (Allred-Rochow)

Effective nuclear charge: 3.00 (Slater); 6.26 (Clementi); 8.72 (Froese–Fischer)

Standard reduction potentials $E^{\ominus}$/V

	III		0
Acid solution	Y^{3+}	$\xrightarrow{-2.37}$	Y
Alkaline solution	$Y(OH)_3$	$\xrightarrow{-2.85}$	Y

Oxidation states

Y^{III} ([Kr])	Y_2O_3, $Y(OH)_3$, $[Y(H_2O)_x]^{3+}$ (aq), Y^{3+} salts, $Y_2(CO_3)_3$, YF_3, YCl_3 etc., YOCl, some complexes
YH_2–YH_3 consists of $Y^{3+}H^-$, complex bonding	

Physical properties

Melting point/K: 1795

Boiling point/K: 3611

ΔH_{fusion}/kJ mol^{-1}: 17.2

ΔH_{vap}/kJ mol^{-1}: 367.4

Thermodynamic properties (298.15 K, 0.1 MPa)

State	$\Delta_f H^{\ominus}$/kJ mol^{-1}	$\Delta_f G^{\ominus}$/kJ mol^{-1}	$S^{\ominus}$/J K^{-1} mol^{-1}	C_p/J K^{-1} mol^{-1}
Solid	0	0	44.43	26.53
Gas	421.3	381.1	1179.48	25.86

Density/kg m^{-3}: 4469 [293 K]

Thermal conductivity/W m^{-1} K^{-1}: 17.2 [300 K]

Electrical resistivity/Ω m: 57.0×10^{-8} [298 K]

Mass magnetic susceptibility/kg^{-1} m^3: $+2.70 \times 10^{-8}$ (s)

Molar volume/cm^3: 19.89

Coefficient of linear thermal expansion/K^{-1}: 10.6×10^{-6}

Lattice structure (cell dimensions/pm), space group

α-Y h.c.p. ($a = 364.74$; $c = 573.06$), $P6_3/mmc$
β-Y b.c.c. ($a = 411$), Im3m
$T(\alpha \rightarrow \beta) = 1763$ K

X-ray diffraction: mass absorption coefficients (μ/ρ)/cm^2 g^{-1}:
CuK_α 134 MoK_α 100

Discovered in 1794 by J. Gadolin at Åbo, Finland

[Named after *Ytterby*, Sweden]

Yttrium

Nuclear properties

Atomic number: 39

Thermal neutron capture cross-section/barns: 1.3±0.1

Number of isotopes (including nuclear isomers): 21

Isotope mass range: 82→96

Key isotopes

Nuclide	Atomic mass	Natural abundance (%)	Half life $T_{1/2}$	Decay mode and energy (MeV)	Nuclear spin I	Nuclear magnetic moment μ	Uses
^{88}Y	87.910	0	106.6d	β^+(3.621); EC, γ			tracer
^{89}Y	88.9054	100	stable		1/2	−0.1373	NMR
^{90}Y		0	64h	β^-(2.27) ; no γ	2	−1.63	tracer

NMR

	^{89}Y
Relative sensitivity (^1H=1.00)	1.18×10^{-4}
Absolute sensitivity (^1H=1.00)	1.18×10^{-4}
Receptivity (^{13}C=1.00)	0.668
Magnetogyric ratio/rad T^{-1} s^{-1}	-1.3108×10^7
Frequency (^1H=100 MHz; 2.3488 T)/MHz	4.899

Reference: $Y(NO_3)_3$ (aq)

Electron shell properties

Ground state electron configuration: $[Kr]4d^15s^2$

Term symbol: $^2D_{3/2}$

Electron affinity ($M\to M^-$)/kJ mol^{-1}: −39

Ionization energies/kJ mol^{-1}

1.	$M \to M^+$	616
2.	$M^+ \to M^{2+}$	1181
3.	$M^{2+} \to M^{3+}$	1980
4.	$M^{3+} \to M^{4+}$	5963
5.	$M^{4+} \to M^{5+}$	7430
6.	$M^{5+} \to M^{6+}$	8970
7.	$M^{6+} \to M^{7+}$	11 200
8.	$M^{7+} \to M^{8+}$	12 400
9.	$M^{8+} \to M^{9+}$	14 137
10.	$M^{9+} \to M^{10+}$	18 400

Principal lines in atomic spectrum

Wavelength/nm	Species	Sensitivity	Application
371.030	II	V1	AE
407.738	I		AA
410.238	I		AA
464.370	I	U2	AE
467.484	I	U1	AE
668.758*	I		

* Lowest energy transition from ground state to nearest empty orbital

Abundance: Earth's crust 31 p.p.m.; seawater 0.0003 p.p.m.

Biological role: None; probably of low toxicity; suspected carcinogen

Zn

Atomic number: 30
Relative atomic mass ($^{12}C = 12.0000$): 65.39

Chemical properties

Bluish-white metal, brittle when cast. Major ores are zinc blende, ZnS, and calamine (smithsonite) $ZnCO_3$. Tarnishes in air, reacts with acids and alkalis. Used ion galvanizing iron, in alloys, e.g. brass, in batteries, as ZnO in rubber, and as polymer stabilizer.

Radii/pm: Zn^{2+} 83; atomic 133.2; covalent 125
Electronegativity: 1.65 (Pauling); 1.66 (Allred-Rochow)
Effective nuclear charge: 4.35 (Slater); 5.97 (Clementi); 8.28 (Froese–Fischer)

Standard reduction potentials $E^\ominus$/V

	II		0
Acid solution	Zn^{2+}	−0.7626	Zn
Alkaline solution	$Zn(OH)_4^{2-}$	−1.285	Zn
	$Zn(OH)_2$	−1.246	Zn

Oxidation states

Zn^I ($d^{10}s^1$)	rare Zn_2^{2+} in Zn/$ZnCl_2$ glass
Zn^{II} (d^{10})	ZnO, ZnS, $Zn(OH)_2$, $[Zn(H_2O)_6]^{2+}$ (aq), $[Zn(OH)_4]^{2-}$ (aq alkali) Zn^{2+} salts, ZnF_2, $ZnCl_2$ etc., many complexes

Physical properties

Melting point/K: 692.73
Boiling point/K: 1180
ΔH_{fusion}/kJ mol^{-1}: 6.67
ΔH_{vap}/kJ mol^{-1}: 114.2

Thermodynamic properties (298.15 K, 0.1 MPa)

State	$\Delta_f H^\ominus$/kJ mol^{-1}	$\Delta_f G^\ominus$/kJ mol^{-1}	$S^\ominus$/J K^{-1} mol^{-1}	C_p/J K^{-1} mol^{-1}
Solid	0	0	41.63	25.40
Gas	130.729	95.145	160.984	20.786

Density/kg m^{-3}: 7133 [293 K]; 6577 [liquid at m.p.]
Thermal conductivity/W m^{-1} K^{-1}: 116 [300 K]
Electrical resistivity/Ω m: 5.916×10^{-8} [293 K]
Mass magnetic susceptibility/kg^{-1} m^3: -2.20×10^{-9} (s)
Molar volume/cm^3: 9.17
Coefficient of linear thermal expansion/K^{-1}: 25.0×10^{-6}

Lattice structure (cell dimensions/pm), space group

h.c.p. ($a = 266.47$; $c = 494.69$), $P6_3/mmc$

X-ray diffraction: mass absorption coefficients (μ/ρ)/cm^2 g^{-1}:
CuK_α 60.3 MoK_α 55.4

Known in India and China before 1500

[German, *zink*]

Zinc

Nuclear properties

Atomic number: 30

Thermal neutron capture cross-section/barns: 1.10 ± 0.04

Number of isotopes (including nuclear isomers): 15

Isotope mass range: $60 \rightarrow 72$

Key isotopes

Nuclide	Atomic mass	Natural abundance (%)	Half life $T_{1/2}$	Decay mode and energy (MeV)	Nuclear spin I	Nuclear magnetic moment μ	Uses
^{64}Zn	63.9291	48.6	$>8 \times 10^{15}$y		0		
^{65}Zn	64.926	0	243.6d	β^+(1.353); EC; γ	5/2	+0.7692	tracer
^{66}Zn	65.9260	27.9	stable		0		
^{67}Zn	66.9271	4.1	stable		5/2	+0.8752	NMR
^{68}Zn	67.9249	18.8	stable		0		
^{69m}Zn		0	13.9h	IT(0.4389); γ			tracer
^{70}Zn	69.9253	0.6	stable				

NMR

	^{67}Zn
Relative sensitivity ($^1H = 1.00$)	2.85×10^{-3}
Absolute sensitivity ($^1H = 1.00$)	1.17×10^{-4}
Receptivity ($^{13}C = 1.00$)	0.665
Magnetogyric ratio/rad T^{-1} s^{-1}	1.6737×10^7
Quadrupole moment/m^2	0.15×10^{-28}
Frequency ($^1H = 100$ MHz; 2.3488 T)/MHz	6.254

References: $Zn(ClO_4)_2$ (aq)

Electron shell properties

Ground state electron configuration: $[Ar]3d^{10}4s^2$

Term symbol: $^1S_0$

Electron affinity ($M \rightarrow M^-$)/kJ mol^{-1}: 9

Ionization energies/kJ mol^{-1}

1. $M \rightarrow M^+$	906.4	6. $M^{5+} \rightarrow M^{6+}$	10 400
2. $M^+ \rightarrow M^{2+}$	1733.3	7. $M^{6+} \rightarrow M^{7+}$	12 900
3. $M^{2+} \rightarrow M^{3+}$	3832.6	8. $M^{7+} \rightarrow M^{8+}$	16 800
4. $M^{3+} \rightarrow M^{4+}$	5730	9. $M^{8+} \rightarrow M^{9+}$	19 600
5. $M^{4+} \rightarrow M^{5+}$	7970	10. $M^{9+} \rightarrow M^{10+}$	23 000

Principal lines in atomic spectrum

Wavelength/nm	Species	Sensitivity	Application
213.856	I	U1	AA, AE
250.200	II	V4	AE
255.795	II	V3	AE
307.590*	I		AA
330.259	I	U3	AE
334.502	I	U2	AE

* Lowest energy transition from ground state to nearest empty orbital

Abundance: Earth's crust 76 p.p.m.; seawater 0.01 p.p.m.

Biological role: Essential trace element; non-toxic except in large excess; carcinogenic

Zr

Atomic number: 40
Relative atomic mass ($^{12}C=12.0000$): 91.224

Chemical properties

Hard, lustrous, silvery metal. Major minerals baddeleyite, ZrO_2, and zircon, $ZrSiO_4$. Very corrosion-resistant due to oxide layer, but will burn in air. Unaffected by acids, (except HF), and alkalis. Used in alloys, coloured glazes, and nuclear reactors.

Radii/pm: Zr^{2+} 109; Zr^{4+} 87; atomic 160; covalent 145

Electronegativity: 1.33 (Pauling); 1.22 (Allred-Rochow)

Effective nuclear charge: 3.15 (Slater); 6.45 (Clementi); 9.20 (Froese–Fischer)

Standard reduction potentials $E^\ominus$/V

IV		0
Zr^{4+}	$\xrightarrow{-1.55}$	Zr

Oxidation states

Zr^0	(d^4)	rare $[Zr(bipyridyl)_3]$
Zr^I	(d^3)	ZrCl?
Zr^{II}	(d^2)	ZrO? $ZrCl_2$
Zr^{III}	(d^1)	$ZrCl_3$, $ZrBr_3$, ZrI_3, Zr^{3+} reduces H_2O
Zr^{IV}	([Kr])	ZrO_2, $Zr(OH)^{3+}$ (aq), ZrF_4, $ZrCl_4$ etc., ZrF_6^{2-}, ZrF_7^{3-}, ZrF_8^{4-}, zirconates, complexes

Physical properties

Melting point/K: 2125

Boiling point/K: 4650

ΔH_{fusion}/kJ mol^{-1}: 23.0

ΔH_{vap}/kJ mol^{-1}: 566.7

Thermodynamic properties (298.15 K, 0.1 MPa)

State	$\Delta_f H^\ominus$/kJ mol^{-1}	$\Delta_f G^\ominus$/kJ mol^{-1}	$S^\ominus$/J K^{-1} mol^{-1}	C_p/J K^{-1} mol^{-1}
Solid	0	0	38.99	25.36
Gas	608.8	566.5	181.36	26.65

Density/kg m^{-3}: 6506 [293 K]; 5800 [liquid at m.p.]

Thermal conductivity/W m^{-1} K^{-1}: 22.7 [300 K]

Electrical resistivity/Ω m: 40.0×10^{-8} [293 K]

Mass magnetic susceptibility/kg^{-1} m^3: $+1.68\times10^{-8}$ (s)

Molar volume/cm^3: 14.02

Coefficient of linear thermal expansion/K^{-1}: 5.78×10^{-6}

Lattice structure (cell dimensions/pm), space group

α-Zr h.c.p. ($a=323.21$; $b=514.77$), $P6_3/mmc$
β-Zr b.c.c. ($a=361.6$), Im3m
$T(\alpha\rightarrow\beta)=1135$ K
High pressure form: ($a=503.6$; $c=310.9$), $P\bar{3}m1$

X-ray diffraction: mass absorption coefficients (μ/ρ)/cm^2 g^{-1}:
CuK_α 143 MoK_α 15.9

Discovered by M. H. Klaproth in 1789 at University of Berlin, Germany; isolated in 1824 by J. J. Berzelius at Stockholm, Sweden

[Arabic, *zargun*, = gold colour]

Zirconium

Nuclear properties

Atomic number: 40

Thermal neutron capture cross-section/barns: 0.182±0.005

Number of isotopes (including nuclear isomers): 20

Isotope mass range: 81→98

Key isotopes

Nuclide	Atomic mass	Natural abundance (%)	Half life $T_{1/2}$	Decay mode and energy (MeV)	Nuclear spin I	Nuclear magnetic moment μ	Uses
^{90}Zr	89.9043	51.45	stable				
^{91}Zr	90.9053	11.32	stable		5/2	−1.303	NMR
^{92}Zr	91.9046	17.19	stable				
^{94}Zr	93.9061	17.28	stable				
^{95}Zr	94.9077	0	65d	β^-(1.121); γ			tracer
^{96}Zr	95.9082	2.76	$>3.6\times10^{17}$y				
^{97}Zr		0	17h	β^-(2.67); γ			tracer

NMR

	^{91}Zr
Relative sensitivity (^{1}H=1.00)	9.48×10^{-3}
Absolute sensitivity (^{1}H=1.00)	1.06×10^{-3}
Receptivity (^{13}C=1.00)	6.04
Magnetogyric ratio/rad T^{-1} s^{-1}	-2.4868×10^7
Quadrupole moment/m^2	-0.21×10^{-28}
Frequency (^{1}H=100 MHz; 2.3488 T)/MHz	9.330

Reference: no agreed standard

Electron shell properties

Ground state electron configuration: [Kr]$4d^25s^2$

Term symbol: $^3F_2$

Electron affinity ($M\rightarrow M^-$)/kJ mol^{-1}: 43

Ionization energies/kJ mol^{-1}

1. $M\rightarrow M^+$	660	6. $M^{5+}\rightarrow M^{6+}$	(9 500)
2. $M^+\rightarrow M^{2+}$	1267	7. $M^{6+}\rightarrow M^{7+}$	(11 200)
3. $M^{2+}\rightarrow M^{3+}$	2218	8. $M^{7+}\rightarrow M^{8+}$	(13 800)
4. $M^{3+}\rightarrow M^{4+}$	3313	9. $M^{8+}\rightarrow M^{9+}$	(15 700)
5. $M^{4+}\rightarrow M^{5+}$	7860	10. $M^{9+}\rightarrow M^{10+}$	(17 500)

Principal lines in atomic spectrum

Wavelength/nm	Species	Sensitivity	Application
303.092	II		AA
333.198	II	V1	AE
351.960	I	U3	AA, AE
354.768	I	U2	AA, AE
360.119	I	U1	AA, AE
676.238*	I		

* Lowest energy transition from ground state to nearest empty orbital

Abundance: Earth's crust 162 p.p.m.; seawater 2.6×10^{-5} p.p.m.

Biological role: None; non-toxic

Tables of properties of the elements in order of element (A tables)

and in ranking order of property (B tables)

Table 1A Abundance of elements in the Earth's crust
p.p.m. (=g per tonne)

Element	p.p.m.	Element	p.p.m.
1 Hydrogen	1520	53 Iodine	0.46
2 Helium	0.003	54 Xenon	nil
3 Lithium	18	55 Caesium	2.6
4 Beryllium	2	56 Barium	390
5 Boron	9	57 Lanthanum	35
6 Carbon	180	58 Cerium	66
7 Nitrogen	19	59 Praseodymium	9.1
8 Oxygen	455 000	60 Neodymium	40
9 Fluorine	544	61 Promethium	trace
10 Neon	0.04	62 Samarium	7.0
11 Sodium	22 700	63 Europium	2.1
12 Magnesium	27 640	64 Gadolinium	6.1
13 Aluminium	83 000	65 Terbium	1.2
14 Silicon	272 000	66 Dysprosium	4.5
15 Phosphorus	1120	67 Holmium	1.3
16 Sulfur	340	68 Erbium	3.5
17 Chlorine	126	69 Thulium	0.5
18 Argon	0.04	70 Ytterbium	3.1
19 Potassium	18 400	71 Lutetium	0.8
20 Calcium	46 600	72 Hafnium	2.8
21 Scandium	25	73 Tantalum	1.7
22 Titanium	6320	74 Tungsten	1.2
23 Vanadium	136	75 Rhenium	0.0007
24 Chromium	122	76 Osmium	0.005
25 Manganese	1060	77 Iridium	0.001
26 Iron	62 000	78 Platinum	0.01
27 Cobalt	29	79 Gold	0.004
28 Nickel	99	80 Mercury	0.08
29 Copper	68	81 Thallium	0.7
30 Zinc	76	82 Lead	13
31 Gallium	19	83 Bismuth	0.008
32 Germanium	1.5	84 Polonium	trace
33 Arsenic	1.8	85 Astatine	trace
34 Selenium	0.05	86 Radon	1.7×10^{-10}
35 Bromine	2.5	87 Francium	trace
36 Krypton	nil	88 Radium	trace
37 Rubidium	78	89 Actinium	trace
38 Strontium	384	90 Thorium	8.1
39 Yttrium	31	91 Protoactinium	trace
40 Zirconium	162	92 Uranium	2.3
41 Niobium	20	93 Neptunium	trace
42 Molybdenum	1.2	94 Plutonium	trace
43 Technetium	trace	95 Americium	nil
44 Ruthenium	0.0001	96 Curium	nil
45 Rhodium	0.0001	97 Berkelium	nil
46 Palladium	0.015	98 Californium	nil
47 Silver	0.08	99 Einsteinium	nil
48 Cadmium	0.16	100 Fermium	nil
49 Indium	0.24	101 Mendelevium	nil
50 Tin	2.1	102 Nobelium	nil
51 Antimony	0.2	103 Lawrencium	nil
52 Tellurium	0.001		

Table 1B Abundance of elements in the Earth's crust in order of abundance p.p.m. (=g per tonne)

Element	p.p.m.	Element	p.p.m.
8 Oxygen	455 000	73 Tantalum	1.7
14 Silicon	272 000	32 Germanium	1.5
13 Aluminium	83 000	67 Holmium	1.3
26 Iron	62 000	42 Molybdenum	1.2
20 Calcium	46 600	74 Tungsten	1.2
12 Magnesium	27 640	65 Terbium	1.2
11 Sodium	22 700	71 Lutetium	0.8
19 Potassium	18 400	81 Thallium	0.7
22 Titanium	6320	69 Thulium	0.5
1 Hydrogen	1520	53 Iodine	0.46
15 Phosphorus	1120	49 Indium	0.24
25 Manganese	1060	51 Antimony	0.2
9 Fluorine	544	48 Cadmium	0.16
56 Barium	390	47 Silver	0.08
38 Strontium	384	80 Mercury	0.08
16 Sulfur	340	34 Selenium	0.05
6 Carbon	180	10 Neon	0.04
40 Zirconium	162	18 Argon	0.04
23 Vanadium	136	46 Palladium	0.015
17 Chlorine	126	78 Platinum	0.01
24 Chromium	122	83 Bismuth	0.008
28 Nickel	99	76 Osmium	0.005
37 Rubidium	78	79 Gold	0.004
30 Zinc	76	2 Helium	0.003
29 Copper	68	77 Iridium	0.001
58 Cerium	66	52 Tellurium	0.001
60 Neodymium	40	75 Rhenium	0.0007
57 Lanthanum	35	44 Ruthenium	0.0001
39 Yttrium	31	45 Rhodium	0.0001
27 Cobalt	29	86 Radon	1.7×10^{-10}
21 Scandium	25	43 Technetium	trace
41 Niobium	20	61 Promethium	trace
7 Nitrogen	19	84 Polonium	trace
31 Gallium	19	85 Astatine	trace
3 Lithium	18	87 Francium	trace
82 Lead	13	88 Radium	trace
59 Praseodymium	9.1	89 Actinium	trace
5 Boron	9	91 Protoactinium	trace
90 Thorium	8.1	93 Neptunium	trace
62 Samarium	7.0	94 Plutonium	trace
64 Gadolinium	6.1	36 Krypton	nil
66 Dysprosium	4.5	54 Xenon	nil
68 Erbium	3.5	95 Americium	nil
70 Ytterbium	3.1	96 Curium	nil
72 Hafnium	2.8	97 Berkelium	nil
55 Caesium	2.6	98 Californium	nil
35 Bromine	2.5	99 Einsteinium	nil
92 Uranium	2.3	100 Fermium	nil
50 Tin	2.1	101 Mendelevium	nil
63 Europium	2.1	102 Nobelium	nil
4 Beryllium	2	103 Lawrencium	nil
33 Arsenic	1.8		

Table 2A Boiling points/K

Element	Boiling point/K	Element	Boiling point/K
1 Hydrogen	20.28	53 Iodine	457.50
2 Helium	4.216	54 Xenon	166.1
3 Lithium	1620	55 Caesium	951.6
4 Beryllium	3243	56 Barium	1910
5 Boron	3931	57 Lanthanum	3730
6 Carbon	5100 (subl.)	58 Cerium	3699
7 Nitrogen	77.4	59 Praseodymium	3785
8 Oxygen	90.19	60 Neodymium	3341
9 Fluorine	85.01	61 Promethium	*c.* 3000
10 Neon	27.10	62 Samarium	2064
11 Sodium	1156.1	63 Europium	1870
12 Magnesium	1363	64 Gadolinium	3539
13 Aluminium	2740	65 Terbium	3396
14 Silicon	2628	66 Dysprosium	2835
15 Phosphorus (P_4)	553	67 Holmium	2968
16 Sulfur	717.824	68 Erbium	3136
17 Chlorine	238.6	69 Thulium	2220
18 Argon	87.3	70 Ytterbium	1466
19 Potassium	1047	71 Lutetium	3668
20 Calcium	1757	72 Hafnium	5470
21 Scandium	3104	73 Tantalum	5698
22 Titanium	3560	74 Tungsten	5930
23 Vanadium	3650	75 Rhenium	5900
24 Chromium	2945	76 Osmium	5300
25 Manganese	2235	77 Iridium	4403
26 Iron	3023	78 Platinum	4100
27 Cobalt	3143	79 Gold	3080
28 Nickel	3005	80 Mercury	629.73
29 Copper	2840	81 Thallium	1730
30 Zinc	1180	82 Lead	2013
31 Gallium	2676	83 Bismuth	1883
32 Germanium	3103	84 Polonium	1235
33 Arsenic	889 (subl.)	85 Astatine	610 (est.)
34 Selenium	958.1	86 Radon	211.4
35 Bromine	331.9	87 Francium	950
36 Krypton	120.85	88 Radium	1413
37 Rubidium	961	89 Actinium	3470
38 Strontium	1657	90 Thorium	5060
39 Yttrium	3611	91 Protactinium	*c.* 4300
40 Zirconium	4650	92 Uranium	4018
41 Niobium	5015	93 Neptunium	4175
42 Molybdenum	4885	94 Plutonium	3505
43 Technetium	5150	95 Americium	2880
44 Ruthenium	4173	96 Curium	n.a.
45 Rhodium	4000	97 Berkelium	n.a.
46 Palladium	3413	98 Californium	n.a.
47 Silver	2485	99 Einsteinium	n.a.
48 Cadmium	1038	100 Fermium	n.a.
49 Indium	2353	101 Mendelevium	n.a.
50 Tin	2543	102 Nobelium	n.a.
51 Antimony	1908	103 Lawrencium	n.a.
52 Tellurium	1263		

Table 2B Boiling points in order of temperature/K

Element		b.p.	Element	b.p.
74 Tungsten		5930	69 Thulium	2220
75 Rhenium		5900	62 Samarium	2064
73 Tantalum		5698	82 Lead	2013
72 Hafnium		5470	56 Barium	1910
76 Osmium		5300	51 Antimony	1908
43 Technetium		5150	83 Bismuth	1883
6 Carbon		5100 (subl.)	63 Europium	1870
90 Thorium		5060	20 Calcium	1757
41 Niobium		5015	81 Thallium	1730
42 Molybdenum		4885	38 Strontium	1657
40 Zirconium		4650	3 Lithium	1620
77 Iridium		4403	70 Ytterbium	1466
91 Protactinium	*c.*	4300	88 Radium	1413
93 Neptunium		4175	12 Magnesium	1363
44 Ruthenium		4173	52 Tellurium	1263
78 Platinum		4100	84 Polonium	1235
92 Uranium		4018	30 Zinc	1180
45 Rhodium		4000	11 Sodium	1156.1
5 Boron		3931	19 Potassium	1047
59 Praseodymium		3785	48 Cadmium	1038
57 Lanthanum		3730	37 Rubidium	961
58 Cerium		3699	34 Selenium	958.1
71 Lutetium		3668	55 Caesium	951.6
23 Vanadium		3650	87 Francium	950
39 Yttrium		3611	33 Arsenic	889 (subl.)
22 Titanium		3560	16 Sulfur	717.824
64 Gadolinium		3539	80 Mercury	629.73
94 Plutonium		3505	85 Astatine	610 (est)
89 Actinium		3470	15 Phosphorus (P_4)	553
46 Palladium		3413	53 Iodine	457.50
65 Terbium		3396	35 Bromine	331.9
60 Neodymium		3341	17 Chlorine	238.6
4 Beryllium		3243	86 Radon	211.4
27 Cobalt		3143	54 Xenon	166.1
68 Erbium		3136	36 Krypton	120.85
21 Scandium		3104	8 Oxygen	90.19
32 Germanium		3103	18 Argon	87.3
79 Gold		3080	9 Fluorine	85.01
26 Iron		3023	7 Nitrogen	77.4
28 Nickel		3005	10 Neon	27.10
61 Promethium	*c.*	3000	1 Hydrogen	20.28
67 Holmium		2968	2 Helium	4.216
24 Chromium		2945		
95 Americium		2880	96 Curium	n.a.
29 Copper		2840	97 Berkelium	n.a.
66 Dysprosium		2835	98 Californium	n.a.
13 Aluminium		2740	99 Einsteinium	n.a.
31 Gallium		2676	100 Fermium	n.a.
14 Silicon		2628	101 Mendelevium	n.a.
50 Tin		2543	102 Nobelium	n.a.
47 Silver		2485	103 Lawrencium	n.a.
49 Indium		2353		
25 Manganese		2235		

Table 3A Coefficient of linear thermal expansion/10^{-6} K^{-1}

1 Hydrogen	n.a./gas	53 Iodine	n.a.
2 Helium	n.a./gas	54 Xenon	n.a./gas
3 Lithium	56	55 Caesium	97
4 Beryllium	11.5	56 Barium	18.1
5 Boron	5	57 Lanthanum	4.9
6 Carbon (diam.)	1.19	58 Cerium	8.5
7 Nitrogen	n.a./gas	59 Praseodymium	6.79
8 Oxygen	n.a./gas	60 Neodymium	6.7
9 Fluorine	n.a./gas	61 Promethium	n.a.
10 Neon	n.a./gas	62 Samarium	10.4
11 Sodium	70.6	63 Europium	32
12 Magnesium	26.1	64 Gadolinium	8.6
13 Aluminium	23.03	65 Terbium	7.0
14 Silicon	4.2	66 Dysprosium	10.0
15 Phosphorus	124.5	67 Holmium	9.5
16 Sulfur	74.33	68 Erbium	9.2
17 Chlorine	n.a./gas	69 Thulium	13.3
18 Argon	n.a./gas	70 Ytterbium	25.0
19 Potassium	83	71 Lutetium	8.12
20 Calcium	22	72 Hafnium	5.9
21 Scandium	10.0	73 Tantalum	6.6
22 Titanium	8.35	74 Tungsten	4.59
23 Vanadium	8.3	75 Rhenium	6.63
24 Chromium	6.2	76 Osmium	6.1‡
25 Manganese	22	77 Iridium	6.8
26 Iron	12.3	78 Platinum	9.0
27 Cobalt	13.36	79 Gold	14.16
28 Nickel	13.3	80 Mercury	n.a./liquid
29 Copper	16.5	81 Thallium	28
30 Zinc	25.0	82 Lead	29.1
31 Gallium	31.5*	83 Bismuth	13.4
32 Germanium	5.57	84 Polonium	23.0
33 Arsenic	4.7	85 Astatine	n.a.
34 Selenium	36.9	86 Radon	n.a./gas
35 Bromine	n.a./liquid	87 Francium	n.a.
36 Krypton	n.a./gas	88 Radium	20.2
37 Rubidium	90	89 Actinium	14.9
38 Strontium	23	90 Thorium	12.5
39 Yttrium	10.6	91 Protactinium	7.3
40 Zirconium	5.78	92 Uranium	12.6
41 Niobium	7.06	93 Neptunium	27.5
42 Molybdenum	5.43	94 Plutonium	55
43 Technetium	8.06	95 Americium	n.a.
44 Ruthenium	9.1	96 Curium	n.a.
45 Rhodium	8.40	97 Berkelium	n.a.
46 Palladium	11.2	98 Californium	n.a.
47 Silver	19.2	99 Einsteinium	n.a.
48 Cadmium	29.8	100 Fermium	n.a.
49 Indium	33	101 Mendelevium	n.a.
50 Tin (α)	5.3†	102 Nobelium	n.a.
51 Antimony	8.5	103 Lawrencium	n.a.
52 Tellurium	16.75		

*This value is along the *b* axis; it is 16.5 along *c* and 11.5 along *a*.
†21.2 for *β* tin.
‡This value is along the *b* axis; it is 6.8 along *c* and 4.3 along *a*.

Table 3B Coefficient of linear thermal expansion in order of expansion/10^{-6} K^{-1}

15 Phosphorus	124.5	22 Titanium	8.35
55 Caesium	97	23 Vanadium	8.3
37 Rubidium	90	71 Lutetium	8.12
19 Potassium	83	43 Technetium	8.06
16 Sulfur	74.33	91 Protactinium	7.3
11 Sodium	70.6	41 Niobium	7.06
3 Lithium	56	65 Terbium	7.0
94 Plutonium	55	77 Iridium	6.8
34 Selenium	36.9	59 Praseodymium	6.79
49 Indium	33	60 Neodymium	6.7
63 Europium	32	75 Rhenium	6.63
31 Gallium	31.5*	73 Tantalum	6.6
48 Cadmium	29.8	24 Chromium	6.2
82 Lead	29.1	76 Osmium	6.1‡
81 Thallium	28	72 Hafnium	5.9
93 Neptunium	27.5	40 Zirconium	5.57
12 Magnesium	26.1	32 Germanium	5.78
30 Zinc	25.0	42 Molybdenum	5.43
70 Ytterbium	25.0	50 Tin (α)	5.3†
13 Aluminium	23.03	5 Boron	5
38 Strontium	23	57 Lanthanum	4.9
84 Polonium	23.0	33 Arsenic	4.7
20 Calcium	22	74 Tungsten	4.59
25 Manganese	22	14 Silicon	4.2
88 Radium	20.2	6 Carbon (diam.)	1.19
47 Silver	19.2		
56 Barium	18.1	1 Hydrogen	n.a./gas
52 Tellurium	16.75	2 Helium	n.a./gas
29 Copper	16.5	7 Nitrogen	n.a./gas
89 Actinium	14.9	8 Oxygen	n.a./gas
79 Gold	14.16	9 Fluorine	n.a./gas
83 Bismuth	13.4	10 Neon	n.a./gas
27 Cobalt	13.36	17 Chlorine	n.a./gas
28 Nickel	13.3	18 Argon	n.a./gas
69 Thulium	13.3	35 Bromine	n.a./liquid
92 Uranium	12.6	36 Krypton	n.a./gas
90 Thorium	12.5	53 Iodine	n.a.
26 Iron	12.3	54 Xenon	n.a./gas
4 Beryllium	11.5	61 Promethium	n.a.
46 Palladium	11.2	80 Mercury	n.a./liquid
39 Yttrium	10.6	85 Astatine	n.a.
62 Samarium	10.4	86 Radon	n.a./gas
21 Scandium	10.0	87 Francium	n.a.
66 Dysprosium	10.0	95 Americium	n.a.
67 Holmium	9.5	96 Curium	n.a.
68 Erbium	9.2	97 Berkelium	n.a.
44 Ruthenium	9.1	98 Californium	n.a.
78 Platinum	9.0	99 Einsteinium	n.a.
64 Gadolinium	8.6	100 Fermium	n.a.
51 Antimony	8.5	101 Mendelevium	n.a.
58 Cerium	8.5	102 Nobelium	n.a.
45 Rhodium	8.40	103 Lawrencium	n.a.

*This value is along the *b* axis; it is 16.5 along *c* and 11.5 along *a*.
†21.2 for *β* tin.
‡This value is along the *b* axis; it is 6.8 along *c* and 4.3 along *a*.

Table 4A Densities ρ for solids at 298 K, unless otherwise specified/kg m^{-3}

Element	ρ	Element	ρ
1 Hydrogen [11 K]	76.0	53 Iodine	4930
2 Helium (liq., 4 K)	124.8	54 Xenon	3540
3 Lithium	534	55 Caesium	1873
4 Beryllium	1847.7	56 Barium	3594
5 Boron	2340	57 Lanthanum	6145
6 Carbon (diam.)	3513	58 Cerium (α)	8240
7 Nitrogen [21 K]	1026	59 Praseodymium	6773
8 Oxygen [55 K]	2000	60 Neodymium	7007
9 Fluorine (liq., 85 K)	1516	61 Promethium	7220
10 Neon [24 K]	1444	62 Samarium	7520
11 Sodium	971	63 Europium	5243
12 Magnesium	1738	64 Gadolinium	7900.4
13 Aluminium	2698	65 Terbium	8229
14 Silicon	2329	66 Dysprosium	8550
15 Phosphorus (P_4)	1820	67 Holmium	8795
16 Sulfur (α)	2070	68 Erbium	9066
17 Chlorine [113 K]	2030	69 Thulium	9321
18 Argon [40 K]	1656	70 Ytterbium	6965
19 Potassium	862	71 Lutetium	9840
20 Calcium	1550	72 Hafnium	13 310
21 Scandium	2989	73 Tantalum	16 654
22 Titanium	4540	74 Tungsten	19 300
23 Vanadium	6110	75 Rhenium	21 020
24 Chromium	7190	76 Osmium	22 590
25 Manganese	7440	77 Iridium	22 420
26 Iron	7874	78 Platinum	21 450
27 Cobalt	8900	79 Gold	19 320
28 Nickel	8902	80 Mercury (liq.)	13 546
29 Copper	8960	81 Thallium	11 850
30 Zinc	7133	82 Lead	11 350
31 Gallium	5907	83 Bismuth	9747
32 Germanium	5323	84 Polonium	9320
33 Arsenic (α)	5780	85 Astatine	n.a.
34 Selenium	4790	86 Radon (liq., 211 K)	4400
35 Bromine [123 K]	4050		
36 Krypton [117 K]	2823	87 Francium	n.a.
37 Rubidium	1532	88 Radium	*c.* 5000
38 Strontium	2540	89 Actinium	10 060
39 Yttrium	4469	90 Thorium	11 720
40 Zirconium	6506	91 Protactinium	15 370 (est.)
41 Niobium	8570	92 Uranium	18 950
42 Molybdenum	10 220	93 Neptunium	20 250
43 Technetium	11 500	94 Plutonium	19 840
44 Ruthenium	12 370	95 Americium	13 670
45 Rhodium	12 410	96 Curium	13 300
46 Palladium	12 020	97 Berkelium	14 790
47 Silver	10 500	98 Californium	n.a.
48 Cadmium	8650	99 Einsteinium	n.a.
49 Indium	7310	100 Fermium	n.a.
50 Tin (β)	7310	101 Mendelevium	n.a.
51 Antimony	6691	102 Nobelium	n.a.
52 Tellurium	6240	103 Lawrencium	n.a.

Table 4B Densities ρ for solids at 298 K, unless otherwise specified, in order of density/kg m^{-3}

76 Osmium	22 590	51 Antimony	6691
77 Iridium	22 420	40 Zirconium	6506
78 Platinum	21 450	52 Tellurium	6240
75 Rhenium	21 020	57 Lanthanum	6145
93 Neptunium	20 250	23 Vanadium	6110
94 Plutonium	19 840	31 Gallium	5907
79 Gold	19 320	33 Arsenic (α)	5780
74 Tungsten	19 300	32 Germanium	5323
92 Uranium	18 950	63 Europium	5243
73 Tantalum	16 654	88 Radium	*c.* 5000
91 Protactinium	15 370 (est)	53 Iodine	4930
97 Berkelium	14 790	34 Selenium	4790
95 Americium	13 670	22 Titanium	4540
80 Mercury (liq.)	13 546	39 Yttrium	4469
72 Hafnium	13 310	86 Radon (liq., 211 K)	4400
96 Curium	13 300	35 Bromine [123 K]	4050
45 Rhodium	12 410	56 Barium	3594
44 Ruthenium	12 370	54 Xenon	3540
46 Palladium	12 020	6 Carbon (diam.)	3513
81 Thallium	11 850	21 Scandium	2989
90 Thorium	11 720	36 Krypton [117 K]	2823
43 Technetium	11 500	13 Aluminium	2698
82 Lead	11 350	38 Strontium	2540
47 Silver	10 500	5 Boron	2340
42 Molybdenum	10 220	14 Silicon	2329
89 Actinium	10 060	16 Sulfur (α)	2070
71 Lutetium	9840	17 Chlorine [113 K]	2030
83 Bismuth	9747	8 Oxygen [55 K]	2000
69 Thulium	9321	55 Caesium	1873
84 Polonium	9320	4 Beryllium	1847.7
68 Erbium	9066	15 Phosphorus (P_4)	1820
29 Copper	8960	12 Magnesium	1738
28 Nickel	8902	18 Argon [40 K]	1656
27 Cobalt	8900	20 Calcium	1550
67 Holmium	8795	37 Rubidium	1532
48 Cadmium	8650	9 Fluorine (liq., 85 K)	1516
41 Niobium	8570	10 Neon [24 K]	1444
66 Dysprosium	8550	7 Nitrogen [21 K]	1026
58 Cerium (α)	8240	11 Sodium	971
65 Terbium	8229	19 Potassium	862
64 Gadolinium	7900.4	3 Lithium	534
26 Iron	7874	2 Helium (liq., 4 K)	124.8
62 Samarium	7520	1 Hydrogen [11 K]	76.0
25 Manganese	7440		
49 Indium	7310	85 Astatine	n.a.
50 Tin (β)	7310	87 Francium	n.a.
61 Promethium	7220	98 Californium	n.a.
24 Chromium	7190	99 Einsteinium	n.a.
30 Zinc	7133	100 Fermium	n.a.
60 Neodymium	7007	101 Mendelevium	n.a.
70 Ytterbium	6965	102 Nobelium	n.a.
59 Praseodymium	6773	103 Lawrencium	n.a.

Table 5A Discovery

Element	Discovery	Element	Discovery
1 Hydrogen	1766	53 Iodine	1811
2 Helium	1895	54 Xenon	1898
3 Lithium	1817	55 Caesium	1860
4 Beryllium	1797	56 Barium	1808
5 Boron	1808	58 Lanthanum	1839
6 Carbon	pre-history	58 Cerium	1803
7 Nitrogen	1772	59 Praseodymium	1885
8 Oxygen	1774	60 Neodymium	1885
9 Fluorine	1886	61 Promethium	1945
10 Neon	1898	62 Samarium	1879
11 Sodium	1807	63 Europium	1901
12 Magnesium	1755	64 Gadolinium	1880
13 Aluminium	1825	65 Terbium	1843
14 Silicon	1824	66 Dysprosium	1886
15 Phosphorus	1669	67 Holmium	1878
16 Sulfur	pre-history	68 Erbium	1842
17 Chlorine	1774	69 Thulium	1879
18 Argon	1894	70 Ytterbium	1878
19 Potassium	1807	71 Lutetium	1907
20 Calcium	1808	72 Hafnium	1923
21 Scandium	1879	73 Tantalum	1802
22 Titanium	1791	74 Tungsten	1783
23 Vanadium	1801	75 Rhenium	1925
24 Chromium	1780	76 Osmium	1803
25 Manganese	1774	77 Iridium	1803
26 Iron	*c.* 2500 BC	78 Platinum	pre 1700
27 Cobalt	1735	79 Gold	*c.* 3000 BC
28 Nickel	1751	80 Mercury	*c.* 1500 BC
29 Copper	*c.* 5000 BC	81 Thallium	1861
30 Zinc	pre 1500*	82 Lead	*c.* 1000 BC
31 Gallium	1875	83 Bismuth	*c.* 1500
32 Germanium	1896	84 Polonium	1898
33 Arsenic	*c.* 1250	85 Astatine	1940
34 Selenium	1817	86 Radon	1900
35 Bromine	1826	87 Francium	1939
36 Krypton	1898	88 Radium	1898
37 Rubidium	1861	89 Actinium	1899
38 Strontium	1808	90 Thorium	1815
39 Yttrium	1794	91 Protactinium	1917
40 Zirconium	1789	92 Uranium	1789
41 Niobium	1801	93 Neptunium	1940
42 Molybdenum	1781	94 Plutonium	1940
43 Technetium	1937	95 Americium	1944
44 Ruthenium	1808	96 Curium	1944
45 Rhodium	1803	97 Berkelium	1949
46 Palladium	1803	98 Californium	1950
47 Silver	*c.* 3000 BC	99 Einsteinium	1952
48 Cadmium	1817	100 Fermium	1952
49 Indium	1863	101 Mendelevium	1955
50 Tin	*c.* 2100 BC	102 Nobelium	1958
51 Antimony	*c.* 1600 BC	103 Lawrencium	1961
52 Tellurium	1783		

*Zinc was known as the copper–zinc alloy, brass, around 20 BC.

Table 5B Discovery in chronological order

Element	Date	Element	Date
6 Carbon	pre-history	48 Cadmium	1817
16 Sulfur	pre-history	14 Silicon	1824
29 Copper	*c.* 5000 BC	13 Aluminium	1825
47 Silver	*c.* 3000 BC	35 Bromine	1826
79 Gold	*c.* 3000 BC	57 Lanthanum	1839
26 Iron	*c.* 2500 BC	68 Erbium	1842
50 Tin	*c.* 2100 BC	65 Terbium	1843
51 Antimony	*c.* 1600 BC	55 Caesium	1860
80 Mercury	*c.* 1500 BC	37 Rubidium	1861
82 Lead	*c.* 1000 BC	81 Thallium	1861
33 Arsenic	*c.* 1250	49 Indium	1863
30 Zinc	pre 1500*	31 Gallium	1875
83 Bismuth	*c.* 1500	67 Holmium	1878
15 Phosphorus	1669	70 Ytterbium	1878
78 Platinum	pre 1700	21 Scandium	1879
		62 Samarium	1879
27 Cobalt	1735	69 Thulium	1879
28 Nickel	1751	64 Gadolinium	1880
12 Magnesium	1755	59 Praseodymium	1885
1 Hydrogen	1766	60 Neodymium	1885
7 Nitrogen	1772	32 Germanium	1886
8 Oxygen	1774	9 Fluorine	1886
17 Chlorine	1774	66 Dysprosium	1886
25 Manganese	1774	18 Argon	1894
24 Chromium	1780	2 Helium	1895
42 Molybdenum	1781	36 Krypton	1898
52 Tellurium	1783	10 Neon	1898
74 Tungsten	1783	54 Xenon	1898
40 Zirconium	1789	84 Polonium	1898
92 Uranium	1789	88 Radium	1898
22 Titanium	1791	89 Actinium	1899
39 Yttrium	1794		
4 Beryllium	1797	86 Radon	1900
		63 Europium	1901
23 Vanadium	1801	71 Lutetium	1907
41 Niobium	1801	91 Protactinium	1917
73 Tantalum	1802	72 Hafnium	1923
45 Rhodium	1803	75 Rhenium	1925
46 Palladium	1803	43 Technetium	1937
76 Osmium	1803	87 Francium	1939
77 Iridium	1803	93 Neptunium	1940
58 Cerium	1803	85 Astatine	1940
19 Potassium	1807	94 Plutonium	1940
11 Sodium	1807	95 Americium	1944
5 Boron	1808	96 Curium	1944
20 Calcium	1808	61 Promethium	1945
38 Strontium	1808	97 Berkelium	1949
44 Ruthenium	1808	98 Californium	1950
56 Barium	1808	99 Einsteinium	1952
53 Iodine	1811	100 Fermium	1952
90 Thorium	1815	101 Mendelevium	1955
3 Lithium	1817	102 Nobelium	1958
34 Selenium	1817	103 Lawrencium	1961

*Zinc was known as the copper–zinc alloy, brass, around 20 BC.

Table 6A Electrical resistivities at 298 K/10^{-8} Ωm

Element	Resistivity	Element	Resistivity
1 Hydrogen	n.a.	53 Iodine	1.3×10^{15}
2 Helium	n.a.	54 Xenon	n.a.
3 Lithium	8.55	55 Caesium	20.0
4 Beryllium	4.0	56 Barium	50
5 Boron	1.8×10^{12}	57 Lanthanum	57
6 Carbon (graph.)	1.375×10^{3}	58 Cerium	73
7 Nitrogen	n.a.	59 Praseodymium	68
8 Oxygen	n.a.	60 Neodymium	64.0
9 Fluorine	n.a.	61 Promethium	50 (est.)
10 Neon	n.a.	62 Samarium	88.0
11 Sodium	4.2	63 Europium	90.0
12 Magnesium	4.45	64 Gadolinium	134.0
13 Aluminium	2.6548	65 Terbium	114
14 Silicon	1×10^{5}	66 Dysprosium	57.0
15 Phosphorus (P_4)	1×10^{17}	67 Holmium	87.0
16 Sulfur	2×10^{23}	68 Erbium	87
17 Chlorine	n.a.	69 Thulium	79.0
18 Argon	n.a.	70 Ytterbium	29.0
19 Potassium	6.15	71 Lutetium	79.0
20 Calcium	3.43	72 Hafnium	35.1
21 Scandium	61.0	73 Tantalum	12.45
22 Titanium	42.0	74 Tungsten	5.65
23 Vanadium	24.8	75 Rhenium	19.3
24 Chromium	12.7	76 Osmium	8.12
25 Manganese	185.0	77 Iridium	5.3
26 Iron	9.71	78 Platinum	10.6
27 Cobalt	6.24	79 Gold	2.35
28 Nickel	6.84	80 Mercury	94.1
29 Copper	1.6730	81 Thallium	18.0
30 Zinc	5.916	82 Lead	20.648
31 Gallium	27	83 Bismuth	106.8
32 Germanium	4.6×10^{5}	84 Polonium	140
33 Arsenic	26	85 Astatine	n.a.
34 Selenium	1×10^{6}	86 Radon	n.a.
35 Bromine	n.a.	87 Francium	n.a.
36 Krypton	n.a.	88 Radium	100
37 Rubidium	12.5	89 Actinium	n.a.
38 Strontium	23.0	90 Thorium	13.0
39 Yttrium	57.0	91 Protactinium	17.7
40 Zirconium	40.0	92 Uranium	30.8
41 Niobium	12.5	93 Neptunium	122
42 Molybdenum	5.2	94 Plutonium	146
43 Technetium	22.6 (373 K)	95 Americium	68
44 Ruthenium	7.6	96 Curium	n.a.
45 Rhodium	4.51	97 Berkelium	n.a.
46 Palladium	10.8	98 Californium	n.a.
47 Silver	1.59	99 Einsteinium	n.a.
48 Cadmium	6.83	100 Fermium	n.a.
49 Indium	8.37	101 Mendelevium	n.a.
50 Tin (α)	11.0	102 Nobelium	n.a.
51 Antimony	39.0	103 Lawrencium	n.a.
52 Tellurium	4.36×10^{5}		

Table 6B Electrical resistivities at 298 K in order of resistivity/10^{-8} Ωm

Element	Resistivity	Element	Resistivity
16 Sulfur	2×10^{23}	37 Rubidium	12.5
15 Phosphorus (P_4)	1×10^{17}	41 Niobium	12.5
53 Iodine	1.3×10^{15}	73 Tantalum	12.45
5 Boron	1.8×10^{12}	50 Tin (α)	11.0
34 Selenium	1×10^{6}	46 Palladium	10.8
32 Germanium	4.6×10^{5}	78 Platinum	10.6
52 Tellurium	4.36×10^{5}	26 Iron	9.71
14 Silicon	1×10^{5}	3 Lithium	8.55
6 Carbon (graph.)	1375	49 Indium	8.37
25 Manganese	185.0	76 Osmium	8.12
94 Plutonium	146	44 Ruthenium	7.6
84 Palonium	140	28 Nickel	6.84
64 Gadolinium	134.0	48 Cadmium	6.83
93 Neptunium	122	27 Cobalt	6.24
65 Terbium	114	19 Potassium	6.15
83 Bismuth	106.8	30 Zinc	5.916
88 Radium	100	74 Tungsten	5.65
80 Mercury	94.1	77 Iridium	5.3
63 Europium	90.0	42 Molybdenum	5.2
62 Samarium	88.0	45 Rhodium	4.51
67 Holmium	87.0	12 Magnesium	4.45
68 Erbium	87	11 Sodium	4.2
69 Thulium	79.0	4 Beryllium	4.0
71 Lutetium	79.0	20 Calcium	3.43
58 Cerium	73	13 Aluminium	2.6548
59 Praseodymium	68	79 Gold	2.35
95 Americium	68	29 Copper	1.6730
60 Neodymium	64.0	47 Silver	1.59
21 Scandium	61.0		
66 Dysprosium	57.0	1 Hydrogen	n.a.
39 Yttrium	57.0	2 Helium	n.a.
57 Lanthanum	57	7 Nitrogen	n.a.
56 Barium	50	8 Oxygen	n.a.
61 Promethium	50 (est.)	9 Fluorine	n.a.
22 Titanium	42.0	10 Neon	n.a.
40 Zirconium	40.0	17 Chlorine	n.a.
51 Antimony	39.0	18 Argon	n.a.
72 Hafnium	35.1	35 Bromine	n.a.
92 Uranium	30.8	36 Krypton	n.a.
70 Ytterbium	29.0	54 Xenon	n.a.
31 Gallium	27	85 Astatine	n.a.
33 Arsenic	26	86 Radon	n.a.
23 Vanadium	24.8	87 Francium	n.a.
38 Strontium	23.0	89 Actinium	n.a.
43 Technetium	22.6 (373 K)	96 Curium	n.a.
82 Lead	20.648	97 Berkelium	n.a.
55 Caesium	20.0	98 Californium	n.a.
75 Rhenium	19.3	99 Einsteinium	n.a.
81 Thallium	18.0	100 Fermium	n.a.
91 Protactinium	17.7	101 Mendelevium	n.a.
90 Thorium	13.0	102 Nobelium	n.a.
24 Chromium	12.7	103 Lawrencium	n.a.

Table 7A Electron affinities $\Delta E(\mathrm{M} \rightarrow \mathrm{M}^-)$*/kJ mol^{-1}

Element	ΔE	Element	ΔE
1 Hydrogen	72.8	53 Iodine	295.3
2 Helium	−21 (calc.)	54 Xenon	−41 (calc.)
3 Lithium	59.8	55 Caesium	45.5
4 Beryllium	−18	56 Barium	−46
5 Boron	23	57 Lanthanum	53
6 Carbon	122.5	58 Cerium	
7 Nitrogen	−7	59 Praseodymium	
8 Oxygen	141	60 Neodymium	
9 Fluorine	322	61 Promethium	
10 Neon	−29 (calc.)	62 Samarium	
11 Sodium	52.9	63 Europium	
12 Magnesium	−21	64 Gadolinium	
13 Aluminium	44	65 Terbium	≤50 (58–71)
14 Silicon	133.6	66 Dysprosium	
15 Phosphorus	71.7	67 Holmium	
16 Sulfur	200.4	68 Erbium	
17 Chlorine	348.7	69 Thulium	
18 Argon	−35 (calc.)	70 Ytterbium	
19 Potassium	43.8	71 Lutetium	
20 Calcium	−186	72 Hafnium	61
21 Scandium	−70	73 Tantalum	14
22 Titanium	−2	74 Tungsten	119
23 Vanadium	61	75 Rhenium	37
24 Chromium	94	76 Osmium	139
25 Manganese	−94	77 Iridium	190
26 Iron	44	78 Platinum	247
27 Cobalt	102	79 Gold	223
28 Nickel	156	80 Mercury	−18
29 Copper	118.3	81 Thallium	30
30 Zinc	9	82 Lead	35.2
31 Gallium	36 (calc.)	83 Bismuth	101
32 Germanium	116	84 Polonium	186
33 Arsenic	77	85 Astatine	270
34 Selenium	195.0	86 Radon	−41 (calc.)
35 Bromine	324.5	87 Francium	44 (calc.)
36 Krypton	−39 (calc.)	88 Radium	n.a.
37 Rubidium	46.9	89 Actinium	n.a.
38 Strontium	−146	90 Thorium	n.a.
39 Yttrium	−39	91 Protactinium	n.a.
40 Zirconium	43	92 Uranium	n.a.
41 Niobium	109	93 Neptunium	n.a.
42 Molybdenum	114	94 Plutonium	n.a.
43 Technetium	96	95 Americium	n.a.
44 Ruthenium	146	96 Curium	n.a.
45 Rhodium	162	97 Berkelium	n.a.
46 Palladium	98.4	98 Californium	n.a.
47 Silver	125.7	99 Einsteinium	n.a.
48 Cadmium	−26	100 Fermium	n.a.
49 Indium	34	101 Mendelevium	n.a.
50 Tin	121	102 Nobelium	n.a.
51 Antimony	101	103 Lawrencium	n.a.
52 Tellurium	190.2		

*By convention electron affinities are positive for energy released.

Table 7B Electron affinities $\Delta E(\mathrm{M}\rightarrow\mathrm{M}^-)$ in order of energies*/kJ mol^{-1}

17 Chlorine	348.7
35 Bromine	324.5
9 Fluorine	322
53 Iodine	295.3
85 Astatine	270
78 Platinum	247
79 Gold	223
16 Sulfur	200.4
34 Selenium	195.0
52 Tellurium	190.2
77 Iridium	190
84 Polonium	186
45 Rhodium	162
28 Nickel	156
44 Ruthenium	146
8 Oxygen	141
76 Osmium	139
14 Silicon	133.6
47 Silver	125.7
6 Carbon	122.5
50 Tin	121
74 Tungsten	119
29 Copper	118.3
32 Germanium	116
42 Molybdenum	114
41 Niobium	109
27 Cobalt	102
51 Antimony	101
83 Bismuth	101
46 Palladium	98.4
43 Technetium	96
24 Chromium	94
33 Arsenic	77
1 Hydrogen	72.8
15 Phosphorus	71.7
23 Vanadium	61
3 Lithium	59.8
57 Lanthanum	53
11 Sodium	52.9
58 Cerium	
59 Praseodymium	
60 Neodymium	
61 Promethium	
62 Samarium	
63 Europium	
64 Gadolinium	≤ 50 (58–70)
65 Terbium	
66 Dysprosium	
67 Holmium	
68 Erbium	
69 Thulium	
70 Ytterbium	
71 Lutetium	$\leqslant 50$
37 Rubidium	46.9
55 Caesium	45.5
13 Aluminium	44
26 Iron	44
87 Francium	44 (calc.)
19 Potassium	43.8
40 Zirconium	43
75 Rhenium	37
31 Gallium	36 (calc.)
82 Lead	35.2
49 Indium	34
81 Thallium	30
5 Boron	23
73 Tantalum	14
30 Zinc	9
22 Titanium	−2
7 Nitrogen	−7
4 Beryllium	−18
80 Mercury	−18
2 Helium	−21 (calc.)
12 Magnesium	−21
48 Cadmium	−26
10 Neon	−29 (calc.)
18 Argon	−35 (calc.)
36 Krypton	−39 (calc.)
39 Yttrium	−39
54 Xenon	−41 (calc.)
86 Radon	−41 (calc.)
56 Barium	−46
72 Hafnium	−61
21 Scandium	−70
25 Manganese	−94
38 Strontium	−146
20 Calcium	−186
88 Radium	n.a.
89 Actinium	n.a.
90 Thorium	n.a.
91 Protactinium	n.a.
92 Uranium	n.a.
93 Neptunium	n.a.
94 Plutonium	n.a.
95 Americium	n.a.
96 Curium	n.a.
97 Berkelium	n.a.
98 Californium	n.a.
99 Einsteinium	n.a.
100 Fermium	n.a.
101 Mendelevium	n.a.
102 Nobelium	n.a.
103 Lawrencium	n.a.

*By convention electron affinities are positive for energy released.

Table 8A Electronegativities (Allred–Rochow values)

Element	Value	Element	Value
1 Hydrogen	2.20*	53 Iodine	2.21
2 Helium	n.a.	54 Xenon	2.6*
3 Lithium	0.97	55 Caesium	0.86
4 Beryllium	1.47	56 Barium	0.97
5 Boron	2.01	57 Lanthanum	1.08
6 Carbon	2.50	58 Cerium	1.06
7 Nitrogen	3.07	59 Praseodymium	1.07
8 Oxygen	3.50	60 Neodymium	1.07
9 Fluorine	4.10	61 Promethium	1.07
10 Neon	n.a.	62 Samarium	1.07
11 Sodium	1.08	63 Europium	1.01
12 Magnesium	1.23	64 Gadolinium	1.11
13 Aluminium	1.47	65 Terbium	1.10
14 Silicon	1.74	66 Dysprosium	1.10
15 Phosphorus	2.06	67 Holmium	1.10
16 Sulfur	2.44	68 Erbium	1.11
17 Chlorine	2.83	69 Thulium	1.11
18 Argon	n.a.	70 Ytterbium	1.06
19 Potassium	0.91	71 Lutetium	1.14
20 Calcium	1.04	72 Hafnium	1.23
21 Scandium	1.20	73 Tantalum	1.33
22 Titanium	1.32	74 Tungsten	1.40
23 Vanadium	1.45	75 Rhenium	1.46
24 Chromium	1.56	76 Osmium	1.52
25 Manganese	1.60	77 Iridium	1.55
26 Iron	1.64	78 Platinum	1.44
27 Cobalt	1.70	79 Gold	1.42
28 Nickel	1.75	80 Mercury	1.44
29 Copper	1.75	81 Thallium	1.44
30 Zinc	1.66	82 Lead	1.55
31 Gallium	1.82	83 Bismuth	1.67
32 Germanium	2.02	84 Polonium	1.76
33 Arsenic	2.20	85 Astatine	1.96
34 Selenium	2.48	86 Radon	*c.* 2.0 (est.)
35 Bromine	2.74	87 Francium	0.86
36 Krypton	2.9*	88 Radium	0.97
37 Rubidium	0.89	89 Actinium	1.00
38 Strontium	0.99	90 Thorium	1.11
39 Yttrium	1.11	91 Protactinium	1.14
40 Zirconium	1.22	92 Uranium	1.22
41 Niobium	1.23	93 Neptunium	1.22
42 Molybdenum	1.30	94 Plutonium	1.22
43 Technetium	1.36	95 Americium	n.a.
44 Ruthenium	1.42	96 Curium	n.a.
45 Rhodium	1.45	97 Berkelium	n.a.
46 Palladium	1.35	98 Californium	n.a.
47 Silver	1.42	99 Einsteinium	n.a.
48 Cadmium	1.46	100 Fermium	n.a.
49 Indium	1.49	101 Mendelevium	n.a.
50 Tin	1.72	102 Nobelium	n.a.
51 Antimony	1.82	103 Lawrencium	n.a.
52 Tellurium	2.01		

*Pauling electronegativity value.

Table 8B Electronegativities (Allred–Rochow values) in ranking order

9 Fluorine	4.10
8 Oxygen	3.50
7 Nitrogen	3.07
36 Krypton	2.9*
17 Chlorine	2.83
35 Bromine	2.74
54 Xenon	2.6*
6 Carbon	2.50
34 Selenium	2.48
16 Sulfur	2.44
53 Iodine	2.21
1 Hydrogen	2.20*
33 Arsenic	2.20
15 Phosphorus	2.06
32 Germanium	2.02
5 Boron	2.01
52 Tellurium	2.01
86 Radon	*c.* 2.0 (est.)
85 Astatine	1.96
31 Gallium	1.82
51 Antimony	1.82
84 Polonium	1.76
28 Nickel	1.75
29 Copper	1.75
14 Silicon	1.74
50 Tin	1.72
27 Cobalt	1.70
83 Bismuth	1.67
30 Zinc	1.66
26 Iron	1.64
25 Manganese	1.60
24 Chromium	1.56
77 Iridium	1.55
82 Lead	1.55
76 Osmium	1.52
49 Indium	1.49
4 Beryllium	1.47
13 Aluminium	1.47
48 Cadmium	1.46
75 Rhenium	1.46
45 Rhodium	1.45
23 Vanadium	1.45
78 Platinum	1.44
80 Mercury	1.44
81 Thallium	1.44
47 Silver	1.42
44 Ruthenium	1.42
79 Gold	1.42
74 Tungsten	1.40
43 Technetium	1.36
46 Palladium	1.35
73 Tantalum	1.33
22 Titanium	1.32
42 Molybdenum	1.30
12 Magnesium	1.23
41 Niobium	1.23
72 Hafnium	1.23
40 Zirconium	1.22
92 Uranium	1.22
93 Neptunium	1.22
94 Plutonium	1.22
21 Scandium	1.20
71 Lutetium	1.14
91 Protactinium	1.14
39 Yttrium	1.11
64 Gadolinium	1.11
68 Erbium	1.11
69 Thulium	1.11
90 Thorium	1.11
65 Terbium	1.10
66 Dysprosium	1.10
67 Holmium	1.10
57 Lanthanum	1.08
11 Sodium	1.08
59 Praseodymium	1.07
60 Neodymium	1.07
61 Promethium	1.07
62 Samarium	1.07
58 Cerium	1.06
70 Ytterbium	1.06
20 Calcium	1.04
63 Europium	1.01
89 Actinium	1.00
38 Strontium	0.99
3 Lithium	0.97
56 Barium	0.97
88 Radium	0.97
19 Potassium	0.91
37 Rubidium	0.89
37 Caesium	0.86
87 Francium	0.86
2 Helium	n.a.
10 Neon	n.a.
18 Argon	n.a.
95 Americium	n.a.
96 Curium	n.a.
97 Berkelium	n.a.
98 Californium	n.a.
99 Einsteinium	n.a.
100 Fermium	n.a.
101 Mendelevium	n.a.
102 Nobelium	n.a.
103 Lawrencium	n.a.

*Pauling electronegativity value.

Table 9A Enthalpy of formation of gaseous atoms $\Delta_f H^{\ominus}$/kJ mol^{-1} at 298.15 K and 0.1 MPa

Element	$\Delta_f H^{\ominus}$	Element	$\Delta_f H^{\ominus}$
1 Hydrogen	218.0	53 Iodine	106.8
2 Helium	0	54 Xenon	0
3 Lithium	159.4	55 Caesium	76.1
4 Beryllium	324.3	56 Barium	180
5 Boron	562.7	57 Lanthanum	431.0
6 Carbon	716.7	58 Cerium	423
7 Nitrogen	472.7	59 Praseodymium	355.6
8 Oxygen	249.2	60 Neodymium	327.6
9 Fluorine	79.0	61 Promethium	n.a.
10 Neon	0	62 Samarium	206.7
11 Sodium	107.3	63 Europium	175.3
12 Magnesium	147.7	64 Gadolinium	397.5
13 Aluminium	326.4	65 Terbium	388.7
14 Silicon	455.6	66 Dysprosium	290.4
15 Phosphorus	314.6	67 Holmium	300.8
16 Sulfur	278.8	68 Erbium	317.1
17 Chlorine	121.7	69 Thulium	232.2
18 Argon	0	70 Ytterbium	152.3
19 Potassium	89.2	71 Lutetium	427.6
20 Calcium	178.2	72 Hafnium	619.2
21 Scandium	377.8	73 Tantalum	782.0
22 Titanium	469.9	74 Tungsten	849.4
23 Vanadium	514.2	75 Rhenium	769.9
24 Chromium	396.6	76 Osmium	791
25 Manganese	280.7	77 Iridium	665.3
26 Iron	416.3	78 Platinum	565.3
27 Cobalt	424.7	79 Gold	366.1
28 Nickel	429.7	80 Mercury	61.3
29 Copper	338.3	81 Thallium	182.2
30 Zinc	130.7	82 Lead	195.0
31 Gallium	277.0	83 Bismuth	207.1
32 Germanium	376.6	84 Polonium	146
33 Arsenic	302.5	85 Astatine	n.a.
34 Selenium	227.1	86 Radon	0
35 Bromine	111.9	87 Francium	72.8
36 Krypton	0	88 Radium	159
37 Rubidium	80.9	89 Actinium	406
38 Strontium	164.4	90 Thorium	598.3
39 Yttrium	421.3	91 Protactinium	607
40 Zirconium	608.8	92 Uranium	535.6
41 Niobium	725.9	93 Neptunium	n.a.
42 Molybdenum	658.1	94 Plutonium	n.a.
43 Technetium	678	95 Americium	n.a.
44 Ruthenium	642.7	96 Curium	n.a.
45 Rhodium	556.9	97 Berkelium	n.a.
46 Palladium	378.2	98 Californium	n.a.
47 Silver	284.6	99 Einsteinium	n.a.
48 Cadmium	112.0	100 Fermium	n.a.
49 Indium	243.3	101 Mendelevium	n.a.
50 Tin	302.1	102 Nobelium	n.a.
51 Antimony	262.3	103 Lawrencium	n.a.
52 Tellurium	196.7		

Table 9B Enthalpy of formation of gaseous atoms of the elements $\Delta_f H^{\ominus}$ at 298.15 K and 0.1 MPa in order of $\Delta_f H^{\ominus}$/kJ mol^{-1}

Element	$\Delta_f H^{\ominus}$	Element	$\Delta_f H^{\ominus}$
74 Tungsten	849.4	51 Antimony	262.3
76 Osmium	791	8 Oxygen	249.2
73 Tantalum	782.0	49 Indium	243.3
75 Rhenium	769.9	69 Thulium	232.2
41 Niobium	725.9	34 Selenium	227.1
6 Carbon	716.7	1 Hydrogen	218.0
43 Technetium	678	83 Bismuth	207.1
77 Iridium	665.3	62 Samarium	206.7
42 Molybdenum	658.1	52 Tellurium	196.7
44 Ruthenium	642.7	82 Lead	195.0
72 Hafnium	619.2	81 Thallium	182.2
40 Zirconium	608.8	56 Barium	180
91 Protactinium	607	20 Calcium	178.2
90 Thorium	598.3	63 Europium	175.3
78 Platinum	565.3	38 Strontium	164.4
5 Boron	562.7	3 Lithium	159.4
45 Rhodium	556.9	88 Radium	159
92 Uranium	535.6	70 Ytterbium	152.3
23 Vanadium	514.2	12 Magnesium	147.7
7 Nitrogen	472.7	84 Polonium	146
22 Titanium	469.9	30 Zinc	130.7
14 Silicon	455.6	17 Chlorine	121.7
57 Lanthanum	431.0	48 Cadmium	112.0
28 Nickel	429.7	35 Bromine	111.9
71 Lutetium	427.6	11 Sodium	107.3
27 Cobalt	424.7	53 Iodine	106.8
58 Cerium	423	19 Potassium	89.2
39 Yttrium	421.3	37 Rubidium	80.9
26 Iron	416.3	9 Fluorine	79.0
89 Actinium	406	55 Caesium	76.1
64 Gadolinium	397.5	87 Francium	72.8
24 Chromium	396.6	80 Mercury	61.3
65 Terbium	388.7	2 Helium	0
46 Palladium	378.2	10 Neon	0
21 Scandium	377.8	18 Argon	0
32 Germanium	376.6	36 Krypton	0
79 Gold	366.1	54 Xenon	0
59 Praseodymium	355.6	86 Radon	0
29 Copper	338.3		
60 Neodymium	327.6	61 Promethium	n.a.
13 Aluminium	326.4	85 Astatine	n.a.
4 Beryllium	324.3	93 Neptunium	n.a.
68 Erbium	317.1	94 Plutonium	n.a.
15 Phosphorus	314.6	95 Americium	n.a.
33 Arsenic	302.5	96 Curium	n.a.
50 Tin	302.1	97 Berkelium	n.a.
67 Holmium	300.8	98 Californium	n.a.
66 Dysprosium	290.4	99 Einsteinium	n.a.
47 Silver	284.6	100 Fermium	n.a.
25 Manganese	280.7	101 Mendelevium	n.a.
16 Sulfur	278.8	102 Nobelium	n.a.
31 Gallium	277.0	103 Lawrencium	n.a.

Table 10A Enthalpies of vaporization ΔH_{vap}/kJ mol^{-1}

Element	ΔH_{vap}	Element	ΔH_{vap}
1 Hydrogen	0.46	53 Iodine	41.67
2 Helium	0.082	54 Xenon	12.65
3 Lithium	147.7	55 Caesium	66.5
4 Beryllium	308.8	56 Barium	150.9
5 Boron	504.5	57 Lanthanum	402.1
6 Carbon	710.9	58 Cerium	398
7 Nitrogen	5.58	59 Praseodymium	357
8 Oxygen	6.82	60 Neodymium	328
9 Fluorine	3.26	61 Promethium	n.a.
10 Neon	1.736	62 Samarium	164.8
11 Sodium	99.2	63 Europium	176
12 Magnesium	127.6	64 Gadolinium	301
13 Aluminium	290.8	65 Terbium	391
14 Silicon	383.3	66 Dysprosium	293
15 Phosphorus (P_4)	51.9	67 Holmium	303
16 Sulfur	9.62	68 Erbium	280
17 Chlorine	20.42	69 Thulium	247
18 Argon	6.53	70 Ytterbium	159
19 Potassium	79.1	71 Lutetium	428
20 Calcium	150.6	72 Hafnium	570.7
21 Scandium	376.1	73 Tantalum	758.22
22 Titanium	425.5	74 Tungsten	824.2
23 Vanadium	459.7	75 Rhenium	704.25
24 Chromium	341.8	76 Osmium	738.06
25 Manganese	220.5	77 Iridium	612.1
26 Iron	340.2	78 Platinum	469
27 Cobalt	382.4	79 Gold	343.1
28 Nickel	374.8	80 Mercury	59.11
29 Copper	306.7	81 Thallium	166.1
30 Zinc	114.2	82 Lead	177.8
31 Gallium	270.3	83 Bismuth	179.1
32 Germanium	327.6	84 Polonium	100.8
33 Arsenic	31.9	85 Astatine	n.a.
34 Selenium	90	86 Radon	18.1
35 Bromine	30.5	87 Francium	n.a.
36 Krypton	9.05	88 Radium	136.7
37 Rubidium	75.7	89 Actinium	293
38 Strontium	154.4	90 Thorium	513.7
39 Yttrium	367.4	91 Protactinium	481
40 Zirconium	566.7	92 Uranium	417.1
41 Niobium	680.19	93 Neptunium	336.6
42 Molybdenum	589.9	94 Plutonium	343.5
43 Technetium	585.2	95 Americium	238.5
44 Ruthenium	567	96 Curium	n.a.
45 Rhodium	494.3	97 Berkelium	n.a.
46 Palladium	361.5	98 Californium	n.a.
47 Silver	257.7	99 Einsteinium	n.a.
48 Cadmium	100.0	100 Fermium	n.a.
49 Indium	231.8	101 Mendelevium	n.a.
50 Tin	296.2	102 Nobelium	n.a.
51 Antimony	165.8	103 Lawrencium	n.a.
52 Tellurium	104.6		

Table 10B Enthalpies of vaporization ΔH_{vap} in order of ΔH_{vap}/kJ mol^{-1}

74 Tungsten	824.2	25 Manganese	220.5
73 Tantalum	758.22	83 Bismuth	179.1
76 Osmium	738.06	82 Lead	177.8
6 Carbon	710.9	63 Europium	176
75 Rhenium	704.25	81 Thallium	166.1
41 Niobium	680.19	51 Antimony	165.8
77 Iridium	612.1	62 Samarium	164.8
42 Molybdenum	589.9	70 Ytterbium	159
43 Technetium	585.2	38 Strontium	154.4
72 Hafnium	570.7	56 Barium	150.9
44 Ruthenium	567	20 Calcium	150.6
40 Zirconium	566.7	3 Lithium	147.7
90 Thorium	513.7	88 Radium	136.7
5 Boron	504.5	12 Magnesium	127.6
45 Rhodium	494.3	30 Zinc	114.2
91 Protactinium	481	52 Tellurium	104.6
78 Platinum	469	84 Polonium	100.8
23 Vanadium	459.7	48 Cadmium	100.0
71 Lutetium	428	11 Sodium	99.2
22 Titanium	425.5	34 Selenium	90
92 Uranium	417.1	19 Potassium	79.1
57 Lanthanum	402.1	37 Rubidium	75.7
58 Cerium	398	55 Caesium	66.5
65 Terbium	391	80 Mercury	59.11
14 Silicon	383.3	15 Phosphorus (P_4)	51.9
27 Cobalt	382.4	53 Iodine	41.67
21 Scandium	376.1	33 Arsenic	31.9
28 Nickel	374.8	35 Bromine	30.5
39 Yttrium	367.4	17 Chlorine	20.42
46 Palladium	361.5	86 Radon	18.1
59 Praseodymium	357	54 Xenon	12.65
94 Plutonium	343.5	16 Sulfur	9.62
79 Gold	343.1	36 Krypton	9.05
24 Chromium	341.8	8 Oxygen	6.82
26 Iron	340.2	18 Argon	6.53
93 Neptunium	336.6	7 Nitrogen	5.58
60 Neodymium	328	9 Fluorine	3.26
32 Germanium	327.6	10 Neon	1.736
4 Beryllium	308.8	1 Hydrogen	0.46
29 Copper	306.7	2 Helium	0.082
67 Holmium	303		
64 Gadolinium	301	61 Promethium	n.a.
50 Tin	296.2	85 Astatine	n.a.
66 Dysprosium	293	87 Francium	n.a.
89 Actinium	293	96 Curium	n.a.
13 Aluminium	290.8	97 Berkelium	n.a.
68 Erbium	280	98 Californium	n.a.
31 Gallium	270.3	99 Einsteinium	n.a.
47 Silver	257.7	100 Fermium	n.a.
69 Thulium	247	101 Mendelevium	n.a.
95 Americium	238.5	102 Nobelium	n.a.
49 Indium	231.8	103 Lawrencium	n.a.

Table 11A Ionization energies of neutral atoms $\Delta E(M \rightarrow M^{+})$ kJ mol^{-1}

Element	ΔE	Element	ΔE
1 Hydrogen	1312.0	53 Iodine	1008.4
2 Helium	2372.3	54 Xenon	1170.4
3 Lithium	513.3	55 Caesium	375.7
4 Beryllium	899.4	56 Barium	502.8
5 Boron	800.6	57 Lanthanum	538.1
6 Carbon	1086.2	58 Cerium	527.4
7 Nitrogen	1402.3	59 Praseodymium	523.1
8 Oxygen	1313.9	60 Neodymium	529.6
9 Fluorine	1681	61 Promethium	535.9
10 Neon	2080.6	62 Samarium	543.3
11 Sodium	495.8	63 Europium	546.7
12 Magnesium	737.7	64 Gadolinium	592.5
13 Aluminium	577.4	65 Terbium	564.6
14 Silicon	786.5	66 Dysprosium	571.9
15 Phosphorus	1011.7	67 Holmium	580.7
16 Sulfur	999.6	68 Erbium	588.7
17 Chlorine	1251.1	69 Thulium	596.7
18 Argon	1520.4	70 Ytterbium	603.4
19 Potassium	418.8	71 Lutetium	523.5
20 Calcium	589.7	72 Hafnium	642
21 Scandium	631	73 Tantalum	761
22 Titanium	658	74 Tungsten	770
23 Vanadium	650	75 Rhenium	760
24 Chromium	652.7	76 Osmium	840
25 Manganese	717.4	77 Iridium	880
26 Iron	759.3	78 Platinum	870
27 Cobalt	760.0	79 Gold	890.1
28 Nickel	736.7	80 Mercury	1007.0
29 Copper	745.4	81 Thallium	589.3
30 Zinc	906.4	82 Lead	715.5
31 Gallium	578.8	83 Bismuth	703.2
32 Germanium	762.1	84 Polonium	812
33 Arsenic	947.0	85 Astatine	930
34 Selenium	940.9	86 Radon	1037
35 Bromine	1139.9	87 Francium	400
36 Krypton	1350.7	88 Radium	509.3
37 Rubidium	403.0	89 Actinium	499
38 Strontium	549.5	90 Thorium	587
39 Yttrium	616	91 Protactinium	568
40 Zirconium	660	92 Uranium	584
41 Niobium	664	93 Neptunium	597
42 Molybdenum	685.0	94 Plutonium	585
43 Technetium	702	95 Americium	578.2
44 Ruthenium	711	96 Curium	581
45 Rhodium	720	97 Berkelium	601
46 Palladium	805	98 Californium	608
47 Silver	731.0	99 Einsteinium	619
48 Cadmium	867.6	100 Fermium	627
49 Indium	558.3	101 Mendelevium	635
50 Tin	708.6	102 Nobelium	642
51 Antimony	833.7	103 Lawrencium	n.a.
52 Tellurium	869.2		

Table 11B Ionization energies of neutral atoms $\Delta E(M \rightarrow M^+)$ in order of energy/kJ mol^{-1}

2 Helium	2372.3	40 Zirconium	660
10 Neon	2080.6	22 Titanium	658
9 Fluorine	1681	24 Chromium	652.7
18 Argon	1520.4	23 Vanadium	650
7 Nitrogen	1402.3	72 Hafnium	642
36 Krypton	1350.7	102 Nobelium	642
8 Oxygen	1313.9	101 Mendelevium	635
1 Hydrogen	1312.0	21 Scandium	631
17 Chlorine	1251.1	100 Fermium	630
54 Xenon	1170.4	99 Einsteinium	619
35 Bromine	1139.9	39 Yttrium	616
6 Carbon	1086.2	98 Californium	608
86 Radon	1037	70 Ytterbium	603.4
15 Phosphorus	1011.7	97 Berkelium	601
53 Iodine	1008.4	93 Neptunium	597
80 Mercury	1007.0	69 Thulium	596.7
16 Sulfur	999.6	64 Gadolinium	592.5
33 Arsenic	947.0	20 Calcium	589.7
34 Selenium	940.9	81 Thallium	589.3
85 Astatine	930	68 Erbium	588.7
30 Zinc	906.4	90 Thorium	587
4 Beryllium	899.4	94 Plutonium	585
79 Gold	890.1	92 Uranium	584
77 Iridium	880	96 Curium	581
78 Platinum	870	67 Holmium	580.7
52 Tellurium	869.2	31 Gallium	578.8
48 Cadmium	867.6	95 Americium	578.2
76 Osmium	840	13 Aluminium	577.4
51 Antimony	833.7	66 Dysprosium	571.9
84 Polonium	812	91 Protactinium	568
46 Palladium	805	65 Terbium	564.6
5 Boron	800.6	49 Indium	558.3
14 Silicon	786.5	38 Strontium	549.5
74 Tungsten	770	63 Europium	546.7
32 Germanium	762.1	62 Samarium	543.3
73 Tantalum	761	57 Lanthanum	538.1
27 Cobalt	760.0	61 Promethium	535.9
75 Rhenium	760	60 Neodymium	529.6
26 Iron	759.3	58 Cerium	527.4
29 Copper	745.4	71 Lutetium	523.5
12 Magnesium	737.7	59 Praseodymium	523.1
28 Nickel	736.7	3 Lithium	513.3
47 Silver	731.0	88 Radium	509.3
45 Rhodium	720	56 Barium	502.8
25 Manganese	717.4	89 Actinium	499
82 Lead	715.5	11 Sodium	495.8
44 Ruthenium	711	19 Potassium	418.8
50 Tin	708.6	37 Rubidium	403.0
83 Bismuth	703.2	87 Francium	400
43 Technetium	702	55 Caesium	375.7
42 Molybdenum	685.0		
41 Niobium	664	103 Lawrencium	n.a.

Table 12A Mass magnetic susceptibilities χ at 298 K/10^{-9} kg^{-1} m^3

Element	χ	Element	χ
1 Hydrogen	−25.0	53 Iodine	−4.40
2 Helium	−5.9	54 Xenon	−4.20
3 Lithium	+25.6	55 Caesium	+2.8
4 Beryllium	−13	56 Barium	+1.9
5 Boron	−7.8	57 Lanthanum	+11
6 Carbon (gra.)	−6.3	58 Cerium	+217
7 Nitrogen	−5.4	59 Praseodymium	+447
8 Oxygen	+1355	60 Neodymium	+490.2
9 Fluorine	n.a.	61 Promethium	n.a.
10 Neon	−4.20	62 Samarium	+152
11 Sodium	+8.8	63 Europium	+2810
12 Magnesium	+6.8	64 Gadolinium	+60 300
13 Aluminium	+7.7	65 Terbium	+11 500
14 Silicon	−1.8	66 Dysprosium	+8000
15 Phosphorus (red)	−8.4	67 Holmium	+5490
16 Sulfur (α)	−6.09	68 Erbium	+3330
17 Chlorine	−7.2	69 Thulium	+1900
18 Argon	−6.16	70 Ytterbium	+18.1
19 Potassium	+6.7	71 Lutetium	+1.3
20 Calcium	+14	72 Hafnium	+5.3
21 Scandium	+88	73 Tantalum	+10.7
22 Titanium	+40.1	74 Tungsten	+4.0
23 Vanadium	+62.8	75 Rhenium	+4.56
24 Chromium	+44.5	76 Osmium	+0.65
25 Manganese (α)	+121	77 Iridium	+1.67
26 Iron	ferromagnetic	78 Platinum	+13.01
27 Cobalt	ferromagnetic	79 Gold	−1.78
28 Nickel	ferromagnetic	80 Mercury	−2.095
29 Copper	−1.081	81 Thallium	−3.13
30 Zinc	−2.20	82 Lead	−1.39
31 Gallium	−3.9	83 Bismuth	−16.84
32 Germanium	−1.328	84 Polonium	n.a.
33 Arsenic (α)	−0.917	85 Astatine	n.a.
34 Selenium	−4.0	86 Radon	n.a.
35 Bromine	−4.44	87 Francium	n.a.
36 Krypton	−4.32	88 Radium	n.a.
37 Rubidium	+2.49	89 Actinium	n.a.
38 Strontium	+13.2	90 Thorium	+7.2
39 Yttrium	+27.0	91 Protactinium	n.a.
40 Zirconium	+16.8	92 Uranium	+21.6
41 Niobium	+27.6	93 Neptunium	n.a.
42 Molybdenum	+12	94 Plutonium	+31.7
43 Technetium	+31	95 Americium	+50
44 Ruthenium	+5.37	96 Curium	n.a.
45 Rhodium	+13.6	97 Berkelium	n.a.
46 Palladium	+67.02	98 Californium	n.a.
47 Silver	−2.27	99 Einsteinium	n.a.
48 Cadmium	−2.21	100 Fermium	n.a.
49 Indium	−7.0	101 Mendelevium	n.a.
50 Tin (α)	−4.0	102 Nobelium	n.a.
51 Antimony	−10	103 Lawrencium	n.a.
52 Tellurium	−3.9		

Table 12B Mass magnetic susceptibilities χ in order of $\chi/10^{-9}\ kg^{-1}\ m^3$

Element	χ	Element	χ
26 Iron	ferromagnetic	33 Arsenic (α)	−0.917
27 Cobalt	ferromagnetic	29 Copper	−1.081
28 Nickel	ferromagnetic	32 Germanium	−1.328
		82 Lead	−1.39
64 Gadolinium	+60 300	79 Gold	−1.78
65 Terbium	+11 500	14 Silicon	−1.8
66 Dysprosium	+8000	80 Mercury	−2.095
67 Holmium	+5490	30 Zinc	−2.20
68 Erbium	+3330	48 Cadmium	−2.21
63 Europium	+2810	47 Silver	−2.27
69 Thulium	+1900	81 Thallium	−3.13
8 Oxygen	+1355	31 Gallium	−3.9
60 Neodymium	+490.2	52 Tellurium	−3.9
59 Praseodymium	+447	34 Selenium	−4.0
58 Cerium	+217	50 Tin (α)	−4.0
62 Samarium	+152	54 Xenon	−4.20
25 Manganese (α)	+121	10 Neon	−4.20
21 Scandium	+88	36 Krypton	−4.32
46 Palladium	+67.02	35 Iodine	−4.40
23 Vanadium	+62.8	35 Bromine	−4.44
95 Americium	+50	7 Nitrogen	−5.4
24 Chromium	+44.5	2 Helium	−5.9
22 Titanium	+40.1	16 Sulfur (α)	−6.09
94 Plutonium	+31.7	18 Argon	−6.16
43 Technetium	+31	6 Carbon (graph.)	−6.3
41 Niobium	+27.6	49 Indium	−7.0
39 Yttrium	+27.0	17 Chlorine	−7.2
3 Lithium	+25.6	5 Boron	−7.8
92 Uranium	+21.6	15 Phosphorus (red)	−8.4
70 Ytterbium	+18.1	51 Antimony	−10
40 Zirconium	+16.8	4 Beryllium	−13
21 Calcium	+14	83 Bismuth	−16.84
45 Rhodium	+13.6	1 Hydrogen	−25.0
38 Strontium	+13.2		
78 Platinum	+13.01	9 Fluorine	n.a.
42 Molybdenum	+12	61 Promethium	n.a.
57 Lanthanum	+11	84 Polonium	n.a.
73 Tantalum	+10.7	85 Astatine	n.a.
11 Sodium	+8.8	86 Radon	n.a.
13 Aluminium	+7.7	87 Francium	n.a.
90 Thorium	+7.2	88 Radium	n.a.
12 Magnesium	+6.8	89 Actinium	n.a.
19 Potassium	+6.7	91 Protactinium	n.a.
44 Ruthenium	+5.37	93 Neptunium	n.a.
72 Hafnium	+5.3	96 Curium	n.a.
75 Rhenium	+4.56	97 Berkelium	n.a.
74 Tungsten	+4.0	98 Californium	n.a.
55 Caesium	+2.8	99 Einsteinium	n.a.
37 Rubidium	+2.49	100 Fermium	n.a.
56 Barium	+1.9	101 Mendelevium	n.a.
77 Iridium	+1.67	102 Nobelium	n.a.
71 Lutetium	+1.3	103 Lawrencium	n.a.
76 Osmium	+0.65		

Table 13A Melting points/K

Element	m.p./K	Element	m.p./K
1 Hydrogen	14.01	53 Iodine	386.7
2 Helium	0.95	54 Xenon	161.3
3 Lithium	453.69	55 Caesium	301.6
4 Beryllium	1551	56 Barium	1002
5 Boron	2573	57 Lanthanum	1194
6 Carbon (diam.)	3820	58 Cerium	1072
7 Nitrogen	63.29	59 Praseodymium	1204
8 Oxygen	54.8	60 Neodymium	1294
9 Fluorine	53.53	61 Promethium	1441
10 Neon	24.48	62 Samarium	1350
11 Sodium	370.96	63 Europium	1095
12 Magnesium	922.0	64 Gadolinium	1586
13 Aluminium	933.5	65 Terbium	1629
14 Silicon	1683	66 Dysprosium	1685
15 Phosphorus (P_4)	317.3	67 Holmium	1747
16 Sulfur (α)	386.0	68 Erbium	1802
17 Chlorine	172.2	69 Thulium	1818
18 Argon	83.8	70 Ytterbium	1097
19 Potassium	336.8	71 Lutetium	1936
20 Calcium	1112	72 Hafnium	2503
21 Scandium	1814	73 Tantalum	3269
22 Titanium	1933	74 Tungsten	3680
23 Vanadium	2160	75 Rhenium	3453
24 Chromium	2130	76 Osmium	3327
25 Manganese	1517	77 Iridium	2683
26 Iron	1808	78 Platinum	2045
27 Cobalt	1768	79 Gold	1337.58
28 Nickel	1726	80 Mercury	234.28
29 Copper	1356.6	81 Thallium	576.6
30 Zinc	692.73	82 Lead	600.65
31 Gallium	302.93	83 Bismuth	544.5
32 Germanium	1210.6	84 Polonium	527
33 Arsenic	1090	85 Astatine	575 (est.)
34 Selenium	490	86 Radon	202
35 Bromine	265.9	87 Francium	300
36 Krypton	116.6	88 Radium	973
37 Rubidium	312.2	89 Actinium	1320
38 Strontium	1042	90 Thorium	2023
39 Yttrium	1795	91 Protactinium	2113
40 Zirconium	2125	92 Uranium	1405.5
41 Niobium	2741	93 Neptunium	913
42 Molybdenum	2890	94 Plutonium	914
43 Technetium	2445	95 Americium	1267
44 Ruthenium	2583	96 Curium	n.a.
45 Rhodium	2239	97 Berkelium	n.a.
46 Palladium	1825	98 Californium	n.a.
47 Silver	1235.1	99 Einsteinium	n.a.
48 Cadmium	594.1	100 Fermium	n.a.
49 Indium	429.32	101 Mendelevium	n.a.
50 Tin (β)	505.118	102 Nobelium	n.a.
51 Antimony	903.9	103 Lawrencium	n.a.
52 Tellurium	722.7		

Table 13B Melting points in order of temperature/K

Element	m.p./K	Element	m.p./K
6 Carbon (diam.)	3820	38 Strontium	1042
74 Tungsten	3680	56 Barium	1002
75 Rhenium	3453	88 Radium	973
76 Osmium	3327	13 Aluminium	933.5
73 Tantalum	3269	12 Magnesium	922.0
42 Molybdenum	2890	94 Plutonium	914
41 Niobium	2741	93 Neptunium	913
77 Iridium	2683	51 Antimony	903.9
44 Ruthenium	2583	52 Tellurium	722.7
5 Boron	2573	30 Zinc	692.73
72 Hafnium	2503	82 Lead	600.65
43 Technetium	2445	48 Cadmium	594.1
45 Rhodium	2239	81 Thallium	576.6
23 Vanadium	2160	85 Astatine	575 (est.)
24 Chromium	2130	83 Bismuth	544.5
40 Zirconium	2125	84 Polonium	527
91 Protactinium	2113	50 Tin (β)	505.118
78 Platinum	2045	34 Selenium	490
90 Thorium	2023	3 Lithium	453.69
71 Lutetium	1936	49 Indium	429.32
22 Titanium	1933	53 Iodine	386.7
46 Palladium	1825	16 Sulfur (α)	386.0
69 Thulium	1818	11 Sodium	370.96
21 Scandium	1814	19 Potassium	336.8
26 Iron	1808	15 Phosphorus (P_4)	317.31
68 Erbium	1802	37 Rubidium	312.2
39 Yttrium	1795	31 Gallium	302.93
27 Cobalt	1768	55 Caesium	301.6
67 Holmium	1747	87 Francium	300
28 Nickel	1726	35 Bromine	265.9
66 Dysprosium	1685	80 Mercury	234.28
14 Silicon	1683	86 Radon	202
65 Terbium	1629	17 Chlorine	172.2
64 Gadolinium	1586	54 Xenon	161.3
4 Beryllium	1551	36 Krypton	116.6
25 Manganese	1517	18 Argon	83.8
61 Promethium	1441	7 Nitrogen	63.29
92 Uranium	1405.5	8 Oxygen	54.8
29 Copper	1356.6	9 Fluorine	53.53
62 Samarium	1350	10 Neon	24.48
79 Gold	1337.58	1 Hydrogen	14.01
89 Actinium	1320	2 Helium	0.95
60 Neodymium	1294		
95 Americium	1267	96 Curium	n.a.
47 Silver	1235.1	97 Berkelium	n.a.
32 Germanium	1210.6	98 Californium	n.a.
59 Praseodymium	1204	99 Einsteinium	n.a.
57 Lanthanum	1194	100 Fermium	n.a.
20 Calcium	1112	101 Mendelevium	n.a.
70 Ytterbium	1097	102 Nobelium	n.a.
63 Europium	1095	103 Lawrencium	n.a.
33 Arsenic	1090		
58 Cerium	1072		

Table 14A Neutron (thermal) capture cross-section/barns

Element	Cross-section	Element	Cross-section
1 Hydrogen	0.332	53 Iodine	6.2
2 Helium	*c.* 0.007	54 Xenon	24.5
3 Lithium	71	55 Caesium	30.0
4 Beryllium	0.0092	56 Barium	1.2
5 Boron	3837	57 Lanthanum	8.9
6 Carbon	0.0034	58 Cerium	0.73
7 Nitrogen	0.075	59 Praseodymium	11.5
8 Oxygen	0.0002	60 Neodymium	49
9 Fluorine	0.098	61 Promethium	8400
10 Neon	0.038	62 Samarium	5820
11 Sodium	0.534	63 Europium	4100
12 Magnesium	0.064	64 Gadolinium	49 000
13 Aluminium	0.232	65 Terbium	30
14 Silicon	0.160	66 Dysprosium	90
15 Phosphorus	0.19	67 Holmium	65
16 Sulfur	0.51	68 Erbium	0.16
17 Chlorine	44	69 Thulium	115
18 Argon	0.65	70 Ytterbium	37
19 Potassium	2.2	71 Lutetium	75
20 Calcium	0.44	72 Hafnium	103
21 Scandium	25	73 Tantalum	22
22 Titanium	6.1	74 Tungsten	18.5
23 Vanadium	5.06	75 Rhenium	85
24 Chromium	3.1	76 Osmium	15.3
25 Manganese	13.3	77 Iridium	425
26 Iron	2.56	78 Platinum	9
27 Cobalt	37.5	79 Gold	98.8
28 Nickel	4.51	80 Mercury	375
29 Copper	3.8	81 Thallium	3.4
30 Zinc	1.10	82 Lead	0.18
31 Gallium	3.1	83 Bismuth	0.034
32 Germanium	2.3	84 Polonium	<0.5
33 Arsenic	4.30	85 Astatine	n.a.
34 Selenium	12.2	86 Radon	0.72
35 Bromine	6.8	87 Francium	n.a.
36 Krypton	24.1	88 Radium	20
37 Rubidium	0.5	89 Actinium	810
38 Strontium	1.21	90 Thorium	7.4
39 Yttrium	1.3	91 Protactinium	200
40 Zirconium	0.182	92 Uranium	7.6
41 Niobium	1.15	93 Neptunium	170
42 Molybdenum	2.65	94 Plutonium	1.8
43 Technetium	22	95 Americium	180
44 Ruthenium	3.0	96 Curium	180
45 Rhodium	150	97 Berkelium	1000
46 Palladium	6.0	98 Californium	2100
47 Silver	63.8	99 Einsteinium	<40
48 Cadmium	2450	100 Fermium	26
49 Indium	194	101 Mendelevium	n.a.
50 Tin	0.63	102 Nobelium	n.a.
51 Antimony	5	103 Lawrencium	n.a.
52 Tellurium	4.7		

Table 14B Neutron (thermal) capture cross-section in order of neutron capture/barns

Element	Cross-section	Element	Cross-section
64 Gadolinium	49 000	23 Vanadium	5.06
61 Promethium	8400	51 Antimony	5
62 Samarium	5820	52 Tellurium	4.7
63 Europium	4100	28 Nickel	4.51
5 Boron	3837	33 Arsenic	4.30
48 Cadmium	2450	29 Copper	3.8
98 Californium	2100	81 Thallium	3.4
97 Berkelium	1000	24 Chromium	3.1
89 Actinium	810	31 Gallium	3.1
77 Iridium	425	44 Ruthenium	3.0
80 Mercury	375	42 Molybdenum	2.65
91 Protactinium	200	26 Iron	2.56
49 Indium	194	32 Germanium	2.3
95 Americium	180	19 Potassium	2.2
96 Curium	180	94 Plutonium	1.8
93 Neptunium	170	39 Yttrium	1.3
45 Rhodium	150	38 Strontium	1.21
69 Thulium	115	56 Barium	1.2
72 Hafnium	103	41 Niobium	1.15
79 Gold	98.8	30 Zinc	1.10
66 Dysprosium	90	74 Tungsten	1
75 Rhenium	85	58 Cerium	0.73
71 Lutetium	75	86 Radon	0.72
3 Lithium	71	18 Argon	0.65
67 Holmium	65	50 Tin	0.63
47 Silver	63.8	11 Sodium	0.534
60 Neodymium	49	16 Sulfur	0.51
17 Chlorine	44	37 Rubidium	0.5
99 Einsteinium	40	84 Polonium	0.5
27 Cobalt	37.5	20 Calcium	0.44
70 Ytterbium	37	1 Hydrogen	0.332
55 Caesium	30.0	13 Aluminium	0.232
65 Terbium	30	15 Phosphorus	0.19
100 Fermium	26	40 Zirconium	0.182
21 Scandium	25	82 Lead	0.18
54 Xenon	24.5	14 Silicon	0.160
36 Krypton	24.1	68 Erbium	0.16
73 Tantalum	22	9 Fluorine	0.098
43 Technetium	22	7 Nitrogen	0.075
88 Radium	20	12 Magnesium	0.064
76 Osmium	15.3	10 Neon	0.038
25 Manganese	13.3	83 Bismuth	0.034
34 Selenium	12.2	4 Beryllium	0.0092
59 Praseodymium	11.5	2 Helium	*c.* 0.007
78 Platinum	9	6 Carbon	0.0034
57 Lanthanum	8.9	8 Oxygen	0.0002
92 Uranium	7.6		
90 Thorium	7.4	85 Astatine	n.a.
35 Bromine	6.8	87 Francium	n.a.
53 Iodine	6.2	101 Mendelevium	n.a.
22 Titanium	6.1	102 Nobelium	n.a.
46 Palladium	6.0	103 Lawrencium	n.a.

Table 15A NMR frequencies of nuclei* at a field of 2.344 T (^1H=100 MHz)/MHz

Element	MHz	Element	MHz
1 Hydrogen	100.000	53 Iodine-127	20.007
2 Helium-3	76.178	54 Xenon-129	27.660
3 Lithium-7	38.863	55 Caesium-133	13.117
4 Beryllium-9	14.053	56 Barium-137	11.113
5 Boron-11	32.084	57 Lanthanum-139	14.126
6 Carbon-13	25.144	58 Cerium-139	10.862
7 Nitrogen-15	10.133	59 Praseodymium-141	29.291
8 Oxygen-17	13.557	60 Neodymium-143	5.437
9 Fluorine-19	94.077	61 Promethium-147	13.51
10 Neon-21	7.894	62 Samarium-147	4.128
11 Sodium-23	26.451	63 Europium-151	24.801
12 Magnesium-25	6.120	64 Gadolinium-155	3.819
13 Aluminium-27	26.057	65 Terbium-159	22.678
14 Silicon-29	19.865	66 Dysprosium-163	4.583
15 Phosphorus-31	40.481	67 Holmium-165	20.513
16 Sulfur-33	7.670	68 Erbium-167	2.890
17 Chlorine-35	9.798	69 Thulium-169	8.271
18 Argon-39	6.6	70 Ytterbium-171	17.613
19 Potassium-39	4.667	71 Lutetium-175	11.407
20 Calcium-43	6.728	72 Hafnium-177	3.120
21 Scandium-45	24.290	73 Tantalum-181	11.970
22 Titanium-49	5.638	74 Tungsten-183	4.161
23 Vanadium-51	26.289	75 Rhenium-187	22.513
24 Chromium-53	5.652	76 Osmium-187	2.282
25 Manganese-55	24.664	77 Iridium-191	1.718
26 Iron-57	3.231	78 Platinum-195	21.499
27 Cobalt-59	23.614	79 Gold-197	1.712
28 Nickel-61	8.936	80 Mercury-199	17.827
29 Copper-63	26.505	81 Thallium-205	57.708
30 Zinc-67	6.254	82 Lead-207	20.921
31 Gallium-71	30.495	83 Bismuth-209	16.069
32 Germanium-73	3.488	84 Polonium-209	28
33 Arsenic-75	17.126	85 Astatine	n.a.
34 Selenium-77	19.092	86 Radon	n.a.
35 Bromine-81	27.006	87 Francium	n.a.
36 Krypton-83	3.847	88 Radium	n.a.
37 Rubidium-87	32.721	89 Actinium-227	13.1
38 Strontium-87	4.333	90 Thorium-229	1.5
39 Yttrium-89	4.899	91 Protactinium-231	12.0
40 Zirconium-91	9.330	92 Uranium-235	1.790
41 Niobium-93	24.442	93 Neptunium-237	11.25
42 Molybdenum-95	6.514	94 Plutonium-239	3.63
43 Technetium-99	22.508	95 Americium-241	5.76
44 Ruthenium-101	4.941	96 Curium-247	0.75
45 Rhodium-103	3.172	97 Berkelium	n.a.
46 Palladium-105	4.576	98 Californium	n.a.
47 Silver-109	4.652	99 Einsteinium	n.a.
48 Cadmium-113	22.182	100 Fermium	n.a.
49 Indium-115	21.914	101 Mendelevium	n.a.
50 Tin-119	37.272	102 Nobelium	n.a.
51 Antimony-121	23.930	103 Lawrencium	n.a.
52 Tellurium-125	31.596		

*Where there are two or more isotopes that can be observed by NMR spectroscopy, only the one that is commonly used is included in this table.

Table 15B NMR frequencies of nuclei* at a field of 2.344 T ($^1H = 100$ MHz) in order of frequency/MHz

1	Hydrogen-1	100.000	58	Cerium-139	10.862
9	Fluorine-19	94.077	7	Nitrogen-15	10.133
2	Helium-3	76.178	17	Chlorine-35	9.798
81	Thallium-205	57.708	40	Zirconium-91	9.330
15	Phosphorus-31	40.481	28	Nickel-61	8.936
3	Lithium-7	38.863	69	Thulium-169	8.271
50	Tin-119	37.272	10	Neon-21	7.894
37	Rubidium-87	32.721	16	Sulfur-33	7.670
5	Boron-11	32.084	20	Calcium-43	6.728
52	Tellurium-125	31.596	18	Argon-39	6.6
31	Gallium-71	30.495	42	Molybdenum-95	6.514
59	Praseodymium-141	29.291	30	Zinc-67	6.254
84	Polonium-209	28	12	Magnesium-25	6.120
54	Xenon-129	27.660	95	Americium-241	5.76
35	Bromine-81	27.006	24	Chromium-53	5.652
29	Copper-63	26.505	22	Titanium-49	5.638
11	Sodium-23	26.451	60	Neodymium-143	5.437
23	Vanadium-51	26.289	44	Ruthenium-101	4.941
13	Aluminium-27	26.057	39	Yttrium-89	4.899
6	Carbon-13	25.144	19	Potassium-39	4.667
63	Europium-151	24.801	47	Silver-109	4.652
25	Manganese-55	24.664	66	Dysprosium-163	4.583
41	Niobium-93	24.442	46	Palladium-105	4.576
21	Scandium-45	24.290	38	Strontium-87	4.333
51	Antimony-121	23.930	74	Tungsten-183	4.161
27	Cobalt-59	23.614	62	Samarium-147	4.128
65	Terbium-159	22.678	36	Krypton-83	3.847
75	Rhenium-187	22.513	64	Gadolinium-155	3.819
43	Technetium-99	22.508	94	Plutonium-239	3.63
48	Cadmium-113	22.182	32	Germanium-73	3.488
49	Indium-115	21.914	26	Iron-57	3.231
78	Platinum-195	21.499	45	Rhodium-103	3.172
82	Lead-207	20.921	72	Hafnium-177	3.120
67	Holmium-165	20.513	68	Erbium-167	2.890
53	Iodine-127	20.007	76	Osmium-187	2.282
14	Silicon-29	19.865	92	Uranium-235	1.790
34	Selenium-77	19.092	77	Iridium-191	1.718
80	Mercury-199	17.827	79	Gold-197	1.712
70	Ytterbium-171	17.613	90	Thorium-229	1.5
33	Arsenic-75	17.126	96	Curium-247	0.75
83	Bismuth-209	16.069			
57	Lanthanum-139	14.126	85	Astatine	n.a.
4	Beryllium-9	14.053	86	Radon	n.a.
8	Oxygen-17	13.557	87	Francium	n.a.
61	Promethium-147	13.51	88	Radium	n.a.
55	Caesium-133	13.117	97	Berkelium	n.a.
89	Actinium-227	13.1	98	Californium	n.a.
91	Protactinium-231	12.0	99	Einsteinium	n.a.
73	Tantalum-181	11.970	100	Fermium	n.a.
71	Lutetium-175	11.407	101	Mendelevium	n.a.
93	Neptunium-237	11.25	102	Nobelium	n.a.
56	Barium-137	11.113	103	Lawrencium	n.a.

*Where there are two or more isotopes that can be observed by NMR spectroscopy, only the one that is commonly used is included in this table.

Table 16A Thermal conductivities at 300 K/W m^{-1} K^{-1}

Element	Thermal conductivity
1 Hydrogen (g)	0.1815
2 Helium (g)	0.152
3 Lithium	84.7
4 Beryllium	200
5 Boron	27.0
6 Carbon (graph.)	1960
7 Nitrogen (g)	0.025 98
8 Oxygen (g)	0.026 74
9 Fluorine (g)	0.0279
10 Neon (g)	0.0493
11 Sodium	141
12 Magnesium	156
13 Aluminium	237
14 Silicon	148
15 Phosphorus (P_4)	0.235
16 Sulfur (α)	0.269
17 Chlorine (g)	0.0089
18 Argon (g)	0.017 72
19 Potassium	102.4
20 Calcium	200
21 Scandium	15.8
22 Titanium	21.9
23 Vanadium	30.7
24 Chromium	93.7
25 Manganese	7.82
26 Iron	80.2
27 Cobalt	100
28 Nickel	90.7
29 Copper	401
30 Zinc	116
31 Gallium	40.6
32 Germanium	59.9
33 Arsenic (α)	50.0
34 Selenium	2.04
35 Bromine (liq.)	0.122
36 Krypton (g)	0.009 49
37 Rubidium	58.2
38 Strontium	35.3
39 Yttrium	17.2
40 Zirconium	22.7
41 Niobium	53.7
42 Molybdenum	138
43 Technetium	50.6
44 Ruthenium	117
45 Rhodium	150
46 Palladium	71.8
47 Silver	429
48 Cadmium	96.8
49 Indium	81.6
50 Tin (α)	66.6
51 Antimony	243
52 Tellurium	2.35
53 Iodine	0.449
54 Xenon (g)	0.005 69
55 Caesium	35.9
56 Barium	18.4
57 Lanthanum	13.5
58 Cerium	11.4
59 Praseodymium	12.5
60 Neodymium	16.5
62 Promethium	17.9 (est.)
62 Samarium	13.3
63 Europium	13.9
64 Gadolinium	10.6
65 Terbium	11.1
66 Dysprosium	10.7
67 Holmium	16.2
68 Erbium	14.3
69 Thulium	16.8
70 Ytterbium	34.9
71 Lutetium	164
72 Hafnium	23.0
73 Tantalum	57.5
74 Tungsten	174
75 Rhenium	47.9
76 Osmium	87.6
77 Iridium	147
78 Platinum	71.6
79 Gold	317
80 Mercury (liq.)	8.34
81 Thallium	46.1
82 Lead	35.3
83 Bismuth	7.87
84 Polonium	20
85 Astatine	1.7
86 Radon (g)	0.003 64 (est.)
87 Francium	15 (est.)
88 Radium	18.6 (est.)
89 Actinium	12
90 Thorium	54.0
91 Protactinium	47 (est.)
92 Uranium	27.6
93 Neptunium	6.3
94 Plutonium	6.74
95 Americium	10 (est.)
96 Curium	10 (est.)
97 Berkelium	10 (est.)
98 Californium	10 (est.)
99 Einsteinium	10 (est.)
100 Fermium	10 (est.)
101 Mendelevium	10 (est.)
102 Nobelium	10 (est.)
103 Lawrencium	10 (est.)

Table 16B Thermal conductivities at 300 K in order of conductivity/W m^{-1} K^{-1}

Z	Element	Conductivity
6	Carbon (graph.)	1960
47	Silver	429
29	Copper	401
79	Gold	317
51	Antimony	243
13	Aluminium	237
4	Beryllium	200
20	Calcium	200
74	Tungsten	174
71	Lutetium	164
12	Magnesium	156
45	Rhodium	150
14	Silicon	148
77	Iridium	147
11	Sodium	141
42	Molybdenum	138
44	Ruthenium	117
30	Zinc	116
27	Cobalt	100
19	Potassium	102.4
48	Cadmium	96.8
24	Chromium	93.7
28	Nickel	90.7
76	Osmium	87.6
3	Lithium	84.7
49	Indium	81.6
26	Iron	80.2
46	Palladium	71.8
78	Platinum	71.6
50	Tin (α)	66.6
32	Germanium	59.9
37	Rubidium	58.2
73	Tantalum	57.5
90	Thorium	54.0
41	Niobium	53.7
43	Technetium	50.6
33	Arsenic (α)	50.0
75	Rhenium	47.9
91	Protactinium	47 (est.)
81	Thallium	46.1
31	Gallium	40.6
55	Caesium	35.9
82	Lead	35.3
38	Strontium	35.3
70	Ytterbium	34.9
23	Vanadium	30.7
92	Uranium	27.6
5	Boron	27.0
72	Hafnium	23.0
40	Zirconium	22.7
22	Titanium	21.9
84	Polonium	20
88	Radium	18.6 (est.)
56	Barium	18.4
61	Promethium	17.9 (est.)
39	Yttrium	17.2
69	Thulium	16.8
60	Neodymium	16.5
67	Holmium	16.2
21	Scandium	15.8
87	Francium	15 (est.)
68	Erbium	14.3
63	Europium	13.9
57	Lanthanum	13.5
62	Samarium	13.3
59	Praseodymium	12.5
89	Actinium	12
58	Cerium	11.4
65	Terbium	11.1
66	Dysprosium	10.7
64	Gadolinium	10.6
95	Americium	10 (est.)
96	Curium	10 (est.)
97	Berkelium	10 (est.)
98	Californium	10 (est.)
99	Einsteinium	10 (est.)
100	Fermium	10 (est.)
101	Mendelevium	10 (est.)
102	Nobelium	10 (est.)
103	Lawrencium	10 (est.)
80	Mercury (liq.)	8.34
83	Bismuth	7.87
25	Manganese	7.82
94	Plutonium	6.74
93	Neptunium	6.3
52	Tellurium	2.35
34	Selenium	2.04
85	Astatine	1.7
53	Iodine	0.449
16	Sulfur (α)	0.269
15	Phosphorus (P_4)	0.235
1	Hydrogen (g)	0.1815
2	Helium (g)	0.152
35	Bromine (liq.)	0.122
10	Neon (g)	0.0493
9	Fluorine (g)	0.0279
8	Oxygen (g)	0.026 74
7	Nitrogen (g)	0.025 98
18	Argon (g)	0.017 72
36	Krypton (g)	0.009 49
17	Chlorine (g)	0.0089
54	Xenon (g)	0.005 69
86	Radon (g)	0.003 64 (est.)

Tables of Elements in alphabetical order by name and by formula, and table of relative atomic masses by atomic number

Table 1 The elements in alphabetical order with formulae and atomic numbers

Actinium	Ac	89	Mercury	Hg	80
Aluminium	Al	13	Molybdenum	Mo	42
Americium	Am	95	Neodymium	Nd	60
Antimony	Sb	51	Neon	Ne	10
Argon	Ar	18	Neptunium	Np	93
Arsenic	As	33	Nickel	Ni	28
Astatine	At	85	Niobium	Nb	41
Barium	Ba	56	Nitrogen	N	7
Berkelium	Bk	97	Nobelium	No	102
Beryllium	Be	4	Osmium	Os	76
Bismuth	Bi	83	Oxygen	O	8
Boron	B	5	Palladium	Pd	46
Bromine	Br	35	Phosphorus	P	15
Cadmium	Cd	48	Platinum	Pt	78
Caesium	Cs	55	Plutonium	Pu	94
Calcium	Ca	20	Polonium	Po	84
Californium	Cf	98	Potassium	K	19
Carbon	C	6	Praseodymium	Pr	59
Cerium	Ce	58	Promethium	Pm	61
Chlorine	Cl	17	Protactinium	Pa	91
Chromium	Cr	24	Radium	Ra	88
Cobalt	Co	27	Radon	Rn	86
Copper	Cu	29	Rhenium	Re	75
Curium	Cm	96	Rhodium	Rh	45
Dysprosium	Dy	66	Rubidium	Rb	37
Einsteinium	Es	99	Ruthenium	Ru	44
Erbium	Er	68	Samarium	Sm	62
Europium	Eu	63	Scandium	Sc	21
Fermium	Fm	100	Selenium	Se	34
Fluorine	F	9	Silicon	Si	14
Francium	Fr	87	Silver	Ag	47
Gadolinium	Gd	64	Sodium	Na	11
Gallium	Ga	31	Strontium	Sr	38
Germanium	Ge	32	Sulfur	S	16
Gold	Au	79	Tantalum	Ta	73
Hafnium	Hf	72	Technetium	Tc	43
Helium	He	2	Tellurium	Te	52
Holmium	Ho	67	Terbium	Tb	65
Hydrogen	H	1	Thallium	Tl	81
Indium	In	49	Thorium	Th	90
Iodine	I	53	Thulium	Tm	69
Iridium	Ir	77	Tin	Sn	50
Iron	Fe	26	Titanium	Ti	22
Krypton	Kr	36	Tungsten	W	74
Lanthanum	La	57	Uranium	U	92
Lawrencium	Lr	103	Vanadium	V	23
Lead	Pb	82	Xenon	Xe	54
Lithium	Li	3	Ytterbium	Yb	70
Lutetium	Lu	71	Yttrium	Y	39
Magnesium	Mg	12	Zinc	Zn	30
Manganese	Mn	25	Zirconium	Zr	40
Mendelevium	Md	101			

Table 2 The elements in alphabetical order of formula with names and atomic numbers

Ac	Actinium	89	Mn	Manganese	25
Ag	Silver	47	Mo	Molybdenum	42
Al	Aluminium	13	N	Nitrogen	7
Am	Americium	95	Na	Sodium	11
Ar	Argon	18	Nb	Niobium	41
As	Arsenic	33	Nd	Neodymium	60
At	Astatine	85	Ne	Neon	10
Au	Gold	79	Ni	Nickel	28
B	Boron	5	No	Nobelium	102
Ba	Barium	56	Np	Neptunium	93
Be	Beryllium	4	O	Oxygen	8
Bi	Bismuth	83	Os	Osmium	76
Bk	Berkelium	97	P	Phosphorus	15
Br	Bromine	35	Pa	Protactinium	91
C	Carbon	6	Pb	Lead	82
Ca	Calcium	20	Pd	Palladium	46
Cd	Cadmium	48	Pm	Promethium	61
Ce	Cerium	58	Po	Polonium	84
Cf	Californium	98	Pr	Praseodymium	59
Cl	Chlorine	17	Pt	Platinum	78
Cm	Curium	96	Pu	Plutonium	94
Co	Cobalt	27	Ra	Radium	88
Cr	Chromium	24	Rb	Rubidium	37
Cs	Caesium	55	Re	Rhenium	75
Cu	Copper	29	Rh	Rhodium	45
Dy	Dysprosium	66	Rn	Radon	86
Er	Erbium	68	Ru	Ruthenium	44
Es	Einsteinium	99	S	Sulfur	16
Eu	Europium	63	Sb	Antimony	51
F	Fluorine	9	Sc	Scandium	21
Fe	Iron	26	Se	Selenium	34
Fm	Fermium	100	Si	Silicon	14
Fr	Francium	87	Sm	Samarium	62
Ga	Gallium	31	Sn	Tin	50
Gd	Gadolinium	64	Sr	Strontium	38
Ge	Germanium	32	Ta	Tantalum	73
H	Hydrogen	1	Tb	Terbium	65
He	Helium	2	Tc	Technetium	43
Hf	Hafnium	72	Te	Tellurium	52
Hg	Mercury	80	Th	Thorium	90
Ho	Holmium	67	Ti	Titanium	22
I	Iodine	53	Tl	Thallium	81
In	Indium	49	Tm	Thulium	69
Ir	Iridium	77	U	Uranium	92
K	Potassium	19	V	Vanadium	23
Kr	Krypton	36	W	Tungsten	74
La	Lanthanum	57	Xe	Xenon	54
Li	Lithium	3	Y	Yttrium	39
Lr	Lawrencium	103	Yb	Ytterbium	70
Lu	Lutetium	71	Zn	Zinc	30
Md	Mendelevium	101	Zr	Zirconium	40
Mg	Magnesium	12			

Table 3 Relative atomic masses (atomic weights) of the elements

1 Hydrogen	1.0079	53 Iodine	126.9045
2 Helium	4.00260	54 Xenon	131.30
3 Lithium	6.941	55 Caesium	132.9054
4 Beryllium	9.01218	56 Barium	137.33
5 Boron	10.81	57 Lanthanum	138.9055
6 Carbon	12.011	58 Cerium	140.12
7 Nitrogen	14.0067	59 Praseodymium	140.9077
8 Oxygen	15.9994	60 Neodymium	144.24
9 Fluorine	18.998403	61 Promethium	(145)
10 Neon	20.179	62 Samarium	150.4
11 Sodium	22.98977	63 Europium	151.96
12 Magnesium	24.305	64 Gadolinium	157.25
13 Aluminium	26.98154	65 Terbium	158.9254
14 Silicon	28.0855	66 Dysprosium	162.50
15 Phosphorus	30.97376	67 Holmium	164.9304
16 Sulfur	32.06	68 Erbium	167.26
17 Chlorine	35.453	69 Thulium	168.9342
18 Argon	39.948	70 Ytterbium	173.04
19 Potassium	39.0983	71 Lutetium	174.967
20 Calcium	40.08	72 Hafnium	178.49
21 Scandium	44.9559	73 Tantalum	180.9479
22 Titanium	47.90	74 Tungsten	183.85
23 Vanadium	50.9415	75 Rhenium	186.207
24 Chromium	51.996	76 Osmium	190.2
25 Manganese	54.9380	77 Iridium	192.22
26 Iron	55.847	78 Platinum	195.09
27 Cobalt	58.9332	79 Gold	196.9665
28 Nickel	58.70	80 Mercury	200.59
29 Copper	63.546	81 Thallium	204.37
30 Zinc	65.38	82 Lead	207.2
31 Gallium	69.735	83 Bismuth	208.9804
32 Germanium	72.59	84 Polonium	(209)
33 Arsenic	74.9216	85 Astatine	(210)
34 Selenium	78.96	86 Radon	(222)
35 Bromine	79.904	87 Francium	(223)
36 Krypton	83.80	88 Radium	226.0254
37 Rubidium	85.4678	89 Actinium	(227)
38 Strontium	87.62	90 Thorium	232.0381
39 Yttrium	88.9059	91 Protactinium	231.0359
40 Zirconium	91.22	92 Uranium	238.029
41 Niobium	92.9064	93 Neptunium	237.0482
42 Molybdenum	95.94	94 Plutonium	(244)
43 Technetium	98.9062	95 Americium	(243)
44 Ruthenium	101.17	96 Curium	(247)
45 Rhodium	102.9055	97 Berkelium	(247)
46 Palladium	106.4	98 Californium	(251)
47 Silver	107.868	99 Einsteinium	(254)
48 Cadmium	112.41	100 Fermium	(257)
49 Indium	114.82	101 Mendelevium	(258)
50 Tin	118.69	102 Nobelium	(259)
51 Antimony	121.75	103 Lawrencium	(260)
52 Tellurium	127.60		